Texts in
Computational Science
and Engineering

2

Editors

Timothy J. Barth
Michael Griebel
David E. Keyes
Risto M. Nieminen
Dirk Roose
Tamar Schlick

Alfio Quarteroni   Fausto Saleri

Scientific Computing with MATLAB and Octave

Second Edition

With 108 Figures and 12 Tables

Alfio Quarteroni
Ecole Polytechnique Fédérale
de Lausanne
CMCS-Modeling and Scientific Computing
1015 Lausanne, Switzerland
and
MOX-Politecnico di Milano
Piazza Leonardo da Vinci 32
20133 Milano, Italy
E-mail: alfio.quarteroni@epfl.ch

Fausto Saleri
MOX-Politecnico di Milano
Piazza Leonardo da Vinci 32
20133 Milano, Italy
E-mail: fausto.saleri@polimi.it

Cover figure by Marzio Sala

Title of the Italian original edition: *Introduzione al Calcolo Scientifico*, Springer-Verlag Italia, Milano, 2006.
ISBN 88-470-0480-2

Library of Congress Control Number: 2006928277

Mathematics Subject Classification: 65-01, 68U01, 68N15

ISBN-10 3-540-32612-X Springer Berlin Heidelberg New York
ISBN-13 978-3-540-32612-0 Springer Berlin Heidelberg New York
ISBN-10 3-540-44363-0 1st Edition Springer Berlin Heidelberg New York

This work is subject to copyright. All rights are reserved, whether the whole or part of the material is concerned, specifically the rights of translation, reprinting, reuse of illustrations, recitation, broadcasting, reproduction on microfilm or in any other way, and storage in data banks. Duplication of this publication or parts thereof is permitted only under the provisions of the German Copyright Law of September 9, 1965, in its current version, and permission for use must always be obtained from Springer. Violations are liable for prosecution under the German Copyright Law.

Springer is a part of Springer Science+Business Media
springer.com
© Springer-Verlag Berlin Heidelberg 2003, 2006
Printed in The Netherlands

The use of general descriptive names, registered names, trademarks, etc. in this publication does not imply, even in the absence of a specific statement, that such names are exempt from the relevant protective laws and regulations and therefore free for general use.

Typesetting: by the authors and techbooks using a Springer LaTeX macro package
Cover design: *design & production* GmbH, Heidelberg

Printed on acid-free paper   SPIN: 11678793   46/techbooks   5 4 3 2 1 0

*This book is dedicated to
Fulvia, Silvia and Marzia,
Paola, Maria and Caterina,
who make our lives
less scientifically computed.*

Preface

Preface to the First Edition

This textbook is an introduction to Scientific Computing. We will illustrate several numerical methods for the computer solution of certain classes of mathematical problems that cannot be faced by paper and pencil. We will show how to compute the zeros or the integrals of continuous functions, solve linear systems, approximate functions by polynomials and construct accurate approximations for the solution of differential equations.

With this aim, in Chapter 1 we will illustrate the rules of the game that computers adopt when storing and operating with real and complex numbers, vectors and matrices.

In order to make our presentation concrete and appealing we will adopt the programming environment MATLAB® [1] as a faithful companion. We will gradually discover its principal commands, statements and constructs. We will show how to execute all the algorithms that we introduce throughout the book. This will enable us to furnish an immediate quantitative assessment of their theoretical properties such as stability, accuracy and complexity. We will solve several problems that will be raised through exercises and examples, often stemming from specific applications.

Several graphical devices will be adopted in order to render the reading more pleasant. We will report in the margin the MATLAB command along side the line where that command is being introduced for the first time. The symbol will be used to indicate the presence of exercises, the symbol to indicate the presence of a MATLAB program, while

---

[1] MATLAB is a trademark of TheMathWorks Inc., 24 Prime Park Way, Natick, MA 01760, Tel: 001+508-647-7000, Fax: 001+508-647-7001.

the symbol ⬛ will be used when we want to attract the attention of the reader on a critical or surprising behavior of an algorithm or a procedure. The mathematical formulae of special relevance are put within a frame. Finally, the symbol ⬛ indicates the presence of a display panel summarizing concepts and conclusions which have just been reported and drawn.

At the end of each chapter a specific section is devoted to mentioning those subjects which have not been addressed and indicate the bibliographical references for a more comprehensive treatment of the material that we have carried out.

Quite often we will refer to the textbook [QSS06] where many issues faced in this book are treated at a deeper level, and where theoretical results are proven. For a more thorough description of MATLAB we refer to [HH05]. All the programs introduced in this text can be downloaded from the web address

mox.polimi.it/qs

No special prerequisite is demanded of the reader, with the exception of an elementary course of Calculus.

However, in the course of the first chapter, we recall the principal results of Calculus and Geometry that will be used extensively throughout this text. The less elementary subjects, those which are not so necessary for an introductory educational path, are highlighted by the special symbol ⬛ .

We express our thanks to Thanh-Ha Le Thi from Springer-Verlag Heidelberg, and to Francesca Bonadei and Marina Forlizzi from Springer-Italia for their friendly collaboration throughout this project. We gratefully thank Prof. Eastham of Cardiff University for editing the language of the whole manuscript and stimulating us to clarify many points of our text.

Milano and Lausanne	*Alfio Quarteroni*
May 2003	*Fausto Saleri*

Preface to the Second Edition

In this second edition we have enriched all the Chapters by introducing several new problems. Moreover, we have added new methods for the numerical solution of linear and nonlinear systems, the eigenvalue computation and the solution of initial-value problems. Another relevant improvement is that we also use the Octave programming environment. Octave is a reimplementation of part of MATLAB which includes many numerical facilities of MATLAB and is freely distributed under the GNU General Public License.

Throughout the book, we shall often make use of the expression "MATLAB command": in this case, MATLAB should be understood as the *language* which is the common subset of both programs MATLAB and Octave. We have striven to ensure a seamless usage of our codes and programs under both MATLAB and Octave. In the few cases where this does not apply, we shall write a short explanation notice at the end of each corresponding section.

For this second edition we would like to thank Paola Causin for having proposed several problems, Christophe Prud'homme, John W. Eaton and David Bateman for their help with Octave, and Silvia Quarteroni for the translation of the new sections. Finally, we kindly acknowledge the support of the Poseidon project of the Ecole Polytechnique Fédérale de Lausanne.

Lausanne and Milano   *Alfio Quarteroni*
May 2006   *Fausto Saleri*

Contents

1 **What can't be ignored** 1
   1.1 Real numbers 2
      1.1.1 How we represent them 2
      1.1.2 How we operate with floating-point numbers 4
   1.2 Complex numbers 6
   1.3 Matrices 8
      1.3.1 Vectors 14
   1.4 Real functions 15
      1.4.1 The zeros 16
      1.4.2 Polynomials 18
      1.4.3 Integration and differentiation 21
   1.5 To err is not only human 23
      1.5.1 Talking about costs 26
   1.6 The MATLAB and Octave environments 28
   1.7 The MATLAB language 29
      1.7.1 MATLAB statements 31
      1.7.2 Programming in MATLAB 32
      1.7.3 Examples of differences between MATLAB
            and Octave languages 36
   1.8 What we haven't told you 37
   1.9 Exercises 37

2 **Nonlinear equations** 39
   2.1 The bisection method 41
   2.2 The Newton method 45
      2.2.1 How to terminate Newton's iterations 47
      2.2.2 The Newton method for systems of nonlinear
            equations 49
   2.3 Fixed point iterations 51
      2.3.1 How to terminate fixed point iterations 55

	2.4	Acceleration using Aitken method	56
	2.5	Algebraic polynomials	60
		2.5.1 Hörner's algorithm	61
		2.5.2 The Newton-Hörner method	63
	2.6	What we haven't told you	65
	2.7	Exercises	67
3	**Approximation of functions and data**		**71**
	3.1	Interpolation	74
		3.1.1 Lagrangian polynomial interpolation	75
		3.1.2 Chebyshev interpolation	80
		3.1.3 Trigonometric interpolation and FFT	81
	3.2	Piecewise linear interpolation	86
	3.3	Approximation by spline functions	88
	3.4	The least-squares method	92
	3.5	What we haven't told you	97
	3.6	Exercises	98
4	**Numerical differentiation and integration**		**101**
	4.1	Approximation of function derivatives	103
	4.2	Numerical integration	105
		4.2.1 Midpoint formula	106
		4.2.2 Trapezoidal formula	108
		4.2.3 Simpson formula	109
	4.3	Interpolatory quadratures	111
	4.4	Simpson adaptive formula	115
	4.5	What we haven't told you	119
	4.6	Exercises	120
5	**Linear systems**		**123**
	5.1	The LU factorization method	126
	5.2	The pivoting technique	134
	5.3	How accurate is the LU factorization?	136
	5.4	How to solve a tridiagonal system	140
	5.5	Overdetermined systems	141
	5.6	What is hidden behind the command \	143
	5.7	Iterative methods	144
		5.7.1 How to construct an iterative method	146
	5.8	Richardson and gradient methods	150
	5.9	The conjugate gradient method	153
	5.10	When should an iterative method be stopped?	156
	5.11	To wrap-up: direct or iterative?	159
	5.12	What we haven't told you	164
	5.13	Exercises	164

6	**Eigenvalues and eigenvectors** 167	
	6.1 The power method 170	
	6.1.1 Convergence analysis 173	
	6.2 Generalization of the power method 174	
	6.3 How to compute the shift............................ 176	
	6.4 Computation of all the eigenvalues.................... 179	
	6.5 What we haven't told you 183	
	6.6 Exercises 183	
7	**Ordinary differential equations** 187	
	7.1 The Cauchy problem................................ 190	
	7.2 Euler methods 191	
	7.2.1 Convergence analysis 194	
	7.3 The Crank-Nicolson method 197	
	7.4 Zero-stability 199	
	7.5 Stability on unbounded intervals 202	
	7.5.1 The region of absolute stability 204	
	7.5.2 Absolute stability controls perturbations 205	
	7.6 High order methods................................. 212	
	7.7 The predictor-corrector methods 216	
	7.8 Systems of differential equations..................... 219	
	7.9 Some examples..................................... 225	
	7.9.1 The spherical pendulum....................... 225	
	7.9.2 The three-body problem 228	
	7.9.3 Some stiff problems........................... 230	
	7.10 What we haven't told you 234	
	7.11 Exercises 234	
8	**Numerical methods for (initial-)boundary-value problems** 237	
	8.1 Approximation of boundary-value problems............ 240	
	8.1.1 Approximation by finite differences 241	
	8.1.2 Approximation by finite elements............... 243	
	8.1.3 Approximation by finite differences of two-dimensional problems 245	
	8.1.4 Consistency and convergence................... 251	
	8.2 Finite difference approximation of the heat equation 253	
	8.3 The wave equation 257	
	8.3.1 Approximation by finite differences 260	
	8.4 What we haven't told you 263	
	8.5 Exercises 264	

9	**Solutions of the exercises**	267
	9.1 Chapter 1	267
	9.2 Chapter 2	270
	9.3 Chapter 3	276
	9.4 Chapter 4	280
	9.5 Chapter 5	285
	9.6 Chapter 6	289
	9.7 Chapter 7	293
	9.8 Chapter 8	301

References 307

Index 311

Listings

2.1	**bisection**: bisection method	43
2.2	**newton**: Newton method	48
2.3	**newtonsys**: Newton method for nonlinear systems	50
2.4	**aitken**: Aitken method	59
2.5	**horner**: synthetic division algorithm	62
2.6	**newtonhorner**: Newton-Hörner method	64
3.1	**cubicspline**: interpolating cubic spline	89
4.1	**midpointc**: composite midpoint quadrature formula	107
4.2	**simpsonc**: composite Simpson quadrature formula	110
4.3	**simpadpt**: adaptive Simpson formula	118
5.1	**lugauss**: Gauss factorization	131
5.2	**itermeth**: general iterative method	148
6.1	**eigpower**: power method	171
6.2	**invshift**: inverse power method with shift	175
6.3	**gershcircles**: Gershgorin circles	177
6.4	**qrbasic**: method of QR iterations	180
7.1	**feuler**: forward Euler method	192
7.2	**beuler**: backward Euler method	193
7.3	**cranknic**: Crank-Nicolson method	198
7.4	**predcor**: predictor-corrector method	218
7.5	**onestep**: one step of forward Euler (`eeonestep`), one step of backward Euler (`eionestep`), one step of Crank-Nicolson (`cnonestep`)	218
7.6	**newmark**: Newmark method	223
7.7	**fvinc**: forcing term for the spherical pendulum problem	227
7.8	**threebody**: forcing term for the simplified three body system	229
8.1	**bvp**: approximation of a two-point boundary-value problem by the finite difference method	242
8.2	**poissonfd**: approximation of the Poisson problem with Dirichlet data by the five-point finite difference method	249

8.3	**heattheta**: θ-method for the heat equation in a square domain	255
8.4	**newmarkwave**: Newmark method for the wave equation	260
9.1	**rk2**: Heun method	295
9.2	**rk3**: explicit Runge-Kutta method of order 3	297
9.3	**neumann**: approximation of a Neumann boundary-value problem	304

1
What can't be ignored

In this book we will systematically use elementary mathematical concepts which the reader should know already, yet he or she might not recall them immediately.

We will therefore use this chapter to refresh them, as well as to introduce new concepts which pertain to the field of Numerical Analysis. We will begin to explore their meaning and usefulness with the help of MATLAB (MATrix LABoratory), an integrated environment for programming and visualization in scientific computing. We shall also use GNU Octave (in short, Octave) which is mostly compatible with MATLAB. In Sections 1.6 and 1.7 we will give a quick introduction to MATLAB and Octave, which is sufficient for the use that we are going to make in this book. We also make some notes about differences between MATLAB and Octave which are relevant for this book. However, we refer the interested readers to the manual [HH05] for a description of the MATLAB language and to the manual [Eat02] for a description of Octave.

Octave is a reimplementation of part of MATLAB which includes a large part of the numerical facilities of MATLAB and is freely distributed under the GNU General Public License.

Through the book, we shall often make use of the expression "MATLAB command": in this case, MATLAB should be understood as the *language* which is the common subset of both programs MATLAB and Octave.

We have striven to ensure a seamless usage of our codes and programs under both MATLAB and Octave. In the few cases where this does not apply, we will write a short explanation notice at the end of each corresponding section.

In the present Chapter we have condensed notions which are typical of courses in Calculus, Linear Algebra and Geometry, yet rephrasing them in a way that is suitable for use in scientific computing.

1.1 Real numbers

While the set $\mathbb{R}$ of real numbers is known to everyone, the way in which computers treat them is perhaps less well known. On one hand, since machines have limited resources, only a subset $\mathbb{F}$ of finite dimension of $\mathbb{R}$ can be represented. The numbers in this subset are called *floating-point numbers*. On the other hand, as we shall see in Section 1.1.2, $\mathbb{F}$ is characterized by properties that are different from those of $\mathbb{R}$. The reason is that any real number x is in principle truncated by the machine, giving rise to a new number (called the *floating-point number*), denoted by $fl(x)$, which does not necessarily coincide with the original number x.

1.1.1 How we represent them

To become acquainted with the differences between $\mathbb{R}$ and $\mathbb{F}$, let us make a few experiments which illustrate the way that a computer deals with real numbers. Note that whether we use MATLAB or Octave rather than another language is just a matter of convenience. The results of our calculation, indeed, depend primarily on the manner in which the computer works, and only to a lesser degree on the programming language. Let us consider the rational number $x = 1/7$, whose decimal representation is $0.\overline{142857}$. This is an infinite representation, since the number of decimal digits is infinite. To get its computer representation, let us introduce after the *prompt* (the symbol >>) the ratio 1/7 and obtain

>> 1/7

```
ans =
    0.1429
```

which is a number with only four decimal digits, the last being different from the fourth digit of the original number.

Should we now consider 1/3 we would find 0.3333, so the fourth decimal digit would now be exact. This behavior is due to the fact that real numbers are *rounded* on the computer. This means, first of all, that only an a priori fixed number of decimal digits are returned, and moreover the last decimal digit which appears is increased by unity whenever the first disregarded decimal digit is greater than or equal to 5.

The first remark to make is that using only four decimal digits to represent real numbers is questionable. Indeed, the internal representation of the number is made of as many as 16 decimal digits, and what we have seen is simply one of several possible MATLAB output formats. The same number can take different expressions depending upon the specific format declaration that is made. For instance, for the number 1/7, some possible output *formats* are:

        `format long`   yields 0.14285714285714,
        `format short e`   "   1.4286e − 01,
        `format long e`   "   1.428571428571428e − 01,
        `format short g`   "   0.14286,
        `format long g`   "   0.142857142857143.   `format`

Some of them are more coherent than others with the internal computer representation. As a matter of fact, in general a computer stores a real number in the following way

$$x = (-1)^s \cdot (0.a_1 a_2 \ldots a_t) \cdot \beta^e = (-1)^s \cdot m \cdot \beta^{e-t}, \quad a_1 \neq 0 \qquad (1.1)$$

where s is either 0 or 1, β (a positive integer larger than or equal to 2) is the *basis* adopted by the specific computer at hand, m is an integer called the *mantissa* whose length t is the maximum number of digits a_i (with $0 \leq a_i \leq \beta - 1$) that are stored, and e is an integral number called the *exponent*. The format `long e` is the one which most resembles this representation, and `e` stands for exponent; its digits, preceded by the sign, are reported to the right of the character `e`. The numbers whose form is given in (1.1) are called floating-point numbers, since the position of the decimal point is not fixed. The digits $a_1 a_2 \ldots a_p$ (with $p \leq t$) are often called the p first significant digits of x.

The condition $a_1 \neq 0$ ensures that a number cannot have multiple representations. For instance, without this restriction the number 1/10 could be represented (in the decimal basis) as $0.1 \cdot 10^0$, but also as $0.01 \cdot 10^1$, etc..

The set $\mathbb{F}$ is therefore fully characterized by the basis β, the number of significant digits t and the range (L, U) (with $L < 0$ and $U > 0$) of variation of the index e. Thus it is denoted as $\mathbb{F}(\beta, t, L, U)$. For instance, in MATLAB we have $\mathbb{F} = \mathbb{F}(2, 53, -1021, 1024)$ (indeed, 53 significant digits in basis 2 correspond to the 15 significant digits that are shown by MATLAB in basis 10 with the `format long`).

Fortunately, the *roundoff error* that is inevitably generated whenever a real number $x \neq 0$ is replaced by its representative $fl(x)$ in $\mathbb{F}$, is small, since

$$\frac{|x - fl(x)|}{|x|} \leq \frac{1}{2} \epsilon_M \qquad (1.2)$$

where $\epsilon_M = \beta^{1-t}$ provides the distance between 1 and its closest floating-point number greater than 1. Note that ϵ_M depends on β and t. For instance, in MATLAB ϵ_M can be obtained through the command `eps`, and we obtain $\epsilon_M = 2^{-52} \simeq 2.22 \cdot 10^{-16}$. Let us point out that in (1.2) we estimate the *relative error* on x, which is undoubtedly more meaningful than the *absolute error* $|x - fl(x)|$. As a matter of fact, the latter doesn't account for the order of magnitude of x whereas the former does.

Number 0 does not belong to $\mathbb{F}$, as in that case we would have $a_1 = 0$ in (1.1): it is therefore handled separately. Moreover, L and U being finite, one cannot represent numbers whose absolute value is either arbitrarily large or arbitrarily small. Precisely, the smallest and the largest positive real numbers of $\mathbb{F}$ are given respectively by

$$x_{min} = \beta^{L-1}, \; x_{max} = \beta^{U}(1 - \beta^{-t}).$$

realmin In MATLAB these values can be obtained through the commands
realmax **realmin** and **realmax**, yielding

$$x_{min} = 2.225073858507201 \cdot 10^{-308},$$
$$x_{max} = 1.7976931348623158 \cdot 10^{+308}.$$

A positive number smaller than x_{min} produces a message of underflow and is treated either as 0 or in a special way (see, e.g., [QSS06], Chapter 2). A positive number greater than x_{max} yields instead a message of overflow and is stored in the variable **Inf** (which is the computer representation of $+\infty$).

Inf

The elements in $\mathbb{F}$ are more dense near x_{min}, and less dense while approaching x_{max}. As a matter of fact, the number in $\mathbb{F}$ nearest to x_{max} (to its left) and the one nearest to x_{min} (to its right) are, respectively

$$x_{max}^{-} = 1.7976931348623157 \cdot 10^{+308},$$
$$x_{min}^{+} = 2.225073858507202 \cdot 10^{-308}.$$

Thus $x_{min}^{+} - x_{min} \simeq 10^{-323}$, while $x_{max} - x_{max}^{-} \simeq 10^{292}$ (!). However, the relative distance is small in both cases, as we can infer from (1.2).

1.1.2 How we operate with floating-point numbers

Since $\mathbb{F}$ is a proper subset of $\mathbb{R}$, elementary algebraic operations on floating-point numbers do not enjoy all the properties of analogous operations on $\mathbb{R}$. Precisely, commutativity still holds for addition (that is $fl(x+y) = fl(y+x)$) as well as for multiplication ($fl(xy) = fl(yx)$), but other properties such as associativity and distributivity are violated. Moreover, 0 is no longer unique. Indeed, let us assign the variable **a** the value 1, and execute the following instructions:

```
>> a = 1; b=1; while a+b ~= a; b=b/2; end
```

The variable b is halved at every step as long as the sum of a and b remains different (~=) from a. Should we operate on real numbers, this program would never end, whereas in our case it ends after a finite number of steps and returns the following value for b: 1.1102e-16= $\epsilon_M/2$. There exists therefore at least one number b different from 0 such that a+b=a. This is possible since $\mathbb{F}$ is made up of isolated numbers; when adding two numbers a and b with b<a and b less than ϵ_M, we always obtain that a+b is equal to a. The MATLAB number a+eps(a) is the smallest number in $\mathbb{F}$ larger than a. Thus the sum a+b will return a for all b < eps(a).

Associativity is violated whenever a situation of overflow or underflow occurs. Take for instance a=1.0e+308, b=1.1e+308 and c=-1.001e+308, and carry out the sum in two different ways. We find that

$$a + (b + c) = 1.0990e + 308, (a + b) + c = \text{Inf}.$$

This is a particular instance of what occurs when one adds two numbers with opposite sign but similar absolute value. In this case the result may be quite inexact and the situation is referred to as *loss*, or *cancellation, of significant digits*. For instance, let us compute $((1+x)-1)/x$ (the obvious result being 1 for any $x \neq 0$):

```
>> x =   1.e-15;  ((1+x)-1)/x

ans = 1.1102
```

This result is rather imprecise, the relative error being larger than 11%!

Another case of numerical cancellation is encountered while evaluating the function

$$f(x) = x^7 - 7x^6 + 21x^5 - 35x^4 + 35x^3 - 21x^2 + 7x - 1 \quad (1.3)$$

at 401 equispaced points with abscissa in $[1 - 2 \cdot 10^{-8}, 1 + 2 \cdot 10^{-8}]$. We obtain the chaotic graph reported in Figure 1.1 (the real behavior is that of $(x-1)^7$, which is substantially constant and equal to the null function in such a tiny neighborhood of $x = 1$). The MATLAB commands that have generated this graph will be illustrated in Section 1.4.

Finally, it is interesting to notice that in $\mathbb{F}$ there is no place for indeterminate forms such as $0/0$ or ∞/∞. Their presence produces what is called *not a number* (NaN in MATLAB or in Octave), for which the   NaN normal rules of calculus do not apply.

Remark 1.1 Whereas it is true that roundoff errors are usually small, when repeated within long and complex algorithms, they may give rise to catastrophic effects. Two outstanding cases concern the explosion of the Arianne missile on June 4, 1996, engendered by an overflow in the computer on board, and the failure of the mission of an American Patriot missile, during the Gulf War in 1991, because of a roundoff error in the computation of its trajectory.

An example with less catastrophic (but still troublesome) consequences is provided by the sequence

$$z_2 = 2, \ z_{n+1} = 2^{n-1/2}\sqrt{1 - \sqrt{1 - 4^{1-n}z_n^2}}, \ n = 2, 3, \ldots \quad (1.4)$$

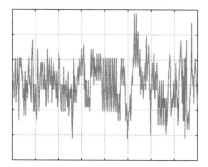

Fig. 1.1. Oscillatory behavior of the function (1.3) caused by cancellation errors

which converges to π when n tends to infinity. When MATLAB is used to compute z_n, the relative error found between π and z_n decreases for the 16 first iterations, then grows because of roundoff errors (as shown in Figure 1.2).

•

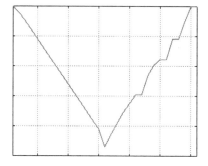

Fig. 1.2. Logarithm of the relative error $|\pi - z_n|/\pi$ versus n

See the Exercises 1.1-1.2.

1.2 Complex numbers

Complex numbers, whose set is denoted by $\mathbb{C}$, have the form $z = x + iy$, where $i = \sqrt{-1}$ is the imaginary unit (that is $i^2 = -1$), while $x = \text{Re}(z)$ and $y = \text{Im}(z)$ are the real and imaginary part of z, respectively. They are generally represented on the computer as pairs of real numbers.

Unless redefined otherwise, MATLAB variables i as well as j denote the imaginary unit. To introduce a complex number with real part x and

1.2 Complex numbers   7

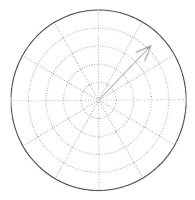

Fig. 1.3. Output of the MATLAB command `compass`

imaginary part y, one can just write x+i*y; as an alternative, one can use the command complex(x,y). Let us also mention the exponential and the trigonometric representations of a complex number z, that are equivalent thanks to the *Euler formula*

$$z = \rho e^{i\theta} = \rho(\cos\theta + i\sin\theta); \qquad (1.5)$$

`complex`

$\rho = \sqrt{x^2 + y^2}$ is the absolute value of the complex number (it can be obtained by setting abs(z)) while θ is its argument, that is the angle between the x axis and the straight line issuing from the origin and passing from the point of coordinate x, y in the complex plane. θ can be found by typing angle(z). The representation (1.5) is therefore:

`abs`

`angle`

$$\text{abs}(z) * (\cos(\text{angle}(z)) + i * \sin(\text{angle}(z))).$$

The graphical polar representation of one or more complex numbers can be obtained through the command compass(z), where z is either a single complex number or a vector whose components are complex numbers. For instance, by typing

`compass`

```
>> z = 3+i*3; compass(z);
```

one obtains the graph reported in Figure 1.3.

For any given complex number z, one can extract its real part with the command real(z) and its imaginary part with imag(z). Finally, the complex conjugate $\bar{z} = x - iy$ of z, can be obtained by simply writing conj(z).

`real`
`imag`

`conj`

In MATLAB all operations are carried out by implicitly assuming that the operands as well as the result are complex. We may therefore find some apparently surprising results. For instance, if we compute the cube root of −5 with the MATLAB command (-5)^(1/3), instead of −1.7099... we obtain the complex number $0.8550 + 1.4809i$. (We anticipate the use of the symbol ^ for the power exponent.) As a matter of fact, all numbers of the form $\rho e^{i(\theta+2k\pi)}$, with k an integer, are indistinguishable from $z = \rho e^{i\theta}$. By computing $\sqrt[3]{z}$ we find $\sqrt[3]{\rho} e^{i(\theta/3+2k\pi/3)}$, that is, the three distinct roots

$$z_1 = \sqrt[3]{\rho} e^{i\theta/3}, \ z_2 = \sqrt[3]{\rho} e^{i(\theta/3+2\pi/3)}, \ z_3 = \sqrt[3]{\rho} e^{i(\theta/3+4\pi/3)}.$$

MATLAB will select the one that is encountered by spanning the complex plane counterclockwise beginning from the real axis. Since the polar representation of $z = -5$ is $\rho e^{i\theta}$ with $\rho = 5$ and $\theta = -\pi$, the three roots are (see Figure 1.4 for their representation in the Gauss plane)

$$z_1 = \sqrt[3]{5}(\cos(-\pi/3) + i\sin(-\pi/3)) \simeq 0.8550 - 1.4809i,$$
$$z_2 = \sqrt[3]{5}(\cos(\pi/3) + i\sin(\pi/3)) \simeq 0.8550 + 1.4809i,$$
$$z_3 = \sqrt[3]{5}(\cos(-\pi) + i\sin(-\pi)) \simeq -1.7100.$$

The second root is the one which is selected.

Finally, by (1.5) we obtain

$$\cos(\theta) = \frac{1}{2}\left(e^{i\theta} + e^{-i\theta}\right), \sin(\theta) = \frac{1}{2i}\left(e^{i\theta} - e^{-i\theta}\right). \quad (1.6)$$

Octave 1.1 The command compass is not available in Octave, however it can be emulated with the following function:

```
function compass(z)
xx = [0 1 .8 1 .8].';
yy = [0 0 .08 0 -.08].';
arrow = xx + yy.*sqrt(-1);
z = arrow * z;
[th,r] = cart2pol(real(z),imag(z));
polar(th,r);
return
```

∎

1.3 Matrices

Let n and m be positive integers. A matrix with m rows and n columns is a set of $m \times n$ elements a_{ij}, with $i = 1, \ldots, m$, $j = 1, \ldots, n$, represented by the following table:

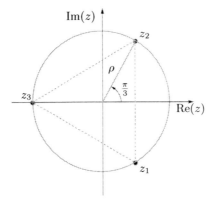

Fig. 1.4. Representation in the complex plane of the three complex cube roots of the real number -5

$$A = \begin{bmatrix} a_{11} & a_{12} & \cdots & a_{1n} \\ a_{21} & a_{22} & \cdots & a_{2n} \\ \vdots & \vdots & & \vdots \\ a_{m1} & a_{m2} & \cdots & a_{mn} \end{bmatrix}. \tag{1.7}$$

In compact form we write $A = (a_{ij})$. Should the elements of A be real numbers, we write $A \in \mathbb{R}^{m \times n}$, and $A \in \mathbb{C}^{m \times n}$ if they are complex.

Square matrices of dimension n are those with $m = n$. A matrix featuring a single column is a *column vector*, whereas a matrix featuring a single row is a *row vector*.

In order to introduce a matrix in MATLAB one has to write the elements from the first to the last row, introducing the character ; to separate the different rows. For instance, the command

```
>> A = [ 1 2 3; 4 5 6]
```

produces

```
A =
     1     2     3
     4     5     6
```

that is, a 2×3 matrix whose elements are indicated above. The $m \times n$ matrix `zeros(m,n)` has all null entries, `eye(m,n)` has all null entries unless a_{ii}, $i = 1, \ldots, \min(m, n)$, on the diagonal that are all equal to 1. The $n \times n$ identity matrix is obtained with the command `eye(n)`: its elements are $\delta_{ij} = 1$ if $i = j$, 0 otherwise, for $i, j = 1, \ldots, n$. Finally, by the command `A=[ ]` we can initialize an empty matrix. `zeros` `eye`

We recall the following matrix operations:

1. if $A = (a_{ij})$ and $B = (b_{ij})$ are $m \times n$ matrices, the *sum* of A and B is the matrix $A + B = (a_{ij} + b_{ij})$;

2. the *product* of a matrix A by a real or complex number λ is the matrix $\lambda A = (\lambda a_{ij})$;
3. the *product* of two matrices is possible only for compatible sizes, precisely if A is $m \times p$ and B is $p \times n$, for some positive integer p. In that case C = AB is an $m \times n$ matrix whose elements are

$$c_{ij} = \sum_{k=1}^{p} a_{ik} b_{kj}, \text{ for } i = 1, \ldots, m, \ j = 1, \ldots, n.$$

Here is an example of the sum and product of two matrices.

```
>> A=[1 2 3; 4 5 6];
>> B=[7 8 9; 10 11 12];
>> C=[13 14; 15 16; 17 18];
>> A+B

ans =
        8    10    12
       14    16    18

>> A*C

ans =
       94   100
      229   244
```

Note that MATLAB returns a diagnostic message when one tries to carry out operations on matrices with incompatible dimensions. For instance:

```
>> A=[1 2 3; 4 5 6];
>> B=[7 8 9; 10 11 12];
>> C=[13 14; 15 16; 17 18];
>> A+C

??? Error using ==> +
Matrix dimensions must agree.

>> A*B

??? Error using ==> *
Inner matrix dimensions must agree.
```

inv If A is a square matrix of dimension n, its *inverse* (provided it exists) is a square matrix of dimension n, denoted by A^{-1}, which satisfies the matrix relation $AA^{-1} = A^{-1}A = I$. We can obtain A^{-1} through the command **inv(A)**. The inverse of A exists iff the *determinant* of A, a number denoted by det(A), is non-zero. The latter condition is satisfied iff the column vectors of A are linearly independent (see Section 1.3.1).

1.3 Matrices

The determinant of a square matrix is defined by the following recursive formula (*Laplace rule*):

$$\det(A) = \begin{cases} a_{11} & \text{if } n = 1, \\ \sum_{j=1}^{n} \Delta_{ij} a_{ij}, & \text{for } n > 1, \forall i = 1, \ldots, n, \end{cases} \quad (1.8)$$

where $\Delta_{ij} = (-1)^{i+j} \det(A_{ij})$ and A_{ij} is the matrix obtained by eliminating the i-th row and j-th column from matrix A. (The result is independent of the row index i.) In particular, if $A \in \mathbb{R}^{2 \times 2}$ one has

$$\det(A) = a_{11} a_{22} - a_{12} a_{21},$$

while if $A \in \mathbb{R}^{3 \times 3}$ we obtain

$$\det(A) = a_{11} a_{22} a_{33} + a_{31} a_{12} a_{23} + a_{21} a_{13} a_{32}$$
$$- a_{11} a_{23} a_{32} - a_{21} a_{12} a_{33} - a_{31} a_{13} a_{22}.$$

We recall that if $A = BC$, then $\det(A) = \det(B)\det(C)$.

To invert a 2×2 matrix and compute its determinant we can proceed as follows:

```
>> A=[1 2; 3 4];
>> inv(A)

ans =
    -2.0000    1.0000
     1.5000   -0.5000

>> det(A)

ans =
    -2
```

Should a matrix be singular, MATLAB returns a diagnostic message, followed by a matrix whose elements are all equal to Inf, as illustrated by the following example:

```
>> A=[1 2; 0 0];
>> inv(A)

Warning: Matrix is singular to working precision.
ans =
    Inf   Inf
    Inf   Inf
```

For special classes of square matrices, the computation of inverses and determinants is rather simple. In particular, if A is a *diagonal matrix*, i.e. one for which only the diagonal elements a_{kk}, $k = 1, \ldots, n$, are non-zero, its determinant is given by $\det(A) = a_{11}a_{22}\cdots a_{nn}$. In particular, A is non-singular iff $a_{kk} \neq 0$ for all k. In such a case the inverse of A is still a diagonal matrix with elements a_{kk}^{-1}.

diag Let v be a vector of dimension n. The command **diag(v)** produces a diagonal matrix whose elements are the components of vector v. The more general command **diag(v,m)** yields a square matrix of dimension n+abs(m) whose m-th upper diagonal (i.e. the diagonal made of elements with indices $i, i + m$) has elements equal to the components of v, while the remaining elements are null. Note that this extension is valid also when m is negative, in which case the only affected elements are those of lower diagonals.

For instance if v = [1 2 3] then:

```
>> A=diag(v,-1)

A =
     0     0     0     0
     1     0     0     0
     0     2     0     0
     0     0     3     0
```

Other special cases are the *upper triangular* and *lower triangular* matrices. A square matrix of dimension n is *lower* (respectively, *upper*) *triangular* if all elements above (respectively, below) the main diagonal are zero. Its determinant is simply the product of the diagonal elements.

tril Through the commands **tril(A)** and **triu(A)**, one can extract from
triu the matrix A of dimension n its lower and upper triangular part. Their extensions **tril(A,m)** or **triu(A,m)**, with m ranging from -n and n, allow the extraction of the triangular part augmented by, or deprived of, m extradiagonals.

For instance, given the matrix A =[3 1 2; -1 3 4; -2 -1 3], by the command L1=tril(A) we obtain

```
L1 =
     3     0     0
    -1     3     0
    -2    -1     3
```

while, by L2=tril(A,1), we obtain

```
L2 =
     3     1     0
    -1     3     4
    -2    -1     3
```

We recall that if A $\in \mathbb{R}^{m \times n}$ its transpose $A^T \in \mathbb{R}^{n \times m}$ is the matrix obtained by interchanging rows and columns of A. When $n = m$ and $A = A^T$ the matrix A is called *symmetric*. Finally, A' denotes the transpose of A if A is real, or its conjugate transpose (that is, A^H) if A is complex. A square complex matrix that coincides with its conjugate transpose A^H is called *hermitian*.

A'

A similar notation, v', is used for the transpose conjugate v^H of the vector v. If v_i denote the components of v, the adjoint vector v^H is a row-vector whose components are the complex conjugate $\bar{v}_i$ of v_i.

v'

Octave 1.2 Also Octave returns a diagnostic message when one tries to carry out operations on matrices having non-compatible dimensions. If we repeat the previous MATLAB examples we obtain:

```
octave:1> A=[1 2 3; 4 5 6];
octave:2> B=[7 8 9; 10 11 12];
octave:3> C=[13 14; 15 16; 17 18];
octave:4> A+C
```

```
error: operator +: nonconformant arguments (op1 is
2x3, op2 is 3x2)
error: evaluating binary operator '+' near line 2,
column 2
```

```
octave:5> A*B
```

```
error: operator *: nonconformant arguments (op1 is
2x3, op2 is 2x3)
error: evaluating binary operator '*' near line 2,
column 2
```

If A is singular, Octave returns a diagnostic message followed by the matrix to be inverted, as illustrated by the following example:

```
octave:1> A=[1 2; 0 0];
octave:2> inv(A)
```

```
warning: inverse: matrix singular to machine
precision, rcond = 0
ans =
   1   2
   0   0
```

∎

1.3.1 Vectors

Vectors will be indicated in boldface; precisely, **v** will denote a column vector whose i-th component is denoted by v_i. When all components are real numbers we can write $\mathbf{v} \in \mathbb{R}^n$.

In MATLAB, vectors are regarded as particular cases of matrices. To introduce a column vector one has to insert between square brackets the values of its components separated by semi-colons, whereas for a row vector it suffices to write the component values separated by blanks or commas. For instance, through the instructions v = [1;2;3] and w = [1 2 3] we initialize the column vector **v** and the row vector **w**, both of dimension 3. The command zeros(n,1) (respectively, zeros(1,n)) produces a column (respectively, row) vector of dimension n with null elements, which we will denote by **0**. Similarly, the command ones(n,1) generates the column vector, denoted with **1**, whose components are all equal to 1.

zeros

ones

A system of vectors $\{\mathbf{y}_1, \ldots, \mathbf{y}_m\}$ is *linearly independent* if the relation

$$\alpha_1 \mathbf{y}_1 + \ldots + \alpha_m \mathbf{y}_m = \mathbf{0}$$

implies that all coefficients $\alpha_1, \ldots, \alpha_m$ are null. A system $\mathcal{B} = \{\mathbf{y}_1, \ldots, \mathbf{y}_n\}$ of n linearly independent vectors in $\mathbb{R}^n$ (or $\mathbb{C}^n$) is a *basis* for $\mathbb{R}^n$ (or $\mathbb{C}^n$), that is, any vector **w** in $\mathbb{R}^n$ can be written as a linear combination of the elements of $\mathcal{B}$,

$$\mathbf{w} = \sum_{k=1}^{n} w_k \mathbf{y}_k,$$

for a unique possible choice of the coefficients $\{w_k\}$. The latter are called the *components* of **w** with respect to the basis $\mathcal{B}$. For instance, the canonical basis of $\mathbb{R}^n$ is the set of vectors $\{\mathbf{e}_1, \ldots, \mathbf{e}_n\}$, where $\mathbf{e}_i$ has its i-th component equal to 1, and all other components equal to 0 and is the one which is normally used.

The *scalar product* of two vectors $\mathbf{v}, \mathbf{w} \in \mathbb{R}^n$ is defined as

$$(\mathbf{v}, \mathbf{w}) = \mathbf{w}^T \mathbf{v} = \sum_{k=1}^{n} v_k w_k,$$

$\{v_k\}$ and $\{w_k\}$ being the components of **v** and **w**, respectively. The corresponding command is w'*v or else dot(v,w), where now the apex denotes transposition of a vector. The length (or modulus) of a vector **v** is given by

dot

$$\|\mathbf{v}\| = \sqrt{(\mathbf{v}, \mathbf{v})} = \sqrt{\sum_{k=1}^{n} v_k^2}$$

norm and can be computed through the command `norm(v)`.

The vector product between two vectors $\mathbf{v}, \mathbf{w} \in \mathbb{R}^n$, $n \geq 3$, $\mathbf{v} \times \mathbf{w}$ or $\mathbf{v} \wedge \mathbf{w}$, is the vector $\mathbf{u} \in \mathbb{R}^n$ orthogonal to both $\mathbf{v}$ and $\mathbf{w}$ whose modulus is $|\mathbf{u}| = |\mathbf{v}|\,|\mathbf{w}|\sin(\alpha)$, where α is the angle formed by $\mathbf{v}$ and $\mathbf{w}$. It can be obtained by the command `cross(v,w)`.   cross

The visualization of a vector can be obtained by the MATLAB command `quiver` in $\mathbb{R}^2$ and `quiver3` in $\mathbb{R}^3$.   quiver

The MATLAB command `x.*y` or `x.^2` indicates that these operations should be carried out component by component. For instance if we   quiver3
define the vectors   .*
```
>> v = [1; 2; 3]; w = [4; 5; 6];
```
   .^
the instruction
```
>> w'*v
```

```
ans =
    32
```
provides their scalar product, while
```
>> w.*v
```

```
ans =
    4
   10
   18
```
returns a vector whose i-th component is equal to $x_i y_i$.

Finally, we recall that a vector $\mathbf{v} \in \mathbb{C}^n$, with $\mathbf{v} \neq \mathbf{0}$, is an *eigenvector* of a matrix $A \in \mathbb{C}^{n \times n}$ associated with the complex number λ if

$$A\mathbf{v} = \lambda \mathbf{v}.$$

The complex number λ is called *eigenvalue* of A. In general, the computation of eigenvalues is quite difficult. Exceptions are represented by diagonal and triangular matrices, whose eigenvalues are their diagonal elements.

See the Exercises 1.3-1.6.

1.4 Real functions

This chapter will deal with manipulation of real functions defined on an interval (a,b). The command `fplot(fun,lims)` plots the graph of the   fplot
function `fun` (which is stored as a string of characters) on the interval (`lims(1)`,`lims(2)`). For instance, to represent $f(x) = 1/(1+x^2)$ on the interval $(-5,5)$, we can write

```
>> fun ='1/(1+x.^2)'; lims=[-5,5]; fplot(fun,lims);
```
or, more directly,
```
>> fplot('1/(1+x.^2)',[-5 5]);
```

In MATLAB the graph is obtained by sampling the function on a set of non-equispaced abscissae and reproduces the true graph of f with a tolerance of 0.2%. To improve the accuracy we could use the command
```
>> fplot(fun,lims,tol,n,'LineSpec',P1,P2,...)
```
where tol indicates the desired tolerance and the parameter n(≥ 1) ensures that the function will be plotted with a minimum of n+1 points. LineSpec is a string specifying the style or the color of the line used for plotting the graph. For example, LineSpec='--' is used for a dashed line, LineSpec='r-.' for a red dashed-dotted line, etc. To use default values for tol, n or LineSpec one can pass empty matrices ([ ]).

eval   To evaluate a function fun at a point x we write y=eval(fun), after having initialized x. The corresponding value is stored in y. Note that x, and correspondingly y, can be a vector. When using this command, the restriction is that the argument of the function fun must be x. When the argument of fun has a different name (this is often the case when this argument is generated at the interior of a program) the command eval would be replaced by feval (see Remark 1.2).

grid   Finally, we point out that if we write grid on after the command fplot, we can obtain the background-grid as that in Figure 1.1.

Octave 1.3 In Octave, using the command fplot(fun,lims,n) the graph is obtained by sampling the function defined in fun (that is the name of a *function* or an expression containing x) on a set of non-equispaced abscissae. The optional parameter n (≥ 1) ensures that the function will be plotted with a minimum of n+1 points. For instance, to represent $f(x) = 1/(1 + x^2)$ we use the following commands:
```
>> fun ='1./(1+x.^2)'; lims=[-5,5];
>> fplot(fun,lims)
```
■

1.4.1 The zeros

We recall that if $f(\alpha) = 0$, α is called *zero* of f or *root* of the equation $f(x) = 0$. A zero is *simple* if $f'(\alpha) \neq 0$, *multiple* otherwise.

From the graph of a function one can infer (within a certain tolerance) which are its real zeros. The direct computation of all zeros of a given function is not always possible. For functions which are polynomials with real coefficients of degree n, that is, of the form

$$p_n(x) = a_0 + a_1 x + a_2 x^2 + \ldots + a_n x^n = \sum_{k=0}^{n} a_k x^k, \quad a_k \in \mathbb{R}, \ a_n \neq 0,$$

we can obtain the only zero $\alpha = -a_0/a_1$, when $n = 1$ (i.e. p_1 represents a straight line), or the two zeros, α_+ and α_-, when $n = 2$ (this time p_2 represents a parabola) $\alpha_\pm = (-a_1 \pm \sqrt{a_1^2 - 4a_0 a_2})/(2a_2)$.

However, there are no explicit formulae for the zeros of an arbitrary polynomial p_n when $n \geq 5$.

In the sequel we will denote with $\mathbb{P}_n$ the space of polynomials of degree less than or equal to n,

$$p_n(x) = \sum_{k=0}^{n} a_k x^k \qquad (1.9)$$

where the a_k are given coefficients, real or complex.

Also the number of zeros of a function cannot in general be determined *a priori*. An exception is provided by polynomials, for which the number of zeros (real or complex) coincides with the polynomial degree. Moreover, should $\alpha = x + iy$ with $y \neq 0$ be a zero of a polynomial with degree $n \geq 2$, its complex conjugate $\bar{\alpha} = x - iy$ is also a zero.

To compute in MATLAB one zero of a function fun, near a given value x0, either real or complex, the command **fzero(fun,x0)** can be used. The result is an approximate value of the desired zero, and also the interval in which the search was made. Alternatively, using the command fzero(fun,[x0 x1]), a zero of fun is searched for in the interval whose extremes are x0,x1, provided f changes sign between x0 and x1.

fzero

Let us consider, for instance, the function $f(x) = x^2 - 1 + e^x$. Looking at its graph we see that there are two zeros in $(-1, 1)$. To compute them we need to execute the following commands:

```
fun=inline('x^2 - 1 + exp(x)','x')
fzero(fun,1)

ans =
    5.4422e-18

fzero(fun,-1)

ans =
    -0.7146
```

Alternatively, after noticing from the function plot that one zero is in the interval $[-1, -0.2]$ and another in $[-0.2, 1]$, we could have written

```
fzero(fun,[-0.2 1])

   ans =
     -5.2609e-17
fzero(fun,[-1 -0.2])

   ans =
     -0.7146
```

The result obtained for the first zero is slightly different than the one obtained previously, due to a different initialization of the algorithm implemented in `fzero`.

In Chapter 2 we will introduce and investigate several methods for the approximate computation of the zeros of an arbitrary function.

Octave 1.4 In Octave, `fzero` accepts only functions defined using the keyword `function` and its corresponding syntax as follows:
```
function y = fun(x)
   y = x.^2 - 1 + exp(x);
end
fzero("fun", 1)

   ans =   2.3762e-17

fzero("fun",-1)

   ans =  -0.71456
```

■

1.4.2 Polynomials

polyval

Polynomials are very special functions and there is a special MATLAB toolbox[1] polyfun for their treatment. The command `polyval` is apt to evaluate a polynomial at one or several points. Its input arguments are a vector p and a vector x, where the components of p are the polynomial coefficients stored in decreasing order, from a_n down to a_0, and the components of x are the abscissae where the polynomial needs to be evaluated. The result can be stored in a vector y by writing
```
>> y = polyval(p,x)
```

---

[1] A toolbox is a collection of special-purpose MATLAB functions

For instance, the values of $p(x) = x^7 + 3x^2 - 1$, at the equispaced abscissae $x_k = -1 + k/4$ for $k = 0, \ldots, 8$, can be obtained by proceeding as follows:

```
>> p = [1 0 0 0 0 3 0 -1]; x = [-1:0.25:1];
>> y = polyval(p,x)

y =
  Columns 1 through 5:

    1.00000    0.55402   -0.25781   -0.81256   -1.00000
  Columns 6 through 9:

   -0.81244   -0.24219    0.82098    3.00000
```

Alternatively, one could use the command `feval`. However, in such case one should provide the entire analytic expression of the polynomial in the input string, and not simply its coefficients.

The program `roots` provides an approximation of the zeros of a polynomial and requires only the input of the vector p.   **roots**

For instance, we can compute the zeros of $p(x) = x^3 - 6x^2 + 11x - 6$ by writing

```
>> p = [1 -6 11 -6]; format long;
>> roots(p)

ans =
   3.00000000000000
   2.00000000000000
   1.00000000000000
```

Unfortunately, the result is not always that accurate. For instance, for the polynomial $p(x) = (x+1)^7$, whose unique zero is $\alpha = -1$ with multiplicity 7, we find (quite surprisingly)

```
>> p = [1 7 21 35 35 21 7 1];
>> roots(p)

ans =
   -1.0101
   -1.0063 + 0.0079i
   -1.0063 - 0.0079i
   -0.9977 + 0.0099i
   -0.9977 - 0.0099i
   -0.9909 + 0.0044i
   -0.9909 - 0.0044i
```

In fact, numerical methods for the computation of the polynomial roots with multiplicity larger than one are particularly subject to round-off errors (see Section 2.5.2).

conv   The command **p=conv(p1,p2)** returns the coefficients of the polynomial given by the product of two polynomials whose coefficients are contained in the vectors p1 and p2.

deconv   Similarly, the command **[q,r]=deconv(p1,p2)** provides the coefficients of the polynomials obtained on dividing p1 by p2, i.e. **p1 = conv(p2,q) + r**. In other words, q and r are the quotient and the remainder of the division.

Let us consider for instance the product and the ratio between the two polynomials $p_1(x) = x^4 - 1$ and $p_2(x) = x^3 - 1$:

```
>> p1 = [1 0 0 0 -1];
>> p2 = [1 0 0 -1];
>> p=conv(p1,p2)

p =
      1    0    0   -1   -1    0    0    1

>> [q,r]=deconv(p1,p2)

q =
      1    0
r =
      0    0    0    1   -1
```

We therefore find the polynomials $p(x) = p_1(x)p_2(x) = x^7 - x^4 - x^3 + 1$, $q(x) = x$ and $r(x) = x - 1$ such that $p_1(x) = q(x)p_2(x) + r(x)$.

polyint   The commands **polyint(p)** and **polyder(p)** provide respectively the
polyder   coefficients of the primitive (vanishing at $x = 0$) and those of the derivative of the polynomial whose coefficients are given by the components of the vector p.

If x is a vector of abscissae and p (respectively, p_1 and p_2) is a vector containing the coefficients of a polynomial p (respectively, p_1 and p_2), the previous commands are summarized in Table 1.1.

command	yields
y=polyval(p,x)	y = values of $p(x)$
z=roots(p)	z = roots of p such that $p(z) = 0$
p=conv(p_1,p_2)	p = coefficients of the polynomial $p_1 p_2$
[q,r]=deconv(p_1,p_2)	q = coefficients of q, r = coefficients of r such that $p_1 = qp_2 + r$
y=polyder(p)	y = coefficients of $p'(x)$
y=polyint(p)	y = coefficients of $\int_0^x p(t)\, dt$

Table 1.1. MATLAB commands for polynomial operations

A further command, `polyfit`, allows the computation of the $n+1$ polynomial coefficients of a polynomial p of degree n once the values attained by p at $n+1$ distinct nodes are available (see Section 3.1.1).

Octave 1.5 The commands `polyderiv` and `polyinteg` have the same functionality of `polyder` and `polyint`, respectively. Notice that the command `polyder` is available as well from the Octave repository, see Section 1.6. ∎

1.4.3 Integration and differentiation

The following two results will often be invoked throughout this book.

1. the *fundamental theorem of integration*: if f is a continuous function in $[a, b)$, then

$$F(x) = \int_a^x f(t)\, dt \qquad \forall x \in [a, b),$$

is a differentiable function, called a *primitive* of f, which satisfies,

$$F'(x) = f(x) \qquad \forall x \in [a, b);$$

2. the *first mean-value theorem for integrals*: if f is a continuous function in $[a, b]$ and $x_1, x_2 \in [a, b]$ with $x_1 < x_2$, then $\exists \xi \in (x_1, x_2)$ such that

$$f(\xi) = \frac{1}{x_2 - x_1} \int_{x_1}^{x_2} f(t)\, dt.$$

Even when it does exist, a primitive might be either impossible to determine or difficult to compute. For instance, knowing that $\ln |x|$ is a primitive of $1/x$ is irrelevant if one doesn't know how to efficiently compute the logarithms. In Chapter 4 we will introduce several methods to compute the integral of an arbitrary continuous function with a desired accuracy, irrespectively of the knowledge of its primitive.

We recall that a function f defined on an interval $[a, b]$ is differentiable in a point $\bar{x} \in (a, b)$ if the following limit exists and is finite

$$f'(\bar{x}) = \lim_{h \to 0} \frac{1}{h}(f(\bar{x} + h) - f(\bar{x})). \tag{1.10}$$

The value of $f'(\bar{x})$ provides the slope of the tangent line to the graph of f at the point $\bar{x}$.

We say that a function which is continuous together with its derivative at any point of $[a,b]$ belongs to the space $C^1([a,b])$. More generally, a function with continuous derivatives up to the order p (a positive integer) is said to belong to $C^p([a,b])$. In particular, $C^0([a,b])$ denotes the space of continuous functions in $[a,b]$.

A result that will be often used is the *mean-value theorem*, according to which, if $f \in C^1([a,b])$, there exists $\xi \in (a,b)$ such that

$$f'(\xi) = (f(b) - f(a))/(b-a).$$

Finally, it is worth recalling that a function that is continuous with all its derivatives up to the order n in a neighborhood of x_0, can be approximated in such a neighborhood by the so-called *Taylor polynomial of degree* n *at the point* x_0:

$$T_n(x) = f(x_0) + (x-x_0)f'(x_0) + \ldots + \frac{1}{n!}(x-x_0)^n f^{(n)}(x_0)$$
$$= \sum_{k=0}^n \frac{(x-x_0)^k}{k!} f^{(k)}(x_0).$$

diff
int
taylor

The MATLAB toolbox `symbolic` provides the commands `diff`, `int` and `taylor` which allow us to obtain the analytical expression of the derivative, the indefinite integral (i.e. a primitive) and the Taylor polynomial, respectively, of a given function. In particular, having defined in the string `f` the function on which we intend to operate, `diff(f,n)` provides its derivative of order `n`, `int(f)` its indefinite integral, and `taylor(f,x,n+1)` the associated Taylor polynomial of degree `n` in a neighborhood of $x_0 = 0$. The variable `x` must be declared *symbolic* by using the command `syms x`. This will allow its algebraic manipulation without specifying its value.

syms

In order to do this for the function $f(x) = (x^2 + 2x + 2)/(x^2 - 1)$, we proceed as follows:

```
>> f = '(x^2+2*x+2)/(x^2-1)';
>> syms x
>> diff(f)
```

(2*x+2)/(x^2-1)-2*(x^2+2*x+2)/(x^2-1)^2*x

```
>> int(f)
```

x+5/2*log(x-1)-1/2*log(1+x)

```
>> taylor(f,x,6)
```

-2-2*x-3*x^2-2*x^3-3*x^4-2*x^5

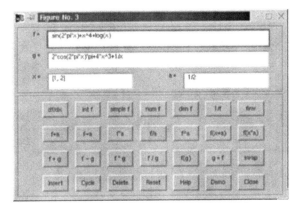

Fig. 1.5. Graphical interface of the command `funtool`

simple   We observe that using the command `simple` it is possible to simplify the expressions generated by `diff`, `int` and `taylor` in order to make
funtool   them as simple as possible. The command `funtool`, by the graphical interface illustrated in Fig. 1.5, allows a very easy symbolic manipulation of arbitrary functions.

Octave 1.6 Symbolic calculations are not yet available in Octave, although it is work in progress.[2]   ∎

See the Exercises 1.7-1.8.

1.5 To err is not only human

As a matter of fact, by re-phrasing the Latin motto *errare humanum est*, we might say that in numerical computation to err is even inevitable.

As we have seen, the simple fact of using a computer to represent real numbers introduces errors. What is therefore important is not to strive to eliminate errors, but rather to be able to control their effect.

Generally speaking, we can identify several levels of errors that occur during the approximation and resolution of a physical problem (see Figure 1.6).

At the highest level stands the error e_m which occurs when forcing the physical reality (PP stands for physical problem and x_{ph} denotes its solution) to obey some mathematical model (MP, whose solution is x). Such errors will limit the applicability of the mathematical model to certain situations and are beyond the control of Scientific Computing.

---

[2] http://www.octave.org

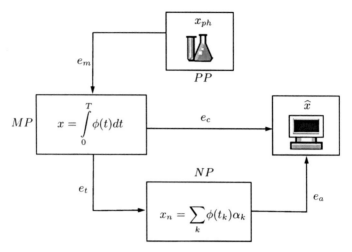

Fig. 1.6. Types of errors in a computational process

The mathematical model (whether expressed by an integral as in the example of Figure 1.6, an algebraic or differential equation, a linear or nonlinear system) is generally not solvable in explicit form. Its resolution by computer algorithms will surely involve the introduction and propagation of roundoff errors at least. Let's call these errors e_a.

On the other hand, it is often necessary to introduce further errors since any procedure of the mathematical model involving an infinite sequence of arithmetic operations cannot be performed by the computer unless approximately. For instance the computation of the sum of a series will necessarily be accomplished in an approximate way by considering a suitable truncation.

It will therefore be necessary to introduce a numerical problem, NP, whose solution x_n differs from x by an error e_t which is called *truncation error*. Such errors do not only occur in mathematical models that are already set in finite dimension (for instance, when solving a linear system). The sum of the errors e_a and e_t constitutes the *computational error* e_c, the quantity we are interested in.

The *absolute* computational error is the difference between x, the exact solution of the mathematical model, and $\widehat{x}$, the solution obtained at the end of the numerical process,

$$e_c^{abs} = |x - \widehat{x}|,$$

while (if $x \neq 0$) the *relative* computational error is

$$e_c^{rel} = |x - \widehat{x}|/|x|,$$

where $|\cdot|$ denotes the modulus, or other measure of size, depending on the meaning of x.

1.5 To err is not only human

The numerical process is generally an approximation of the mathematical model obtained as a function of a discretization parameter, which we will refer to as h and suppose positive. If, as h tends to 0, the numerical process returns the solution of the mathematical model, we will say that the numerical process is *convergent*. Moreover, if the (absolute or relative) error can be bounded as a function of h as

$$\boxed{e_c \leq Ch^p} \qquad (1.11)$$

where C is independent of h and p is a positive number, we will say that the method is *convergent of order p*. It is sometimes even possible to replace the symbol $\leq$ with $\simeq$, in the case where, besides the upper bound (1.11), a lower bound $C'h^p \leq e_c$ is also available (C' being another constant independent from h and p).

Example 1.1 Suppose we approximate the derivative of a function f at a point $\bar{x}$ with the incremental ratio that appears in (1.10). Obviously, if f is differentiable at $\bar{x}$, the error committed by replacing f' by the incremental ratio tends to 0 as $h \to 0$. However, as we will see in Section 4.1, the error can be considered as Ch only if $f \in C^2$ in a neighborhood of $\bar{x}$. ∎

While studying the convergence properties of a numerical procedure we will often deal with graphs reporting the error as a function of h in a logarithmic scale, which shows $\log(h)$ on the abscissae axis and $\log(e_c)$ on the ordinates axis. The purpose of this representation is easy to see: if $e_c = Ch^p$ then $\log e_c = \log C + p \log h$. In logarithmic scale therefore p represents the slope of the straight line $\log e_c$, so if we must compare two methods, the one presenting the greater slope will be the one with a higher order. To obtain graphs in a logarithmic scale one just needs to type `loglog(x,y)`, x and y being the vectors containing the abscissae and the ordinates of the data to be represented.

`loglog`

As an instance, in Figure 1.7 we report the straight lines relative to the behavior of the errors in two different methods. The continuous line represents a first-order approximation, while the dashed line represents a second-order one.

There is an alternative to the graphical way of establishing the order of a method when one knows the errors e_i relative to some given values h_i of the parameter of discretization, with $i = 1, \ldots, N$: it consists in supposing that e_i is equal to Ch_i^p, where C does not depend on i. One can then approach p with the values:

$$p_i = \log(e_i/e_{i-1})/\log(h_i/h_{i-1}), \qquad i = 2, \ldots, N. \qquad (1.12)$$

Actually the error is not a computable quantity since it depends on the unknown solution. Therefore it is necessary to introduce computable quantities that can be used to estimate the error itself, the so called *error estimator*. We will see some examples in Sections 2.2.1, 2.3 and 4.4.

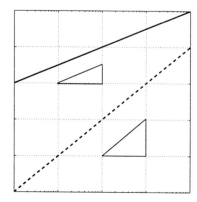

Fig. 1.7. Plot in logarithmic scales

1.5.1 Talking about costs

In general a problem is solved on the computer by an algorithm, which is a precise directive in the form of a finite text specifying the execution of a finite series of elementary operations. We are interested in those algorithms which involve only a finite number of steps.

The *computational cost* of an algorithm is the number of floating-point operations that are required for its execution. Often, the speed of a computer is measured by the maximum number of floating-point operations which the computer can execute in one second (*flops*). In particular, the following abridged notations are commonly used: Mega-flops, equal to 10^6 $flops$, Giga-flops equal to 10^9 $flops$, Tera-flops equal to 10^{12} $flops$. The fastest computers nowadays reach as many as 40 of Tera-flops.

In general, the exact knowledge of the number of operations required by a given algorithm is not essential. Rather, it is useful to determine its order of magnitude as a function of a parameter d which is related to the problem dimension. We therefore say that an algorithm has *constant* complexity if it requires a number of operations independent of d, i.e. $\mathcal{O}(1)$ operations, *linear* complexity if it requires $\mathcal{O}(d)$ operations, or, more generally, *polynomial* complexity if it requires $\mathcal{O}(d^m)$ operations, for a positive integer m. Other algorithms may have *exponential* ($\mathcal{O}(c^d)$ operations) or even *factorial* ($\mathcal{O}(d!)$ operations) complexity. We recall that the symbol $\mathcal{O}(d^m)$ means "it behaves, for large d, like a constant times d^m".

Example 1.2 (matrix-vector product) Le A be a square matrix of order n and let **v** be a vector of $\mathbb{R}^n$. The $j - th$ component of the product A**v** is given by

$$a_{j1}v_1 + a_{j2}v_2 + \ldots + a_{jn}v_n,$$

and requires n products and $n-1$ additions. One needs therefore $n(2n-1)$ operations to compute all the components. Thus this algorithm requires $\mathcal{O}(n^2)$ operations, so it has a quadratic complexity with respect to the parameter n. The same algorithm would require $\mathcal{O}(n^3)$ operations to compute the product of two matrices of order n. However, there is an algorithm, due to Strassen, which requires "only" $\mathcal{O}(n^{\log_2 7})$ operations and another, due to Winograd and Coppersmith, requiring $\mathcal{O}(n^{2.376})$ operations. ■

Example 1.3 (computation of a matrix determinant) As already mentioned, the determinant of a square matrix of order n can be computed using the recursive formula (1.8). The corresponding algorithm has a factorial complexity with respect to n and would be usable only for matrices of small dimension. For instance, if $n = 24$, a computer capable of performing as many as 1 Peta-flops (i.e. 10^{15} floating-point operations per second) would require 20 years to carry out this computation. One has therefore to resort to more efficient algorithms. Indeed, there exists an algorithm allowing the computation of determinants through matrix-matrix products, with henceforth a complexity of $\mathcal{O}(n^{\log_2 7})$ operations by applying the Strassen algorithm previously mentioned (see [BB96]). ■

The number of operations is not the sole parameter which matters in the analysis of an algorithm. Another relevant factor is represented by the time that is needed to access the computer memory (which depends on the way the algorithm has been coded). An indicator of the performance of an algorithm is therefore the CPU time (CPU stands for *central processing unit*), and can be obtained using the MATLAB command **cputime**. The total elapsed time between the *input* and *output* phases can be obtained by the command **etime**.

cputime
etime

Example 1.4 In order to compute the time needed for a matrix-vector multiplication we set up the following program:
```
>> n = 4000; step = 50; A = rand(n,n); v = rand(n); T=[];
>> sizeA = [ ]; count = 1;
>> for k = 50:step:n
   AA = A(1:k,1:k); vv = v(1:k)';
   t = cputime;  b = AA*vv; tt = cputime - t;
   T = [T, tt]; sizeA = [sizeA,k];
end
```
The instruction **a:step:b** appearing in the **for** cycle generates all numbers having the form a+step*k where k is an integer ranging from 0 to the largest value kmax for which a+step*kmax is not greater than b (in the case at hand, a=50, b=4000 and step=50). The command **rand(n,m)** defines an n×m matrix of random entries. Finally, T is the vector whose components contain the CPU time needed to carry out every single matrix-vector product, whereas cputime returns the CPU time in seconds that has been used by the MATLAB process since MATLAB started. The time necessary to execute a single program is therefore the difference between the actual CPU time and the one computed before the execution of the current program which is stored in the variable t. Figure 1.8, which is obtained by the command **plot(sizeA,T,'o')**, shows that the CPU time grows like the square of the matrix order **n**. ■

rand

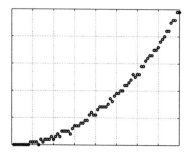

Fig. 1.8. Matrix-vector product: the CPU time (in seconds) versus the dimension n of the matrix (on a PC at 2.53 GHz)

1.6 The MATLAB and Octave environments

MATLAB and Octave, the programs, are integrated environments for scientific computing and visualization. They are written in C and C++ languages.

MATLAB is distributed by The MathWorks (see the website www.mathworks.com). The name stands for *MATrix LABoratory* since originally it was developed for matrix computation.

Octave, also known as GNU Octave (see the website www.octave.org), is a freely redistributable software. You may redistribute it and/or modify it under the terms of the GNU General Public License (GPL) as published by the Free Software Foundation.

As mentioned in the introduction of this chapter, there are differences between MATLAB and Octave environments, languages and toolboxes. However, there is a level of compatibility that allows us to write most programs of this book and run them seamlessly both in MATLAB and Octave. When this is not possible, either because some commands are spelt differently, or because they operate in a different way, or merely because they are just not implemented, a note has been and will be written at the end of each section; it provides an explanation and indicates what could be done.

Just as MATLAB has its toolboxes, Octave has a rich set of functions available through a project called Octave-forge (see the website octave.sourceforge.net). This function repository grows steadily in many different areas such as linear algebra, sparse matrices support or optimization, to name but a few. In order to run properly all programs and examples in this book under Octave, it is mandatory to install Octave-forge.

Once installed, the execution of MATLAB and Octave allow access to a working environment characterized by the *prompt* >> and octave:1>, respectively. For instance, when executing MATLAB on our personal computer we see

>>
octave:1>

```
            < M A T L A B >
        Copyright 1984-2004 The MathWorks, Inc.
              Version 7.0.0.19901 (R14)
                    May 06, 2004

   To get started, select MATLAB Help or Demos from the Help
   menu.
>>
```

When executing Octave on our personal computer we see

```
GNU Octave, version 2.1.72 (x86_64-pc-linux-gnu).
Copyright (C) 2005 John W. Eaton.
This is free software; see the source code for copying conditions.
There is ABSOLUTELY NO WARRANTY; not even for MERCHANTIBILITY or
FITNESS FOR A PARTICULAR PURPOSE.  For details, type 'warranty'.

Additional information about Octave is available at
http://www.octave.org.

Please contribute if you find this software useful.
For more information, visit http://www.octave.org/help-wanted.html

Report bugs to <bug@octave.org> (but first, please read
http://www.octave.org/bugs.html to learn how to write a helpful
report).

octave:1>
```

1.7 The MATLAB language

After the introductory remarks of the previous section, we are now ready to work in either the MATLAB or Octave environments. And from now on MATLAB should be understood as the subset of commands which are common to both MATLAB and Octave.

After pressing the *enter* key (or else *return*), all what is written after the *prompt* will be interpreted.[3] Precisely, MATLAB will first check whether what is written corresponds either to variables which have already been defined or to the name of one of the programs or commands defined in MATLAB. Should all those checks fail, MATLAB returns an error warning. Otherwise, the command is executed and an *output* will possibly be displayed. In all cases, the system eventually returns the *prompt* to acknowledge that it is ready for a new command. To close a MATLAB session one should write the command quit (or else exit)

quit
exit

---
[3] Thus a MATLAB program does not necessarily have to be compiled as other languages do, e.g. Fortran or C.

and press the *enter* key. From now it will be understood that to execute a program or a command one has to press the *enter* key. Moreover, the terms program, function or command will be used in an equivalent manner. When our command coincides with one of the elementary structures characterizing MATLAB (e.g. a number or a string of characters that are put between apices) they are immediately returned in *output* in the
ans *default* variable **ans** (abbreviation of *answer*). Here is an example:
>> 'home'

ans =
home

If we now write a different string (or number), **ans** will assume this new value.

We can turn off the automatic display of the *output* by writing a semicolon after the string. Thus if we write 'home'; MATLAB will simply return the *prompt* (yet assigning the value 'home' to the variable **ans**).

= More generally, the command **=** allows the assignment of a value (or a string of characters) to a given variable. For instance, to assign the string 'Welcome to Milan' to the variable a we can write
>> a='Welcome to Milan';

Thus there is no need to declare the *type* of a variable, MATLAB will do it automatically and dynamically. For instance, should we write a=5, the variable a will now contain a number and no longer a string of characters. This flexibility is not cost-free. If we set a variable named quit equal to the number 5 we are inhibiting the use of the MATLAB command quit. We should therefore try to avoid using variables having
clear the name of MATLAB commands. However, by the command **clear** followed by the name of a variable (e.g. quit), it is possible to cancel this assignment and restore the original meaning of the command quit.
save By the command **save** all the session variables (that are stored in the so-called *base workspace*) are saved in the binary file matlab.mat.
load Similarly, the command **load** restores in the current session all variables stored in matlab.mat. A file name can be specified after save or load. One can also save only selected variables, say v1, v2 and v3, in a given file named, e.g., area.mat, using the command save area v1 v2 v3.
help By the command **help** one can see the whole family of commands and pre-defined variables, including the so-called *toolboxes* which are sets of specialized commands. Among them let us recall those which define
sin cos the elementary functions such as sine (**sin(a)**), cosine (**cos(a)**), square
sqrt exp root (**sqrt(a)**), exponential (**exp(a)**).

There are special characters that cannot appear in the name of a
+ - variable or in a command, for instance the algebraic operators (+, -,
* / & | * and /), the logical operators *and* (&), *or* (|), *not* (˜), the relational

operators *greater than* (>), *greater than or equal to* (>=), *less than* (<), *less than or equal to* (<=), *equal to* (==). Finally, a name can never begin with a digit, a bracket or with any punctuation mark.

~ > >= <
<= ==

1.7.1 MATLAB statements

A special programming language, the MATLAB language, is also available enabling the users to write new programs. Although its knowledge is not required for understanding how to use the several programs which we will introduce throughout this book, it may provide the reader with the capability of modifying them as well as producing new ones.

The MATLAB language features standard statements, such as conditionals and loops.

The *if-elseif-else* conditional has the following general form:

```
if condition(1)
   statement(1)
elseif condition(2)
   statement(2)
   .
   .
   .
else
   statement(n)
end
```

where condition(1), condition(2), ... represent MATLAB sets of logical expressions, with values 0 or 1 (false or true) and the entire construction allows the execution of that statement corresponding to the condition taking value equal to 1. Should all conditions be false, the execution of statement(n) will take place. In fact, if the value of condition(k) is zero, the control moves on.

For instance, to compute the roots of a quadratic polynomial $ax^2 + bx + c$ one can use the following instructions (the command disp(.) simply displays what is written between brackets):

```
>> if   a    ~= 0
     sq = sqrt(b*b - 4*a*c);
     x(1) = 0.5*(-b + sq)/a;
     x(2) = 0.5*(-b - sq)/a;
   elseif  b   ~= 0
     x(1) = -c/b;
   elseif  c   ~= 0
     disp(' Impossible  equation');
   else
     disp(' The  given  equation  is  an  identity');
   end
```
(1.13)

Note that MATLAB does not execute the entire construction until the statement end is typed.

MATLAB allows two types of loops, a *for-loop* (comparable to a Fortran *do-loop* or a C *for-loop*) and a *while-loop*. A for-loop repeats the statements in the loop as the loop index takes on the values in a given row vector. For instance, to compute the first six terms of the Fibonacci sequence $f_i = f_{i-1} + f_{i-2}$, for $i \geq 3$, with $f_1 = 0$ and $f_2 = 1$, one can use the following instructions:

```
>> f(1) = 0; f(2) = 1;
>> for i = [3 4 5 6]
     f(i) = f(i-1) + f(i-2);
   end
```

Note that a semicolon can be used to separate several MATLAB instructions typed on the same line. Also, note that we can replace the second instruction by the equivalent >> for i = 3:6. The while-loop repeats as long as the given condition is true. For instance, the following set of instructions can be used as an alternative to the previous set:

```
>> f(1) = 0; f(2) = 1; k = 3;
>> while k <= 6
     f(k) = f(k-1) + f(k-2); k = k + 1;
   end
```

Other statements of perhaps less frequent use exist, such as *switch, case, otherwise*. The interested reader can have access to their meaning by the help command.

1.7.2 Programming in MATLAB

Let us now explain briefly how to write MATLAB programs. A new program must be put in a file with a given name with extension m, which is called *m-file*. They must be located in one of the directories in which MATLAB automatically searches for m-files; their list can be obtained by the command **path** (see help path to learn how to add a directory to this list). The first directory scanned by MATLAB is the current working directory.

path

It is important at this level to distinguish between *scripts* and *functions*. A script is simply a collection of MATLAB commands in an *m-file* and can be used interactively. For instance, the set of instructions (1.13) can give rise to a script (which we could name equation) by copying it in the file equation.m. To launch it, one can simply write the instruction equation after the MATLAB prompt >>. We report two examples below:

```
>> a = 1; b = 1; c = 1;
>> equation

   ans =
     -0.5000 + 0.8660i  -0.5000 - 0.8660i
```

1.7 The MATLAB language

```
>> a = 0; b = 1; c = 1;
>> equation

  ans =
      -1
```

Since we have no input/output interface, all variables used in a *script* are also the variables of the working session and are therefore cleared only upon an explicit command (clear). This is not at all satisfactory when one intends to write complex programs involving many temporary variables and comparatively fewer input and output variables, which are the only ones that can be effectively saved once the execution of the program is terminated. Much more flexible than scripts are *functions*.

A *function* is still defined in a m-file, e.g. name.m, but it has a well defined input/output interface that is introduced by the command **function**

```
function [out1,...,outn]=name(in1,...,inm)
```

where out1,...,outn are the output variables and in1,...,inm are the input variables.

The following file, called det23.m, defines a new function called det23 which computes, according to the formulae given in Section 1.3, the determinant of a matrix whose dimension could be either 2 or 3:

```
function det=det23(A)
%DET23 computes the determinant of a square matrix
% of dimension 2 or 3
[n,m]=size(A);
if n==m
  if n==2
    det = A(1,1)*A(2,2)-A(2,1)*A(1,2);
  elseif n == 3
    det = A(1,1)*det23(A([2,3],[2,3]))-...
          A(1,2)*det23(A([2,3],[1,3]))+...
          A(1,3)*det23(A([2,3],[1,2]));
  else
    disp(' Only 2x2 or 3x3 matrices ');
  end
else
  disp(' Only square matrices ');
end
return
```

Notice the use of the continuation characters ... meaning that the instruction is continuing on the next line and the character % to begin comments. The instruction A([i,j],[k,l]) allows the construction of a 2×2 matrix whose elements are the elements of the original matrix A lying at the intersections of the i-th and j-th rows with the k-th and l-th columns.

When a function is invoked, MATLAB creates a local workspace (the *function's workspace*). The commands in the function cannot refer to

function

...

%

variables from the global (interactive) workspace unless they are passed as input. In particular, variables used in a function are erased when the execution terminates, unless they are returned as output parameters.

Functions usually terminate when the end of the function is reached, however a **return** statement can be used to force an early return (upon the fulfillment of a certain condition).

For instance, in order to approximate the golden section number $\alpha = 1.6180339887\ldots$, which is the limit for $k \to \infty$ of the quotient of two consecutive Fibonacci numbers f_k/f_{k-1}, by iterating until the difference between two consecutive ratios is less than 10^{-4}, we can construct the following function:

```
function [golden,k]=fibonacci0
f(1) = 0; f(2) = 1; goldenold = 0;
kmax = 100; tol = 1.e-04;
for k = 3:kmax
   f(k) = f(k-1) + f(k-2);
   golden = f(k)/f(k-1);
   if abs(golden - goldenold) <= tol
      return
   end
   goldenold = golden;
end
return
```

Its execution is interrupted either after `kmax=100` iterations or when the absolute value of the difference between two consecutive iterates is smaller than `tol=1.e-04`. Then, we can write

`[alpha,niter]=fibonacci0`

```
alpha =
    1.61805555555556
niter =
    14
```

After 14 iterations the function has returned an approximate value which shares with α the first 5 significant digits.

The number of input and output parameters of a MATLAB function can vary. For instance, we could modify the Fibonacci function as follows:

```
function [golden,k]=fibonacci1(tol,kmax)
if nargin == 0
  kmax = 100; tol = 1.e-04; % default values
elseif nargin == 1
  kmax = 100; % default value only for kmax
end
f(1) = 0; f(2) = 1; goldenold = 0;
for k = 3:kmax
   f(k) = f(k-1) + f(k-2);
   golden = f(k)/f(k-1);
   if abs(golden - goldenold) <= tol
      return
   end
   goldenold = golden;
end
return
```

The **nargin** function counts the number of input parameters. In the new version of the **fibonacci** function we can prescribe the maximum number of inner iterations allowed (**kmax**) and a specific tolerance **tol**. When this information is missing the function must provide default values (in our case, **kmax** = 100 and **tol** = 1.e-04). A possible use of it is as follows:

nargin

```
[alpha,niter]=fibonacci1(1.e-6,200)

   alpha =
       1.61803381340013
   niter =
       19
```

Note that using a stricter tolerance we have obtained a new approximate value that shares with α as many as 8 significant digits.

The **nargin** function can be used externally to a given function to obtain the number of input parameters. Here is an example:

```
nargin('fibonacci1')

   ans =
       2
```

Remark 1.2 (inline functions) The command **inline**, whose most simple syntax reads g=inline(expr,arg1,arg2,...,argn), declares a function g which depends on the strings arg1,arg2,...,argn. The string expr contains the expression of g. For instance, g=inline('sin(r)','r') declares the function $g(r) = \sin(r)$. The shorthand command g=inline(expr) implicitly assumes that expr is a function of the default variable x. Once an inline function has been declared, it can be evaluated at any set of variables through the command feval. For instance, to evaluate g at the points z=[0 1] we can write

inline

```
>> feval('g',z);
```

We note that, contrarily to the case of the **eval** command, with **feval** the name of the variable (z) needs not coincide with the symbolic name (r) assigned by the **inline** command. •

After this quick introduction, our suggestion is to explore MATLAB using the command *help*, and get acquainted with the implementation of various algorithms by the programs described throughout this book. For instance, by typing **help for** we get not only a complete description on the command **for** but also an indication on instructions similar to **for**, such as **if**, **while**, **switch**, **break** and **end**. By invoking their *help* we can progressively improve our knowledge of MATLAB.

Octave 1.7 Generally speaking, one area with little commonalities is that of the plotting facilities of MATLAB and Octave. We checked that most plotting commands in the book are reproducible in both programs, but there are in fact many fundamental differences. By default, Octave's plotting framework is gnuplot; however the plotting command set is different and operates differently than MATLAB does. At the time of writing this section, there are other plotting libraries in Octave such as octaviz (see, the website http://octaviz.sourceforge.net/), epstk (http://www.epstk.de/) and octplot (http://octplot.sourceforge.net). The last is an attempt to reproduce MATLAB plotting commands in Octave. ■

See Exercises 1.9-1.14.

1.7.3 Examples of differences between MATLAB and Octave languages

As already mentioned, what has been written in the previous section about the MATLAB language applies to both MATLAB and Octave environments without changes. However, some differences exist for the language itself. So programs written in Octave may not run in MATLAB and viceversa. For example, Octave supports strings with single and double quotes

```
octave:1> a="Welcome to Milan"
a = Welcome to Milan
octave:2> a='Welcome to Milan'
a = Welcome to Milan
```

whereas MATLAB supports only single quotes, double quotes will result in parsing errors.

Here we provide a list of few other incompatibilities between the two languages:

- MATLAB does not allow a blank before the transpose operator. For instance, [0 1]' works in MATLAB, but [0 1] ' does not. Octave properly parses both cases;
- MATLAB always requires ...,
  ```
  rand (1, ...
        2)
  ```
  while both
  ```
  rand (1,
        2)
  ```
  and
  ```
  rand (1, \
        2)
  ```
  work in Octave in addition to ...;

- for exponentiation, Octave can use ^ or **; MATLAB requires ^;
- for ends, Octave can use end but also endif, endfor, ...; MATLAB requires end.

1.8 What we haven't told you

A systematic discussion on floating-point numbers can be found in [Übe97], [Hig02] and in [QSS06].

For matters concerning the issue of complexity, we refer, e.g., to [Pan92].

For a more systematic introduction to MATLAB the interested reader can refer to the MATLAB manual [HH05] as well as to specific books such as [HLR01], [Pra02], [EKM05], [Pal04] or [MH03].

For Octave we recommend the manual book mentioned at the beginning of this chapter.

1.9 Exercises

Exercise 1.1 How many numbers belong to the set $\mathbb{F}(2, 2, -2, 2)$? What is the value of ϵ_M for such set?

Exercise 1.2 Show that the set $\mathbb{F}(\beta, t, L, U)$ contains precisely $2(\beta-1)\beta^{t-1}(U-L+1)$ elements.

Exercise 1.3 Prove that i^i is a real number, then check this result using MATLAB.

Exercise 1.4 Write the MATLAB instructions to build an upper (respectively, lower) triangular matrix of dimension 10 having 2 on the main diagonal and -3 on the upper (respectively, lower) diagonal.

Exercise 1.5 Write the MATLAB instructions which allow the interchange of the third and seventh row of the matrices built up in Exercise 1.3, and then the instructions allowing the interchange between the fourth and eighth column.

Exercise 1.6 Verify whether the following vectors in $\mathbb{R}^4$ are linearly independent:
$$\mathbf{v}_1 = [0\ 1\ 0\ 1],\ \mathbf{v}_2 = [1\ 2\ 3\ 4],\ \mathbf{v}_3 = [1\ 0\ 1\ 0],\ \mathbf{v}_4 = [0\ 0\ 1\ 1].$$

Exercise 1.7 Write the following functions and compute their first and second derivatives, as well as their primitives, using the symbolic toolbox of MATLAB:
$$f(x) = \sqrt{x^2 + 1}, \qquad g(x) = \sin(x^3) + \cosh(x).$$

Exercise 1.8 For any given vector v of dimension n, using the command poly c=poly(v) one can construct the $n+1$ coefficients of the polynomial $p(x) = \sum_{k=1}^{n+1} c(k) x^{n+1-k}$ which is equal to $\Pi_{k=1}^{n}(x - v(k))$. In exact arithmetics, one should find that v = roots(poly(c)). However, this cannot occur due to roundoff errors, as one can check by using the command roots(poly([1:n])), where n ranges from 2 to 25.

Exercise 1.9 Write a program to compute the following sequence:

$$I_0 = \frac{1}{e}(e-1),$$

$$I_{n+1} = 1 - (n+1)I_n, \text{ for } n = 0, 1, \ldots.$$

Compare the numerical result with the exact limit $I_n \to 0$ for $n \to \infty$.

Exercise 1.10 Explain the behavior of the sequence (1.4) when computed in MATLAB.

Exercise 1.11 Consider the following algorithm to compute π. Generate n couples $\{(x_k, y_k)\}$ of random numbers in the interval $[0, 1]$, then compute the number m of those lying inside the first quarter of the unit circle. Obviously, π turns out to be the limit of the sequence $\pi_n = 4m/n$. Write a MATLAB program to compute this sequence and check the error for increasing values of n.

Exercise 1.12 Since π is the sum of the series

$$\pi = \sum_{m=0}^{\infty} 16^{-m} \left(\frac{4}{8m+1} - \frac{2}{8m+4} + \frac{1}{8m+5} + \frac{1}{8m+6} \right)$$

we can compute an approximation of π by summing up to the n-th term, for a sufficiently large n. Write a MATLAB *function* to compute finite sums of the above series. How large should n be in order to obtain an approximation of π at least as accurate as the one stored in the variable π?

Exercise 1.13 Write a program for the computation of the binomial coefficient $\binom{n}{k} = n!/(k!(n-k)!)$, where n and k are two natural numbers with $k \leq n$.

Exercise 1.14 Write a recursive MATLAB *function* that computes the n-th element f_n of the Fibonacci sequence. Noting that

$$\begin{bmatrix} f_i \\ f_{i-1} \end{bmatrix} = \begin{bmatrix} 1 & 1 \\ 1 & 0 \end{bmatrix} \begin{bmatrix} f_{i-1} \\ f_{i-2} \end{bmatrix} \quad (1.14)$$

write another *function* that computes f_n based on this new recursive form. Finally, compute the related CPU-time.

2
Nonlinear equations

Computing the *zeros* of a real function f (equivalently, the *roots* of the equation $f(x) = 0$) is a problem that we encounter quite often in scientific computing. In general, this task cannot be accomplished in a finite number of operations. For instance, we have already seen in Section 1.4.1 that when f is a generic polynomial of degree greater than four, there do not exist explicit formulae for the zeros. The situation is even more difficult when f is not a polynomial.

Iterative methods are therefore adopted. Starting from one or several initial data, the methods build up a sequence of values $x^{(k)}$ that hopefully will converge to a zero α of the function f at hand.

Problem 2.1 (Investment fund) At the beginning of every year a bank customer deposits v euros in an investment fund and withdraws, at the end of the n-th year, a capital of M euros. We want to compute the average yearly rate of interest r of this investment. Since M is related to r by the relation

$$M = v\sum_{k=1}^{n}(1+r)^k = v\frac{1+r}{r}\left[(1+r)^n - 1\right],$$

we deduce that r is the root of the algebraic equation:

$$f(r) = 0, \quad \text{where } f(r) = M - v\frac{1+r}{r}[(1+r)^n - 1].$$

This problem will be solved in Example 2.1. ∎

Problem 2.2 (State equation of a gas) We want to determine the volume V occupied by a gas at temperature T and pressure p. The state equation (i.e. the equation that relates p, V and T) is

$$\left[p + a(N/V)^2\right](V - Nb) = kNT, \qquad (2.1)$$

where a and b are two coefficients that depend on the specific gas, N is the number of molecules which are contained in the volume V and k is the Boltzmann constant. We need therefore to solve a nonlinear equation whose root is V (see Exercise 2.2). ∎

Problem 2.3 (Rods system) Let us consider the mechanical system represented by the four rigid rods $\mathbf{a}_i$ of Figure 2.1. For any admissible value of the angle β, let us determine the value of the corresponding angle α between the rods $\mathbf{a}_1$ and $\mathbf{a}_2$. Starting from the vector identity

$$\mathbf{a}_1 - \mathbf{a}_2 - \mathbf{a}_3 - \mathbf{a}_4 = \mathbf{0}$$

and noting that the rod $\mathbf{a}_1$ is always aligned with the x-axis, we can deduce the following relationship between β and α:

$$\frac{a_1}{a_2}\cos(\beta) - \frac{a_1}{a_4}\cos(\alpha) - \cos(\beta - \alpha) = -\frac{a_1^2 + a_2^2 - a_3^2 + a_4^2}{2a_2 a_4}, \quad (2.2)$$

where a_i is the known length of the i-th rod. This is called the Freudenstein equation, and we can rewrite it as $f(\alpha) = 0$, where

$$f(x) = (a_1/a_2)\cos(\beta) - (a_1/a_4)\cos(x) - \cos(\beta - x) + \frac{a_1^2 + a_2^2 - a_3^2 + a_4^2}{2a_2 a_4}.$$

A solution in explicit form is available only for special values of β. We would also like to mention that a solution does not exist for all values of β, and may not even be unique. To solve the equation for any given β lying between 0 and π we should invoke numerical methods (see Exercise 2.9). ∎

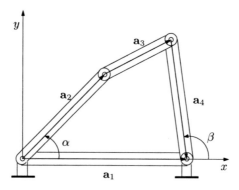

Fig. 2.1. System of four rods of Problem 2.3

Problem 2.4 (Population dynamics) In the study of populations (e.g. bacteria), the equation $x^+ = \phi(x) = xR(x)$ establishes a link between the number of individuals in a generation x and the number of individuals in the following generation. Function $R(x)$ models the variation rate of the considered population and can be chosen in different ways. Among the most known, we can mention:

1. Malthus's model (Thomas Malthus, 1766-1834),

$$R(x) = R_M(x) = r, \qquad r > 0;$$

2. the growth with limited resources model (by Pierre Francois Verhulst, 1804-1849),

$$R(x) = R_V(x) = \frac{r}{1 + xK}, \qquad r > 0, K > 0, \qquad (2.3)$$

which improves on Malthus's model in considering that the growth of a population is limited by the available resources;

3. the predator/prey model with saturation,

$$R(x) = R_P = \frac{rx}{1 + (x/K)^2}, \qquad (2.4)$$

which represents the evolution of Verhulst's model in the presence of an antagonist population.

The dynamics of a population is therefore defined by the iterative process

$$x^{(k)} = \phi(x^{(k-1)}), \qquad k \geq 1, \qquad (2.5)$$

where $x^{(k)}$ represents the number of individuals present k generations later than the initial generation $x^{(0)}$. Moreover, the stationary (or equilibrium) states x^* of the considered population are the solutions of problem

$$x^* = \phi(x^*),$$

or, equivalently, $x^* = x^* R(x^*)$ i.e. $R(x^*) = 1$. Equation (2.5) is an instance of a fixed point method (see Section 2.3). ∎

2.1 The bisection method

Let f be a continuous function in $[a, b]$ which satisfies $f(a)f(b) < 0$. Then necessarily f has at least one zero in (a, b). Let us assume for simplicity that it is unique, and let us call it α.

(In the case of several zeros, by the help of the command fplot we can locate an interval which contains only one of them.)

2 Nonlinear equations

The strategy of the bisection method is to halve the given interval and select that subinterval where f features a sign change. More precisely, having named $I^{(0)} = (a,b)$ and, more generally, $I^{(k)}$ the sub-interval selected at step k, we choose as $I^{(k+1)}$ the sub-interval of $I^{(k)}$ at whose end-points f features a sign change. Following such procedure, it is guaranteed that every $I^{(k)}$ selected this way will contain α. The sequence $\{x^{(k)}\}$ of the midpoints of these subintervals $I^{(k)}$ will inevitably tend to α since the length of the subintervals tends to zero as k tends to infinity.

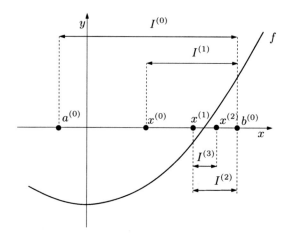

Fig. 2.2. A few iterations of the bisection method

Precisely, the method is started by setting

$$a^{(0)} = a,\ b^{(0)} = b,\ I^{(0)} = (a^{(0)}, b^{(0)}),\ x^{(0)} = (a^{(0)} + b^{(0)})/2.$$

At each step $k \geq 1$ we select the subinterval $I^{(k)} = (a^{(k)}, b^{(k)})$ of the interval $I^{(k-1)} = (a^{(k-1)}, b^{(k-1)})$ as follows:

given $x^{(k-1)} = (a^{(k-1)} + b^{(k-1)})/2$, if $f(x^{(k-1)}) = 0$ then $\alpha = x^{(k-1)}$ and the method terminates;

otherwise,

if $f(a^{(k-1)})f(x^{(k-1)}) < 0$ set $a^{(k)} = a^{(k-1)}$, $b^{(k)} = x^{(k-1)}$;

if $f(x^{(k-1)})f(b^{(k-1)}) < 0$ set $a^{(k)} = x^{(k-1)}$, $b^{(k)} = b^{(k-1)}$.

Then we define $x^{(k)} = (a^{(k)} + b^{(k)})/2$ and increase k by 1.

For instance, in the case represented in Figure 2.2, which corresponds to the choice $f(x) = x^2 - 1$, by taking $a^{(0)} = -0.25$ and $b^{(0)} = 1.25$, we would obtain

$$I^{(0)} = (-0.25, 1.25), \quad x^{(0)} = 0.5,$$
$$I^{(1)} = (0.5, 1.25), \quad x^{(1)} = 0.875,$$
$$I^{(2)} = (0.875, 1.25), \quad x^{(2)} = 1.0625,$$
$$I^{(3)} = (0.875, 1.0625), \quad x^{(3)} = 0.96875.$$

Notice that each subinterval $I^{(k)}$ contains the zero α. Moreover, the sequence $\{x^{(k)}\}$ necessarily converges to α since at each step the length $|I^{(k)}| = b^{(k)} - a^{(k)}$ of $I^{(k)}$ halves. Since $|I^{(k)}| = (1/2)^k |I^{(0)}|$, the error at step k satisfies

$$|e^{(k)}| = |x^{(k)} - \alpha| < \frac{1}{2}|I^{(k)}| = \left(\frac{1}{2}\right)^{k+1}(b-a).$$

In order to guarantee that $|e^{(k)}| < \varepsilon$, for a given tolerance ε it suffices to carry out k_{min} iterations, k_{min} being the smallest integer satisfying the inequality

$$k_{min} > \log_2\left(\frac{b-a}{\varepsilon}\right) - 1 \qquad (2.6)$$

Obviously, this inequality makes sense in general, and is not confined to the specific choice of f that we have made previously.

The bisection method is implemented in Program 2.1: `fun` is a *function* (or an inline function) specifying the function f, `a` and `b` are the endpoints of the search interval, `tol` is the tolerance ε and `nmax` is the maximum number of allotted iterations. Besides the first argument which represents the independent variable, the function `fun` can accept other auxiliary parameters.

Output parameters are `zero`, which contains the approximate value of α, the residual `res` which is the value of f in `zero` and `niter` which is the total number of iterations that are carried out. The command `find(fx==0)` finds those indices of the vector `fx` corresponding to null components.

find

Program 2.1. bisection: bisection method

```
function [zero,res,niter]=bisection(fun,a,b,tol,...
                            nmax,varargin)
%BISECTION Find function zeros.
% ZERO=BISECTION(FUN,A,B,TOL,NMAX) tries to find a zero
% ZERO of the continuous function FUN in the interval
% [A,B] using the bisection method. FUN accepts real
% scalar input x and returns a real scalar value. If
% the search fails an errore message is displayed. FUN
% can also be an inline object.
% ZERO=BISECTION(FUN,A,B,TOL,NMAX,P1,P2,...) passes
% parameters P1,P2,... to the function FUN(X,P1,P2,...).
% [ZERO,RES,NITER]=BISECTION(FUN,...) returns the value
% of the residual in ZERO and the iteration number at
```

```
% which ZERO was computed.
x = [a, (a+b)*0.5, b]; fx = feval(fun,x,varargin{:});
if fx(1)*fx(3) > 0
    error([' The sign of the function at the ',...
        'endpoints of the interval must be different']);
elseif fx(1) == 0
    zero = a; res = 0;   niter = 0;  return
elseif fx(3) == 0
    zero = b; res = 0;   niter = 0;  return
end
niter = 0;
I = (b - a)*0.5;
while I >= tol & niter <= nmax
 niter = niter + 1;
 if fx(1)*fx(2) < 0
    x(3) = x(2);    x(2) = x(1)+(x(3)-x(1))*0.5;
    fx = feval(fun,x,varargin{:}); I = (x(3)-x(1))*0.5;
 elseif fx(2)*fx(3) < 0
    x(1) = x(2);    x(2) = x(1)+(x(3)-x(1))*0.5;
    fx = feval(fun,x,varargin{:}); I = (x(3)-x(1))*0.5;
 else
    x(2) = x(find(fx==0)); I = 0;
 end
end
if niter > nmax
  fprintf(['bisection stopped without converging ',...
        'to the desired tolerance because the ',...
        'maximum number of iterations was ',...
        'reached\n']);
end
zero = x(2); x = x(2); res = feval(fun,x,varargin{:});
return
```

Example 2.1 (Investment fund) Let us apply the bisection method to solve Problem 2.1, assuming that v is equal to 1000 euros and that after 5 years M is equal to 6000 euros. The graph of the function f can be obtained by the following instructions

```
f=inline('M-v*(1+r).*((1+r).^5 - 1)./r','r','M','v');
plot([0.01,0.3],feval(f,[0.01,0.3],6000,1000));
```

We see that f has a unique zero in the interval $(0.01, 0.1)$, which is approximately equal to 0.06. If we execute Program 2.1 with tol= 10^{-12}, a= 0.01 and b= 0.1 as follows

```
[zero,res,niter]=bisection(f,0.01,0.1,1.e-12,1000,...
        6000,1000);
```

after 36 iterations the method converges to the value 0.06140241153618, in perfect agreement with the estimate (2.6) according to which $k_{min} = 36$. Thus, we conclude that the interest rate r is approximately equal to 6.14%. ∎

In spite of its simplicity, the bisection method does not guarantee a monotone reduction of the error, but simply that the search interval is halved from one iteration to the next. Consequently, if the only stopping criterion adopted is the control of the length of $I^{(k)}$, one might discard approximations of α which are quite accurate.

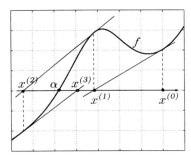

Fig. 2.3. The first iterations generated by the Newton method with initial guess $x^{(0)}$ for the function $f(x) = x + e^x + 10/(1+x^2) - 5$

As a matter of fact, this method does not take into proper account the actual behavior of f. A striking fact is that it does not converge in a single iteration even if f is a linear function (unless the zero α is the midpoint of the initial search interval).

See Exercises 2.1-2.5.

2.2 The Newton method

The sign of the given function f at the endpoints of the subintervals is the only information exploited by the bisection method. A more efficient method can be constructed by exploiting the values attained by f and its derivative (in the case that f is differentiable). In that case,

$$y(x) = f(x^{(k)}) + f'(x^{(k)})(x - x^{(k)})$$

provides the equation of the tangent to the curve $(x, f(x))$ at the point $x^{(k)}$.

If we pretend that $x^{(k+1)}$ is such that $y(x^{(k+1)}) = 0$, we obtain:

$$\boxed{x^{(k+1)} = x^{(k)} - \frac{f(x^{(k)})}{f'(x^{(k)})}, \quad k \geq 0} \qquad (2.7)$$

provided $f'(x^{(k)}) \neq 0$. This formula allows us to compute a sequence of values $x^{(k)}$ starting from an initial guess $x^{(0)}$. This method is known as Newton's method and corresponds to computing the zero of f by locally replacing f by its tangent line (see Figure 2.3).

As a matter of fact, by developing f in Taylor series in a neighborhood of a generic point $x^{(k)}$ we find

$$f(x^{(k+1)}) = f(x^{(k)}) + \delta^{(k)} f'(x^{(k)}) + \mathcal{O}((\delta^{(k)})^2), \qquad (2.8)$$

where $\delta^{(k)} = x^{(k+1)} - x^{(k)}$. Forcing $f(x^{(k+1)})$ to be zero and neglecting the term $\mathcal{O}((\delta^{(k)})^2)$, we can obtain $x^{(k+1)}$ as a function of $x^{(k)}$ as stated in (2.7). In this respect (2.7) can be regarded as an approximation of (2.8).

Obviously, (2.7) converges in a single step when f is linear, that is when $f(x) = a_1 x + a_0$.

Example 2.2 Let us solve Problem 2.1 by Newton's method, taking as initial data $x^{(0)} = 0.3$. After 6 iterations the difference between two subsequent iterates is less than or equal to 10^{-12}. ∎

The Newton method in general does not converge for all possible choices of $x^{(0)}$, but only for those values of $x^{(0)}$ which are *sufficiently close* to α. At first glance, this requirement looks meaningless: indeed, in order to compute α (which is unknown), one should start from a value sufficiently close to α!

In practice, a possible initial value $x^{(0)}$ can be obtained by resorting to a few iterations of the bisection method or, alternatively, through an investigation of the graph of f. If $x^{(0)}$ is properly chosen and α is a simple zero (that is, $f'(\alpha) \neq 0$) then the Newton method converges. Furthermore, in the special case where f is continuously differentiable up to its second derivative one has the following convergence result (see Exercise 2.8),

$$\lim_{k \to \infty} \frac{x^{(k+1)} - \alpha}{(x^{(k)} - \alpha)^2} = \frac{f''(\alpha)}{2f'(\alpha)} \qquad (2.9)$$

Consequently, if $f'(\alpha) \neq 0$ Newton's method is said to converge *quadratically*, or with order 2, since for sufficiently large values of k the error at step $(k+1)$ behaves like the square of the error at step k multiplied by a constant which is independent of k.

In the case of zeros with multiplicity m larger than 1, the order of convergence of Newton's method downgrades to 1 (see Exercise 2.15). In such case one could recover the order 2 by modifying the original method (2.7) as follows:

$$x^{(k+1)} = x^{(k)} - m\frac{f(x^{(k)})}{f'(x^{(k)})}, \qquad k \geq 0 \qquad (2.10)$$

provided that $f'(x^{(k)}) \neq 0$. Obviously, this requires the *a-priori* knowledge of m. If this is not the case, one could develop an *adaptive Newton method*, still of order 2, as described in [QSS06, Section 6.6.2].

Example 2.3 The function $f(x) = (x-1)\log(x)$ has a single zero $\alpha = 1$ of multiplicity $m = 2$. Let us compute it by both Newton's method (2.7) and by its modified version (2.10). In Figure 2.4 we report the error obtained using the

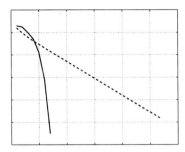

Fig. 2.4. Error versus iteration number for the function of Example 2.3. The dashed line corresponds to Newton's method (2.7), solid line to the modified Newton's method (2.10) (with $m = 2$)

two methods versus the iteration number. Note that for the classical version of Newton's method the convergence is only linear. ∎

2.2.1 How to terminate Newton's iterations

In theory, a convergent Newton's method returns the zero α only after an infinite number of iterations. In practice, one requires an approximation of α up to a prescribed tolerance ε. Thus the iterations can be terminated at the smallest value of k_{min} for which the following inequality holds:

$$|e^{(k_{min})}| = |\alpha - x^{(k_{min})}| < \varepsilon.$$

This is a test on the error. Unfortunately, since the error is unknown, one needs to adopt in its place a suitable *error estimator*, that is, a quantity that can be easily computed and through which we can estimate the real error. At the end of Section 2.3, we will see that a suitable error estimator for Newton's method is provided by the difference between two successive iterates. This means that one terminates the iterations at step k_{min} as soon as

$$\boxed{|x^{(k_{min})} - x^{(k_{min}-1)}| < \varepsilon} \qquad (2.11)$$

This is a test on the increment.

We will see in Section 2.3.1 that the test on the increment is satisfactory when α is a simple zero of f. Alternatively, one could use a test on the *residual* at step k, $r^{(k)} = f(x^{(k)})$ (note that the residual is null when $x^{(k)}$ is a zero of the function f).

Precisely, we could stop the iteration at the first k_{min} for which

$$\boxed{|r^{(k_{min})}| = |f(x^{(k_{min})})| < \varepsilon} \qquad (2.12)$$

The test on the residual is satisfactory only when $|f'(x)| \simeq 1$ in a neighborhood I_α of the zero α (see Figure 2.5). Otherwise, it will produce an over estimation of the error if $|f'(x)| \gg 1$ for $x \in I_\alpha$ and an under estimation if $|f'(x)| \ll 1$ (see also Exercise 2.6).

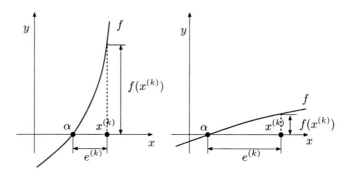

Fig. 2.5. Two situations in which the residual is a poor error estimator: $|f'(x)| \gg 1$ (*left*), $|f'(x)| \ll 1$ (*right*), with x belonging to a neighborhood of α

In Program 2.2 we implement Newton's method (2.7). Its modified form can be obtained simply by replacing f' with f'/m. The input parameters fun and dfun are the strings which define function f and its first derivative, while x0 is the initial guess. The method will be terminated when the absolute value of the difference between two subsequent iterates is less than the prescribed tolerance tol, or when the maximum number of iterations nmax has been reached.

Program 2.2. newton: Newton method

```
function [zero,res,niter]=newton(fun,dfun,x0,tol,...
                        nmax,varargin)
%NEWTON Find function zeros.
%   ZERO=NEWTON(FUN,DFUN,X0,TOL,NMAX) tries to find the
%   zero ZERO of the continuous and differentiable
%   function FUN nearest to X0 using the Newton method.
%   FUN and its derivative DFUN accept real scalar input
%   x and returns a real scalar value. If the search fails
%   an errore message is displayed. FUN and DFUN can also
%   be inline objects.
%   ZERO=NEWTON(FUN,DFUN,X0,TOL,NMAX,P1,P2,...) passes
%   parameters P1,P2,... to functions: FUN(X,P1,P2,...)
%   and DFUN(X,P1,P2,...).
%   [ZERO,RES,NITER]=NEWTON(FUN,...) returns the value of
%   the residual in ZERO and the iteration number at which
%   ZERO was computed.
x = x0;
fx = feval(fun,x,varargin{:});
dfx = feval(dfun,x,varargin{:});
niter = 0; diff = tol+1;
while diff >= tol & niter <= nmax
```

```
       niter = niter + 1;      diff = - fx/dfx;
       x = x + diff;           diff = abs(diff);
       fx  = feval(fun ,x,varargin{:});
       dfx = feval(dfun,x,varargin{:});
end
if niter > nmax
      fprintf(['newton stopped without converging to ',...
        'the desired tolerance because the maximum ',...
        'number of iterations was reached\n']);
end
zero = x; res = fx;
return
```

2.2.2 The Newton method for systems of nonlinear equations

Let us consider a system of nonlinear equations of the form

$$\begin{cases} f_1(x_1, x_2, \ldots, x_n) = 0, \\ f_2(x_1, x_2, \ldots, x_n) = 0, \\ \vdots \\ f_n(x_1, x_2, \ldots, x_n) = 0, \end{cases} \quad (2.13)$$

where $f_1, \ldots, f_n$ are nonlinear functions. Setting $\mathbf{f} = (f_1, \ldots, f_n)^T$ and $\mathbf{x} = (x_1, \ldots, x_n)^T$, system (2.13) can be written in a compact way as

$$\mathbf{f}(\mathbf{x}) = \mathbf{0}. \quad (2.14)$$

An example is given by the following nonlinear system

$$\begin{cases} f_1(x_1, x_2) = x_1^2 + x_2^2 = 1, \\ f_2(x_1, x_2) = \sin(\pi x_1/2) + x_2^3 = 0. \end{cases} \quad (2.15)$$

In order to extend Newton's method to the case of a system, we replace the first derivative of the scalar function f with the *Jacobian matrix* $J_\mathbf{f}$ of the vectorial function $\mathbf{f}$ whose components are

$$(J_\mathbf{f})_{ij} = \frac{\partial f_i}{\partial x_j}, \quad i, j = 1, \ldots, n.$$

The symbol $\partial f_i/\partial x_j$ represents the partial derivative of f_i with respect to x_j (see definition 8.3). With this notation, Newton's method for (2.14) then becomes: given $\mathbf{x}^{(0)} \in \mathbb{R}^n$, for $k = 0, 1, \ldots$, until convergence

$$\boxed{\begin{array}{l} \text{solve } J_\mathbf{f}(\mathbf{x}^{(k)})\delta\mathbf{x}^{(k)} = -\mathbf{f}(\mathbf{x}^{(k)}) \\ \text{set} \quad \mathbf{x}^{(k+1)} = \mathbf{x}^{(k)} + \delta\mathbf{x}^{(k)} \end{array}} \quad (2.16)$$

Therefore, Newton's method applied to a system requires at each step the solution of a linear system with matrix $J_\mathbf{f}(\mathbf{x}^{(k)})$.

Program 2.3 implements this method by using the MATLAB command \ (see Section 5.6) to solve the linear system with the jacobian matrix. In input we must define a column vector x0 representing the initial datum and two *functions*, Ffun and Jfun, which compute (respectively) the column vector F containing the evaluations of $\mathbf{f}$ for a generic vector x and the jacobian matrix J, also evaluated for a generic vector x. The method stops when the difference between two consecutive iterates has an euclidean norm smaller than tol or when nmax, the maximal number of allowed iterations, has been reached.

Program 2.3. newtonsys: Newton method for nonlinear systems

```
function [x,F,iter] = newtonsys(Ffun,Jfun,x0,tol,...
                                nmax, varargin)
%NEWTONSYS find a zero of a nonlinear system
% [ZERO,F,ITER]=NEWTONSYS(FFUN,JFUN,X0,TOL,NMAX)
% tries to find the vector ZERO, zero of a nonlinear
% system defined in FFUN with jacobian matrix defined
% in the function JFUN, nearest to the vector X0.
iter = 0; err = tol + 1; x = x0;
while err > tol & iter <= nmax
    J = feval(Jfun,x,varargin{:});
    F = feval(Ffun,x,varargin{:});
    delta = - J\F;
    x = x + delta;
    err = norm(delta);
    iter = iter + 1;
end
F = norm(feval(Ffun,x,varargin{:}));
if iter >= nmax
 fprintf(' Fails to converge within maximum ',...
         'number of iterations\n ');
 fprintf(' The iterate returned has relative ',...
         'residual %e\n',F);
else
 fprintf(' The method converged at iteration ',...
         '%i with a residual %e\n',iter,F);
end
return
```

Example 2.4 Let us consider the nonlinear system (2.15) which allows the two (graphically detectable) solutions $(0.4761, -0.8794)$ and $(-0.4761, 0.8794)$ (where we only report the four first significant digits). In order to use Program 2.3 we define the following *functions*

```
function J=Jfun(x)
pi2 = 0.5*pi;
J(1,1) = 2*x(1);
J(1,2) = 2*x(2);
J(2,1) = pi2*cos(pi2*x(1));
J(2,2) = 3*x(2)^2;
return
```

2.3 Fixed point iterations

```
function F=Ffun(x)
F(1,1) = x(1)^2 + x(2)^2 - 1;
F(2,1) = sin(pi*x(1)/2) + x(2)^3;
return
```

Starting from an initial datum of x0=[1;1] Newton's method, launched with the command

```
x0=[1;1]; tol=1e-5; maxiter=10;
[x,F,iter] = newtonsys(@Ffun,@Jfun,x0,tol,maxiter);
```

converges in 8 iterations to the values

    4.760958225338114e-01
    -8.793934089897496e-01

(The special character @ tells newtonsys that Ffun and Jfun are *functions*.)

Notice that the method converges to the other root starting from x0=[-1,-1]. In general, exactly as in the case of scalar functions, convergence of Newton's method will actually depend on the choice of the initial datum $\mathbf{x}^{(0)}$ and in particular we should guarantee that $\det(J_f(\mathbf{x}^{(0)})) \neq 0$. ∎

Let us summarize

1. Methods for the computation of the zeros of a function f are usually of iterative type;
2. the bisection method computes a zero of a function f by generating a sequence of intervals whose length is halved at each iteration. This method is convergent provided that f is continuous in the initial interval and has opposite signs at the endpoints of this interval;
3. Newton's method computes a zero α of f by taking into account the values of f and of its derivative. A necessary condition for convergence is that the initial datum belongs to a suitable (sufficiently small) neighborhood of α;
4. Newton's method is quadratically convergent only when α is a simple zero of f, otherwise convergence is linear;
5. the Newton method can be extended to the case of a nonlinear system of equations.

See Exercises 2.6-2.14.

2.3 Fixed point iterations

Playing with a pocket calculator, one may verify that by applying repeatedly the cosine key to the real value 1, one gets the following sequence of real numbers:

$$x^{(1)} = \cos(1) = 0.54030230586814,$$
$$x^{(2)} = \cos(x^{(1)}) = 0.85755321584639,$$
$$\vdots$$
$$x^{(10)} = \cos(x^{(9)}) = 0.74423735490056,$$
$$\vdots$$
$$x^{(20)} = \cos(x^{(19)}) = 0.73918439977149,$$

which should tend to the value $\alpha = 0.73908513\ldots$. Since, by construction, $x^{(k+1)} = \cos(x^{(k)})$ for $k = 0, 1, \ldots$ (with $x^{(0)} = 1$), the limit α satisfies the equation $\cos(\alpha) = \alpha$. For this reason α is called a fixed point of the cosine function. We may wonder how such iterations could be exploited in order to compute the zeros of a given function. In the previous example, α is not only a fixed point for the cosine function, but also a zero of the function $f(x) = x - \cos(x)$, hence the previously proposed method can be regarded as a method to compute the zeros of f. On the other hand, not every function has fixed points. For instance, by repeating the previous experiment using the exponential function and $x^{(0)} = 1$ one encounters a situation of overflow after 4 steps only (see Figure 2.6).

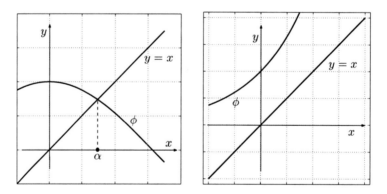

Fig. 2.6. The function $\phi(x) = \cos x$ admits one and only one fixed point (*left*), whereas the function $\phi(x) = e^x$ does not have any (*right*)

Let us clarify the intuitive idea above by considering the following problem. Given a function $\phi : [a, b] \to \mathbb{R}$, find $\alpha \in [a, b]$ such that

$$\alpha = \phi(\alpha).$$

If such an α exists it will be called a *fixed point* of ϕ and it could be computed by the following algorithm:

$$\boxed{x^{(k+1)} = \phi(x^{(k)}), \quad k \geq 0} \qquad (2.17)$$

where $x^{(0)}$ is an initial guess. This algorithm is called *fixed point iterations* and ϕ is said to be the *iteration function*. The introductory example is therefore an instance of fixed point iterations with $\phi(x) = \cos(x)$.

A geometrical interpretation of (2.17) is provided in Figure 2.7 (*left*). One can guess that if ϕ is a continuous function and the limit of the sequence $\{x^{(k)}\}$ exists, then such limit is a fixed point of ϕ. We will make this result more precise in Propositions 2.1 and 2.2.

Example 2.5 The Newton method (2.7) can be regarded as an algorithm of fixed point iterations whose iteration function is

$$\phi(x) = x - \frac{f(x)}{f'(x)}. \qquad (2.18)$$

From now on this function will be denoted by ϕ_N (where N stands for Newton). This is not the case for the bisection method since the generic iterate $x^{(k+1)}$ depends not only on $x^{(k)}$ but also on $x^{(k-1)}$. ∎

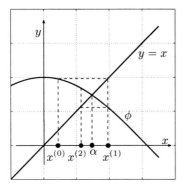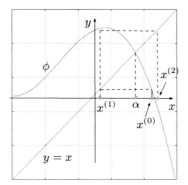

Fig. 2.7. Representation of a few fixed point iterations for two different iteration functions. To the left, the iterations converge to the fixed point α, whereas the iterations on the right produce a divergence sequence

As shown in Figure 2.7 (*right*), fixed point iterations may not converge. Indeed, the following result holds.

Proposition 2.1 *Assume that the iteration function in (2.17) satisfies the following properties:*

1. $\phi(x) \in [a, b]$ *for all* $x \in [a, b]$;
2. ϕ *is differentiable in* $[a, b]$;
3. $\exists K < 1$ *such that* $|\phi'(x)| \leq K$ *for all* $x \in [a, b]$.

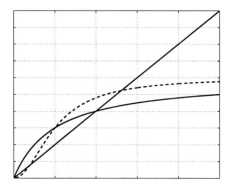

Fig. 2.8. Two fixed points for two different population dynamics: Verhulst's model (*solid line*) and predator/prey model (*dashed line*)

> Then ϕ has a unique fixed point $\alpha \in [a, b]$ and the sequence defined in (2.17) converges to α, whatever choice is made for the initial datum $x^{(0)}$ in $[a, b]$. Moreover
>
> $$\lim_{k \to \infty} \frac{x^{(k+1)} - \alpha}{x^{(k)} - \alpha} = \phi'(\alpha) \qquad (2.19)$$

From (2.19) one deduces that the fixed point iterations converge at least linearly, that is, for k sufficiently large the error at step $k+1$ behaves like the error at step k multiplied by a constant $\phi'(\alpha)$ which is independent of k and whose absolute value is strictly less than 1.

Example 2.6 The function $\phi(x) = \cos(x)$ satisfies all the assumptions of Proposition 2.1. Indeed, $|\phi'(\alpha)| = |\sin(\alpha)| \simeq 0.67 < 1$, and thus by continuity there exists a neighborhood I_α of α such that $|\phi'(x)| < 1$ for all $x \in I_\alpha$. The function $\phi(x) = x^2 - 1$ has two fixed points $\alpha_\pm = (1 \pm \sqrt{5})/2$, however it does not satisfy the assumption for either since $|\phi'(\alpha_\pm)| = |1 \pm \sqrt{5}| > 1$. The corresponding fixed point iterations will not converge. ∎

Example 2.7 (Population dynamics) Let us apply the fixed point iterations to the function $\phi_V(x) = rx/(1 + xK)$ of Verhulst's model (2.3) and to the function $\phi_P(x) = rx^2/(1 + (x/K)^2)$, for $r = 3$ and $K = 1$, of the predator/prey model (2.4). Starting from the initial point $x^{(0)} = 1$, we find the fixed point $\alpha = 2$ in the first case and $\alpha = 2.6180$ in the second case (see Figure 2.8). The fixed point $\alpha = 0$, common to either ϕ_V and ϕ_P, can be obtained using the fixed point iterations on ϕ_P but not those on ϕ_V. In fact, $\phi'_P(\alpha) = 0$, while $\phi'_V(\alpha) = r > 1$. The third fixed point of ϕ_P, $\alpha = 0.3820\ldots$, cannot be obtained by fixed point iterations since $\phi'_P(\alpha) > 1$. ∎

2.3 Fixed point iterations

The Newton method is not the only iterative procedure featuring quadratic convergence. Indeed, the following general property holds.

> **Proposition 2.2** *Assume that all hypotheses of Proposition 2.1 are satisfied. In addition assume that ϕ is differentiable twice and that*
>
> $$\phi'(\alpha) = 0, \ \phi''(\alpha) \neq 0.$$
>
> *Then the fixed point iterations (2.17) converge with order 2 and*
>
> $$\lim_{k \to \infty} \frac{x^{(k+1)} - \alpha}{(x^{(k)} - \alpha)^2} = \frac{1}{2}\phi''(\alpha) \qquad (2.20)$$

Example 2.5 shows that the fixed point iterations (2.17) could also be used to compute the zeros of the function f. Clearly for any given f the function ϕ defined in (2.18) is not the only possible iteration function. For instance, for the solution of the equation $\log(x) = \gamma$, after setting $f(x) = \log(x) - \gamma$, the choice (2.18) could lead to the iteration function

$$\phi_N(x) = x(1 - \log(x) + \gamma).$$

Another fixed point iteration algorithm could be obtained by adding x to both sides of the equation $f(x) = 0$. The associated iteration function is now $\phi_1(x) = x + \log(x) - \gamma$. A further method could be obtained by choosing the iteration function $\phi_2(x) = x\log(x)/\gamma$. Not all these methods are convergent. For instance, if $\gamma = -2$, the methods corresponding to the iteration functions ϕ_N and ϕ_2 are both convergent, whereas the one corresponding to ϕ_1 is not since $|\phi_1'(x)| > 1$ in a neighborhood of the fixed point α.

2.3.1 How to terminate fixed point iterations

In general, fixed point iterations are terminated when the absolute value of the difference between two consecutive iterates is less than a prescribed tolerance ε.

Since $\alpha = \phi(\alpha)$ and $x^{(k+1)} = \phi(x^{(k)})$, using the mean value theorem (see Section 1.4.3) we find

$$\alpha - x^{(k+1)} = \phi(\alpha) - \phi(x^{(k)}) = \phi'(\xi^{(k)}) \left(\alpha - x^{(k)} \right) \text{ with } \xi^{(k)} \in I_{\alpha, x^{(k)}},$$

$I_{\alpha, x^{(k)}}$ being the interval with endpoints α and $x^{(k)}$. Using the identity

$$\alpha - x^{(k)} = (\alpha - x^{(k+1)}) + (x^{(k+1)} - x^{(k)}),$$

it follows that

56   2 Nonlinear equations

$$\alpha - x^{(k)} = \frac{1}{1 - \phi'(\xi^{(k)})}(x^{(k+1)} - x^{(k)}). \qquad (2.21)$$

Consequently, if $\phi'(x) \simeq 0$ in a neighborhood of α, the difference between two consecutive iterates provides a satisfactory error estimator. This is the case for methods of order 2, including Newton's method. This estimate becomes the more unsatisfactory the more ϕ' approaches 1.

Example 2.8 Let us compute with Newton's method the zero $\alpha = 1$ of the function $f(x) = (x-1)^{m-1} \log(x)$ for $m = 11$ and $m = 21$, whose multiplicity is equal to m. In this case Newton's method converges with order 1; moreover, it is possible to prove (see Exercise 2.15) that $\phi'_N(\alpha) = 1 - 1/m$, ϕ_N being the iteration function of the method, regarded as a fixed point iteration algorithm. As m increases, the accuracy of the error estimate provided by the difference between two consecutive iterates decreases. This is confirmed by the numerical results in Figure 2.9 where we compare the behavior of the true error with that of our estimator for both $m = 11$ and $m = 21$. The difference between these two quantities is greater for $m = 21$. ∎

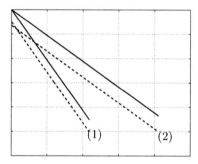

Fig. 2.9. Absolute values of the errors (*solid line*) and absolute values of the difference between two consecutive iterates (*dashed line*), plotted versus the number of iterations for the case of Example 2.8. Graphs (1) refer to $m = 11$, graphs (2) to $m = 21$

2.4 Acceleration using Aitken method

In this paragraph we will illustrate a technique which allows to accelerate the convergence of a sequence obtained via fixed point iterations. Therefore, we suppose that $x^{(k)} = \phi(x^{(k-1)})$, $k \geq 1$. If the sequence $\{x^{(k)}\}$ converges *linearly* to a fixed point α of ϕ, we have from (2.19) that, for a given k, there must be a value λ (to be determined) such that

$$\phi(x^{(k)}) - \alpha = \lambda(x^{(k)} - \alpha), \qquad (2.22)$$

2.4 Acceleration using Aitken method

where we have deliberately avoided to identify $\phi(x^{(k)})$ with $x^{(k+1)}$. Indeed, the idea underlying Aitken's method consists in defining a new value for $x^{(k+1)}$ (and thus a new sequence) which is a better approximation for α than that given by $\phi(x^{(k)})$. As a matter of fact, from (2.22) we have that

$$\alpha = \frac{\phi(x^{(k)}) - \lambda x^{(k)}}{1-\lambda} = \frac{\phi(x^{(k)}) - \lambda x^{(k)} + x^{(k)} - x^{(k)}}{1-\lambda}$$

or

$$\alpha = x^{(k)} + (\phi(x^{(k)}) - x^{(k)})/(1-\lambda) \qquad (2.23)$$

We must now compute λ. To do so, we introduce the following sequence

$$\lambda^{(k)} = \frac{\phi(\phi(x^{(k)})) - \phi(x^{(k)})}{\phi(x^{(k)}) - x^{(k)}} \qquad (2.24)$$

and verify that the following property holds:

Lemma 2.1 *If the sequence of elements $x^{(k+1)} = \phi(x^{(k)})$ converges to α, then $\lim_{k \to \infty} \lambda^{(k)} = \phi'(\alpha)$.*

Proof 2.1 If $x^{(k+1)} = \phi(x^{(k)})$, then $x^{(k+2)} = \phi(\phi(x^{(k)}))$ and from (2.24), we obtain that $\lambda^{(k)} = (x^{(k+2)} - x^{(k+1)})/(x^{(k+1)} - x^{(k)})$ or

$$\lambda^{(k)} = \frac{x^{(k+2)} - \alpha - (x^{(k+1)} - \alpha)}{x^{(k+1)} - \alpha - (x^{(k)} - \alpha)} = \frac{\frac{x^{(k+2)} - \alpha}{x^{(k+1)} - \alpha} - 1}{1 - \frac{x^{(k)} - \alpha}{x^{(k+1)} - \alpha}}$$

from which, computing the limit and recalling (2.19), we find

$$\lim_{k \to \infty} \lambda^{(k)} = \frac{\phi'(\alpha) - 1}{1 - 1/\phi'(\alpha)} = \phi'(\alpha).$$

Thanks to Lemma 2.1 we can conclude that, for a given k, $\lambda^{(k)}$ can be considered as an approximation of the previously introduced unknown value λ. Thus, we use (2.24) in (2.23) and define a new $x^{(k+1)}$ as follows:

$$x^{(k+1)} = x^{(k)} - \frac{(\phi(x^{(k)}) - x^{(k)})^2}{\phi(\phi(x^{(k)})) - 2\phi(x^{(k)}) + x^{(k)}}, \quad k \geq 0 \qquad (2.25)$$

This expression is known as *Aitken's extrapolation formula* and, by (2.25), it can be considered as a *new* fixed point iteration for the new iteration function

$$\phi_\Delta(x) = \frac{x\phi(\phi(x)) - [\phi(x)]^2}{\phi(\phi(x)) - 2\phi(x) + x}.$$

This method is sometimes called *Steffensen's method*. Clearly, function ϕ_Δ is undetermined for $x = \alpha$ as the numerator and denominator vanish. However, by applying de l'Hôpital's formula and assuming that ϕ is differentiable with $\phi'(\alpha) \neq 1$ one finds

$$\lim_{x \to \alpha} \phi_\Delta(x) = \frac{\phi(\phi(\alpha)) + \alpha\phi'(\phi(\alpha))\phi'(\alpha) - 2\phi(\alpha)\phi'(\alpha)}{\phi'(\phi(\alpha))\phi'(\alpha) - 2\phi'(\alpha) + 1}$$
$$= \frac{\alpha + \alpha[\phi'(\alpha)]^2 - 2\alpha\phi'(\alpha)}{[\phi'(\alpha)]^2 - 2\phi'(\alpha) + 1} = \alpha.$$

Consequently, $\phi_\Delta(x)$ can be extended by continuity to $x = \alpha$ by setting $\phi_\Delta(\alpha) = \alpha$.

When $\phi(x) = x - f(x)$, the case $\phi'(\alpha) = 1$ corresponds to a root with multiplicity of at least 2 for f (since $\phi'(\alpha) = 1 - f'(\alpha)$). In such situation however, we can once again prove by evaluating the limit that $\phi_\Delta(\alpha) = \alpha$. Moreover, we can also verify that the fixed points of ϕ_Δ are all and exclusively the fixed points of ϕ.

Aitken's method can thus be applied for any fixed point method. Indeed, the following theorem holds:

Theorem 2.1 *Consider the fixed point iterations (2.17) with $\phi(x) = x - f(x)$ for computing the roots of f. Then if f is sufficiently regular we have:*

- *if the fixed point iterations converge linearly to a simple root of f, then Aitken's method converges quadratically to the same root;*
- *if the fixed point iterations converge with order $p \geq 2$ to a simple root of f, then Aitken's method converges to the same root with order $2p - 1$;*
- *if the fixed point iterations converge linearly to a root with multiplicity $m \geq 2$ of f, then Aitken's method converges linearly to the same root with an asymptotic convergence factor of $C = 1 - 1/m$.*

In particular, if $p = 1$ and the root of f is simple, Aitken's extrapolation method converges even if the corresponding fixed point iterations diverge.

In Program 2.4 we report an implementation of Aitken's method. Here phi is a *function* (or an *inline function*) which defines the expression of the iteration function of the fixed point method to which Aitken's extrapolation technique is applied. The initial datum is defined by the variable x0, while tol and nmax are the stopping criterion tolerance (on

2.4 Acceleration using Aitken method

the absolute value of the difference between two consecutive iterates) and the maximal number of iterations allowed, respectively. If undefined, *default* values nmax=100 and tol=1.e-04 are assumed.

Program 2.4. aitken: Aitken method

```
function [x,niter]=aitken(phi,x0,tol,nmax,varargin)
%AITKEN Aitken's method.
%   [ALPHA,NITER]=AITKEN(PHI,X0) computes an
%   approximation of a fixed point ALPHA of function PHI
%   starting from the initial datum X0 using Aitken's
%   extrapolation method. The method stops after 100
%   iterations or after the absolute value of the
%   difference between two consecutive iterates is
%   smaller than 1.e-04. PHI must be defined as a
%   function or an inline function.
%   [ALPHA,NITER]=AITKEN(PHI,X0,TOL,NMAX) allows to
%   define the tolerance on the stopping criterion and
%   the maximum number of iterations.
if nargin == 2
    tol = 1.e-04;    nmax = 100;
elseif nargin == 3
    nmax = 100;
end
x = x0;
diff = tol + 1;
niter = 0;
while niter <= nmax & diff >= tol
    gx = feval(phi,x,varargin{:});
    ggx = feval(phi,gx,varargin{:});
    xnew = (x*ggx-gx^2)/(ggx-2*gx+x);
    diff = abs(x-xnew);
    x = xnew;
    niter = niter + 1;
end
if niter >= nmax
    fprintf(' Fails to converge within maximum ',...
            'number of iterations\n ');
end
return
```

Example 2.9 In order to compute the single root $\alpha = 1$ for function $f(x) = e^x(x-1)$ we apply Aitken's method starting from the two following iteration functions

$$\phi_0(x) = \log(xe^x), \quad \phi_1(x) = \frac{e^x + x}{e^x + 1}.$$

We use Program 2.4 with tol=1.e-10, nmax=100, x0=2 and we define the two iteration functions as follows:

```
phi0 = inline('log(x*exp(x))','x');
phi1 = inline('(exp(x)+x)/(exp(x)+1)','x');
```

We now run Program 2.4 as follows:

```
[alpha,niter]=aitken(phi0,x0,tol,nmax)
```

```
alpha =
   1.0000 + 0.0000i
niter =
   10

[alpha,niter]=aitken(phi1,x0,tol,nmax)

alpha =
   1
niter =
   4
```

As we can see, the convergence is extremely rapid. For comparison the fixed point method with iteration function ϕ_1 and the same stopping criterion would have required 18 iterations, while the method corresponding to ϕ_0 would not have been convergent as $|\phi'_0(1)| = 2$. ∎

Let us summarize

1. A number α satisfying $\phi(\alpha) = \alpha$ is called a fixed point of ϕ. For its computation we can use the so-called fixed point iterations: $x^{(k+1)} = \phi(x^{(k)})$;
2. fixed point iterations converge under suitable assumptions on the iteration function ϕ and its first derivative. Typically, convergence is linear, however, in the special case when $\phi'(\alpha) = 0$, the fixed point iterations converge quadratically;
3. fixed point iterations can also be used to compute the zeros of a function;
4. given a fixed point iteration $x^{(k+1)} = \phi(x^{(k)})$, it is always possible to construct a new sequence using Aitken's method, which in general converges faster.

See Exercises 2.15-2.18.

2.5 Algebraic polynomials

In this section we will consider the case where f is a polynomial of degree $n \geq 0$ of the form (1.9). As already anticipated, the space of all polynomials (1.9) is denoted by the symbol $\mathbb{P}_n$. When $n \geq 2$ and all the coefficients a_k are real, if $\alpha \in \mathbb{C}$ is a complex root of $p_n \in \mathbb{P}_n$ (i.e. with $\text{Im}(\alpha) \neq 0$), then $\bar{\alpha}$ (the complex conjugate of α) is a root of p_n too.

Abel's theorem guarantees that there does not exist an explicit form to compute all the zeros of a generic polynomial p_n, when $n \geq 5$. This fact further motivates the use of numerical methods for computing the roots of p_n.

As we have previously seen for such methods it is important to choose an appropriate initial datum $x^{(0)}$ or a suitable search interval $[a, b]$ for the root. In the case of polynomials this is sometimes possible on the basis of the following results.

> **Theorem 2.2 (Descartes's sign rule)** *Let us denote by ν the number of sign changes of the coefficients $\{a_j\}$ and with k the number of real positive roots of p_n, each counted with its own multiplicity. Then $k \leq \nu$ and $\nu - k$ is even.*

Example 2.10 The polynomial $p_6(x) = x^6 - 2x^5 + 5x^4 - 6x^3 + 2x^2 + 8x - 8$ has zeros $\{\pm 1, \pm 2i, 1 \pm i\}$ and thus has 1 real positive root ($k = 1$). Indeed, the number of sign changes ν of its coefficients is 5 and thereafter $k \leq \nu$ and $\nu - k = 4$ is even. ∎

> **Theorem 2.3 (Cauchy)** *All of the zeros of p_n are included in the circle Γ in the complex plane*
>
> $$\Gamma = \{z \in \mathbb{C} : |z| \leq 1 + \eta\}, \text{ where } \eta = \max_{0 \leq k \leq n-1} |a_k/a_n|. \quad (2.26)$$

This property is barely useful when $\eta \gg 1$ (for polynomial p_6 in Example 2.10 for instance, we have $\eta = 8$, while all of the roots are in circles with clearly smaller radii).

2.5.1 Hörner's algorithm

In this paragraph we will illustrate a method for the effective evaluation of a polynomial (and its derivative) in a given point z. Such algorithm allows to generate an automatic procedure, called *deflation method*, for the progressive approximation of *all* the roots of a polynomial.

From an algebraic point of view, (1.9) is equivalent to the following representation

$$p_n(x) = a_0 + x(a_1 + x(a_2 + \ldots + x(a_{n-1} + a_n x) \ldots)). \quad (2.27)$$

However, while (1.9) requires n sums and $2n - 1$ products to evaluate $p_n(x)$ (for a given x), (2.27) only requires n sums and n products. The expression (2.27), also known as the nested product algorithm, is the basis for Hörner's algorithm. This method allows to effectively evaluate the polynomial p_n in a point z by using the following *synthetic division algorithm*

$$\boxed{\begin{aligned} b_n &= a_n, \\ b_k &= a_k + b_{k+1}z, \; k = n-1, n-2, ..., 0 \end{aligned}} \quad (2.28)$$

In (2.28) all of the coefficients b_k with $k \leq n-1$ depend on z and we can verify that $b_0 = p_n(z)$. The polynomial

$$q_{n-1}(x;z) = b_1 + b_2 x + ... + b_n x^{n-1} = \sum_{k=1}^{n} b_k x^{k-1}, \quad (2.29)$$

of degree $n-1$ in x, depends on the z parameter (via the b_k coefficients) and is called the *associated polynomial* of p_n. Algorithm (2.28) is implemented in Program 2.5. The a_j coefficients of the polynomial to be evaluated are stored in vector a starting from a_n up to a_0.

Program 2.5. horner: synthetic division algorithm

```
function [y,b] = horner(a,z)
%HORNER Horner algorithm
%   Y=HORNER(A,Z) computes
%   Y = A(1)*Z^N + A(2)*Z^(N-1) + ... + A(N)*Z + A(N+1)
%   using Horner's synthetic division algorithm.
n = length(a)-1;
b = zeros(n+1,1);
b(1) = a(1);
for j=2:n+1
    b(j) = a(j)+b(j-1)*z;
end
y = b(n+1);
b = b(1:end-1);
return
```

We now want to introduce an effective algorithm which, knowing the root of a polynomial (or its approximation), is able to remove it and then to allow the computation of the following one until all roots are determinated.

In order to do this we should recall the following property of *polynomial division*:

Proposition 2.3 *Given two polynomials $h_n \in \mathbb{P}_n$ and $g_m \in \mathbb{P}_m$ with $m \leq n$, there are a unique polynomial $\delta \in \mathbb{P}_{n-m}$ and a unique polynomial $\rho \in \mathbb{P}_{m-1}$ such that*

$$h_n(x) = g_m(x)\delta(x) + \rho(x). \quad (2.30)$$

Thus, by dividing a polynomial $p_n \in \mathbb{P}_n$ by $x - z$, one deduces by (2.30) that

$$p_n(x) = b_0 + (x-z)q_{n-1}(x;z),$$

having denoted by q_{n-1} the quotient and by b_0 the remainder of the division. If z is a root of p_n, then we have $b_0 = p_n(z) = 0$ and therefore $p_n(x) = (x-z)q_{n-1}(x;z)$. In this case the algebric equation $q_{n-1}(x;z) = 0$ provides the $n-1$ remaining roots of $p_n(x)$. This remark suggests to adopt the following *deflation criterion* to compute *all* the roots of p_n.

For $m = n, n-1, \ldots, 1$:

1. find a root r_m for p_m with an appropriate approximation method;
2. compute $q_{m-1}(x;r_m)$ using (2.28)-(2.29) (having set $z = r_m$);
3. set $p_{m-1} = q_{m-1}$.

In the following paragraph we propose the most widely known method in this group, which uses Newton's method for the approximation of the roots.

2.5.2 The Newton-Hörner method

As its name suggests, the *Newton-Hörner method* implements the deflation procedure using Newton's method to compute the roots r_m. The advantage lies in the fact that the implementation of Newton's method conveniently exploits Hörner's algorithm (2.28).

As a matter of fact, if q_{n-1} is the polynomial associated to p_n defined in (2.29), since

$$p'_n(x) = q_{n-1}(x;z) + (x-z)q'_{n-1}(x;z),$$

one has

$$p'_n(z) = q_{n-1}(z;z).$$

Thanks to this identity, the Newton-Hörner method for the approximation of a (real or complex) root r_j of p_n ($j = 1, \ldots, n$) takes the following form:
given an initial estimation $r_j^{(0)}$ of the root, compute for each $k \geq 0$ until convergence

$$r_j^{(k+1)} = r_j^{(k)} - \frac{p_n(r_j^{(k)})}{p'_n(r_j^{(k)})} = r_j^{(k)} - \frac{p_n(r_j^{(k)})}{q_{n-1}(r_j^{(k)};r_j^{(k)})} \qquad (2.31)$$

We now use the deflation technique, exploiting the fact that $p_n(x) = (x-r_j)p_{n-1}(x)$. We can then proceed to the approximation of a zero of p_{n-1} and so on until all the roots of p_n are processed.

Consider that when $r_j \in \mathbb{C}$, it is necessary to perform the computation in complex arithmetics, taking $r_j^{(0)}$ as the non-null imaginary part. Otherwise, the Newton-Hörner method would generate a sequence $\{r_j^{(k)}\}$ of real numbers.

The Newton-Hörner method is implemented in Program 2.6. The coefficients a_j of the polynomial for which we intend to compute the roots are stored in vector a starting from a_n up to a_0. The other input parameters, tol and nmax, are the stopping criterion tolerance (on the absolute value of the difference between two consecutive iterates) and the maximal number of iterations allowed, respectively. If undefined, the *default* values nmax=100 and tol=1.e-04 are assumed. As an output, the program returns in vectors roots and iter the computed roots and the number of iterations required to compute each of the values, respectively.

Program 2.6. newtonhorner: Newton-Hörner method

```
function [roots,iter]=newtonhorner(a,x0,tol,nmax)
%NEWTONHORNER Newton-Horner method
% [roots,ITER]=NEWTONHORNER(A,X0) computes the roots of
% polynomial
% P(X) = A(1)*X^N + A(2)*X^(N-1) + ... + A(N)*X +
% A(N+1)
% using the Newton-Horner method starting from the
% initial datum X0. The method stops for each root
% after 100 iterations or after the absolute value of
% the difference between two consecutive iterates is
% smaller than 1.e-04.
% [roots,ITER]=NEWTONHORNER(A,X0,TOL,NMAX) allows to
% define the tolerance on the stopping criterion and
% the maximal number of iterations.
if nargin == 2
   tol = 1.e-04; nmax = 100;
elseif nargin == 3
   nmax = 100;
end
n=length(a)-1; roots = zeros(n,1); iter = zeros(n,1);
for k = 1:n
   % Newton iterations
   niter = 0; x = x0; diff = tol + 1;
   while niter <= nmax & diff >= tol
       [pz,b] = horner(a,x);    [dpz,b] = horner(b,x);
       xnew = x - pz/dpz;       diff = abs(xnew-x);
       niter = niter + 1;       x = xnew;
   end
   if niter >= nmax
       fprintf(' Fails to converge within maximum ',...
               'number of iterations\n ');
   end
   % Deflation
   [pz,a] = horner(a,x); roots(k) = x; iter(k) = niter;
end
return
```

Remark 2.1 In order to minimize the propagation of roundoff errors, during the deflation process it is better to first approximate the root r_1 with minimal absolute value and then to proceed to the computation of the following roots $r_2, r_3, \ldots$, until the one with the maximal absolute value is reached (to learn more, see for instance [QSS06]). •

Example 2.11 To compute the roots $\{1,2,3\}$ of the polynomial $p_3(x) = x^3 - 6x^2 + 11x - 6$ we use Program 2.6
```
a=[1 -6 11 -6]; [x,niter]=newtonhorner(a,0,1.e-15,100)

x =
    1
    2
    3
niter =
    8
    8
    2
```
The method computes all three roots accurately and in few iterations. As pointed out in Remark 2.1 however, the method is not always so effective. For instance, if we consider the polynomial $p_4(x) = x^4 - 7x^3 + 15x^2 - 13x + 4$ (which has the root 1 of multiplicity 3 and a single root with value 4) we find the following results
```
a=[1 -7 15 -13 4]; format long;
[x,niter]=newtonhorner(a,0,1.e-15,100)

x =
   1.00000693533737
   0.99998524147571
   1.00000782324144
   3.99999999994548
niter =
   61
   101
   6
   2
```
The loss of accuracy is quite evident for the computation of the multiple root, and becomes as more relevant as the multiplicity increases (see [QSS06]). ∎

2.6 What we haven't told you

The most sophisticated methods for the computation of the zeros of a function combine different algorithms. In particular, the MATLAB function `fzero` (see Section 1.4.1) adopts the so called Dekker-Brent method (see [QSS06], Section 6.2.3). In its basic form `fzero(fun,x0)` computes the zero of the function `fun`, where `fun` can be either a string which is a function of x, the name of an inline function, or the name of a m-file.

`fzero`

For instance, we could solve the problem in Example 2.1 also by `fzero`, using the initial value `x0=0.3` (as done by Newton's method) via the following instructions:

```
function y=Rfunc(r)
y=6000 - 1000*(1+r)/r*((1+r)^5 - 1);
end

x0=0.3;
[alpha,res,flag]=fzero('Rfunc',x0);
```

We obtain alpha=0.06140241153653 with residual res=9.0949e-13 in iter=29 iterations. When flag is negative it means that fzero cannot find the zero. The Newton method converges in 6 iterations to the value 0.06140241153652 with a residual equal to 2.3646e-11.

In order to compute the zeros of a polynomial, in addition to the Newton-Hörner method, we can cite the methods based on Sturm sequences, Müller's method, (see [Atk89] or [QSS06]) and Bairstow's method ([RR85], page 371 and following). A different approach consists in characterizing the zeros of a function as the eigenvalues of a special matrix (called the *companion matrix*) and then using appropriate techniques for their computation. This approach is adopted by the MATLAB function roots which has been introduced in Section 1.4.2.

We have mentioned in Section 2.2.2 how to set up a Newton method for a nonlinear system, like (2.13). More in general, any fixed point iteration can be easily extended to compute the roots of nonlinear systems. Other methods exist as well, such as the Broyden and quasi-Newton methods, which can be regarded as generalizations of Newton's method (see [DS83], [Deu04], [SM03] and [QSS06, Chapter 7]).

fsolve The MATLAB instruction

```
zero=fsolve('fun',x0)
```

allows the computation of one zero of a nonlinear system defined via the user function fun starting from the vector x0 as initial guess. The function fun returns the n values $f_i(\bar{x}_1,\ldots,\bar{x}_n)$, $i = 1,\ldots,n$, for any given input vector $(\bar{x}_1,\ldots,\bar{x}_n)^T$.

For instance, in order to solve the nonlinear system (2.15) using fsolve the corresponding MATLAB user function, which we call systemnl, is defined as follows:

```
function fx=systemnl(x)
fx(1) = x(1)^2+x(2)^2-1;
fx(2) = sin(pi*0.5*x(1))+x(2)^3;
```

The MATLAB instructions to solve this system are therefore:

```
x0 = [1 1];
alpha=fsolve('systemnl',x0)

alpha =
    0.4761    -0.8794
```

Using this procedure we have found only one of the two roots. The other can be computed starting from the initial datum -x0.

Octave 2.1 The commands `fzero` and `fsolve` have exactly the same purpose in MATLAB and Octave, however there interface differ slightly between MATLAB and Octave in the optional arguments. We encourage the reader to study the `help` documentation of both commands in each environment. ∎

2.7 Exercises

Exercise 2.1 Given the function $f(x) = \cosh x + \cos x - \gamma$, for $\gamma = 1, 2, 3$ find an interval that contains the zero of f. Then compute the zero by the bisection method with a tolerance of 10^{-10}.

Exercise 2.2 (State equation of a gas) For carbon dioxide (CO_2) the coefficients a and b in (2.1) take the following values: $a = 0.401$Pa m^6, $b = 42.7 \cdot 10^{-6}$m^3 (Pa stands for Pascal). Find the volume occupied by 1000 molecules of CO_2 at a temperature $T = 300$K and a pressure $p = 3.5 \cdot 10^7$ Pa by the bisection method, with a tolerance of 10^{-12} (the Boltzmann constant is $k = 1.3806503 \cdot 10^{-23}$ Joule K^{-1}).

Exercise 2.3 Consider a plane whose slope varies with constant rate ω, and a dimensionless object which is steady at the initial time $t = 0$. At time $t > 0$ its position is

$$s(t, \omega) = \frac{g}{2\omega^2}[\sinh(\omega t) - \sin(\omega t)],$$

where $g = 9.8$ m/s^2 denotes the gravity acceleration. Assuming that this object has moved by 1 meter in 1 second, compute the corresponding value of ω with a tolerance of 10^{-5}.

Exercise 2.4 Prove inequality (2.6).

Exercise 2.5 Motivate why in Program 2.1 the instruction `x(2) = x(1)+ (x(3)- x(1))*0.5` has been used instead of the more natural one `x(2)=(x(1)+ x(3))*0.5` in order to compute the midpoint.

Exercise 2.6 Apply Newton's method to solve Exercise 2.1. Why is this method not accurate when $\gamma = 2$?

Exercise 2.7 Apply Newton's method to compute the square root of a positive number a. Proceed in a similar manner to compute the cube root of a.

Exercise 2.8 Assuming that Newton's method converges, show that (2.9) is true when α is a simple root of $f(x) = 0$ and f is twice continuously differentiable in a neighborhood of α.

Exercise 2.9 (Rods system) Apply Newton's method to solve Problem 2.3 for $\beta \in [0, 2\pi/3]$ with a tolerance of 10^{-5}. Assume that the lengths of the rods are $a_1 = 10$ cm, $a_2 = 13$ cm, $a_3 = 8$ cm and $a_4 = 10$ cm. For each value of β consider two possible initial data, $x^{(0)} = -0.1$ and $x^{(0)} = 2\pi/3$.

Exercise 2.10 Notice that the function $f(x) = e^x - 2x^2$ has 3 zeros, $\alpha_1 < 0$, α_2 and α_3 positive. For which value of $x^{(0)}$ does Newton's method converge to α_1?

Exercise 2.11 Use Newton's method to compute the zero of $f(x) = x^3 - 3x^2 2^{-x} + 3x 4^{-x} - 8^{-x}$ in $[0, 1]$ and explain why convergence is not quadratic.

Exercise 2.12 A projectile is ejected with velocity v_0 and angle α in a tunnel of height h and reaches its maximum range when α is such that $\sin(\alpha) = \sqrt{2gh/v_0^2}$, where $g = 9.8$ m/s^2 is the gravity acceleration. Compute α using Newton's method, assuming that $v_0 = 10$ m/s and $h = 1$ m.

Exercise 2.13 (Investment fund) Solve Problem 2.1 by Newton's method with a tolerance of 10^{-12}, assuming $M = 6000$ euros, $v = 1000$ euros and $n = 5$. As an initial guess take the result obtained after 5 iterations of the bisection method applied on the interval $(0.01, 0.1)$.

Exercise 2.14 A corridor has the form indicated in Figure 2.10. The maximum length L of a rod that can pass from one extreme to the other by sliding on the ground is given by

$$L = l_2/(\sin(\pi - \gamma - \alpha)) + l_1/\sin(\alpha),$$

where α is the solution of the nonlinear equation

$$l_2 \frac{\cos(\pi - \gamma - \alpha)}{\sin^2(\pi - \gamma - \alpha)} - l_1 \frac{\cos(\alpha)}{\sin^2(\alpha)} = 0. \qquad (2.32)$$

Compute α by Newton's method when $l_2 = 10$, $l_1 = 8$ and $\gamma = 3\pi/5$.

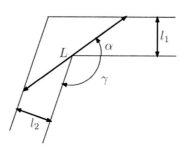

Fig. 2.10. The problem of a rod sliding in a corridor

Exercise 2.15 Let ϕ_N be the iteration function of Newton's method when regarded as a fixed point iteration. Show that $\phi_N'(\alpha) = 1 - 1/m$ where α is a zero of f with multiplicity m. Deduce that Newton's method converges quadratically if α is a simple root of $f(x) = 0$, and linearly otherwise.

Exercise 2.16 Deduce from the graph of $f(x) = x^3 + 4x^2 - 10$ that this function has a unique real zero α. To compute α use the following fixed point iterations: given $x^{(0)}$, define $x^{(k+1)}$ such that

$$x^{(k+1)} = \frac{2(x^{(k)})^3 + 4(x^{(k)})^2 + 10}{3(x^{(k)})^2 + 8x^{(k)}}, \qquad k \geq 0$$

and analyze its convergence to α.

Exercise 2.17 Analyze the convergence of the fixed point iterations

$$x^{(k+1)} = \frac{x^{(k)}[(x^{(k)})^2 + 3a]}{3(x^{(k)})^2 + a}, \qquad k \geq 0,$$

for the computation of the square root of a positive number a.

Exercise 2.18 Repeat the computations carried out in Exercise 2.11 this time using the stopping criterion based on the residual. Which result is the more accurate?

3
Approximation of functions and data

Approximating a function f consists of replacing it by another function $\tilde{f}$ of simpler form that may be used as its surrogate. This strategy is used frequently in numerical integration where, instead of computing $\int_a^b f(x)dx$, one carries out the exact computation of $\int_a^b \tilde{f}(x)dx$, $\tilde{f}$ being a function simple to integrate (e.g. a polynomial), as we will see in the next chapter. In other instances the function f may be available only partially through its values at some selected points. In these cases we aim at constructing a continuous function $\tilde{f}$ that could represent the empirical law which is behind the finite set of data. We provide some examples which illustrate this kind of approach.

Problem 3.1 (Climatology) The air temperature near the ground depends on the concentration K of the carbon acid (H_2CO_3) therein. In Table 3.1 (taken from Philosophical Magazine 41, 237 (1896)) we report for different latitudes on the Earth and for four different values of K, the variation $\delta_K = \theta_K - \theta_{\bar{K}}$ of the average temperature with respect to the average temperature corresponding to a reference value $\bar{K}$ of K. Here $\bar{K}$ refers to the value measured in 1896, and is normalized to one. In this case we can generate a function that, on the basis of the available data, provides an approximate value of the average temperature at any possible latitude and for other values of K (see Example 3.1). ∎

Problem 3.2 (Finance) In Figure 3.1 we report the price of a stock at the Zurich stock exchange over two years. The curve was obtained by joining with a straight line the prices reported at every day's closure. This simple representation indeed implicitly assumes that the prices change linearly in the course of the day (we anticipate that this approximation is called composite linear interpolation). We ask whether from this graph one could predict the stock price for a short time interval beyond the time of the last quotation. We will see in Section 3.4 that this kind of

72   3 Approximation of functions and data

Latitude	$K = 0.67$	δ_K $K = 1.5$	$K = 2.0$	$K = 3.0$
65	-3.1	3.52	6.05	9.3
55	-3.22	3.62	6.02	9.3
45	-3.3	3.65	5.92	9.17
35	-3.32	3.52	5.7	8.82
25	-3.17	3.47	5.3	8.1
15	-3.07	3.25	5.02	7.52
5	-3.02	3.15	4.95	7.3
-5	-3.02	3.15	4.97	7.35
-15	-3.12	3.2	5.07	7.62
-25	-3.2	3.27	5.35	8.22
-35	-3.35	3.52	5.62	8.8
-45	-3.37	3.7	5.95	9.25
-55	-3.25	3.7	6.1	9.5

Table 3.1. Variation of the average yearly temperature on the Earth for four different values of the concentration K of carbon acid at different latitudes

prediction could be guessed by resorting to a special technique known as *least-squares* approximation of data (see Example 3.9). ■

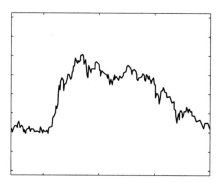

Fig. 3.1. Price variation of a stock over two years

Problem 3.3 (Biomechanics) We consider a mechanical test to establish the link between stresses (MPa= 100 N/cm^2) and deformations of a sample of biological tissue (an intervertebral disc, see Figure 3.2). Starting from the data collected in Table 3.2 (taken from P.Komarek, Chapt. 2 of *Biomechanics of Clinical Aspects of Biomedicine*, 1993, J.Valenta ed., Elsevier) in Example 3.10 we will estimate the deformation corresponding to a stress $\sigma = 0.9$ MPa. ■

3 Approximation of functions and data   73

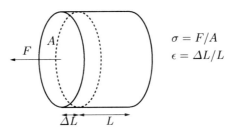

Fig. 3.2. A schematic representation of an intervertebral disc

test	stress σ	stress ϵ	test	stress σ	stress ϵ
1	0.00	0.00	5	0.31	0.23
2	0.06	0.08	6	0.47	0.25
3	0.14	0.14	7	0.60	0.28
4	0.25	0.20	8	0.70	0.29

Table 3.2. Values of the deformation for different values of a stress applied on an intervertebral disc

Problem 3.4 (Robotics) We want to approximate the planar trajectory followed by a robot (idealized as a material point) during a working cycle in an industry. The robot should satisfy a few constraints: it must be steady at the point $(0,0)$ in the plane at the initial time (say, $t = 0$), transit through the point $(1,2)$ at $t = 1$, get the point $(4,4)$ at $t = 2$, stop and restart immediately and reach the point $(3,1)$ at $t = 3$, return to the initial point at time $t = 5$, stop and restart a new working cycle. In Example 3.7 we will solve this problem using the *splines* functions. ∎

A function f can be replaced in a given interval by its Taylor polynomial, which was introduced in Section 1.4.3. This technique is computationally expensive since it requires the knowledge of f and its derivatives up to the order n (the polynomial degree) at a given point x_0. Moreover, the Taylor polynomial may fail to accurately represent f far enough from the point x_0. For instance, in Figure 3.3 we compare the behavior of $f(x) = 1/x$ with that of its Taylor polynomial of degree 10 built around the point $x_0 = 1$. This picture also shows the graphical interface of the MATLAB function taylortool which allows the computation of taylortool
Taylor's polynomial of arbitrary degree for any given function f. The agreement between the function and its Taylor polynomial is very good in a small neighborhood of $x_0 = 1$ while it becomes unsatisfactory when $x - x_0$ gets large. Fortunately, this is not the case of other functions such as the exponential function which is approximated quite nicely for all $x \in \mathbb{R}$ by its Taylor polynomial related to $x_0 = 0$, provided that the degree n is sufficiently large.

In the course of this chapter we will introduce approximation methods that are based on alternative approaches.

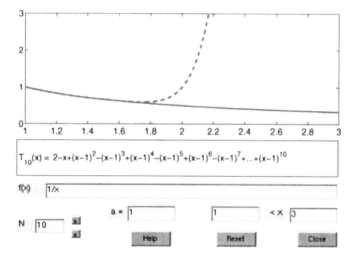

Fig. 3.3. Comparison between the function $f(x) = 1/x$ (*solid line*) and its Taylor polynomial of degree 10 related to the point $x_0 = 1$ (*dashed line*). The explicit form of the Taylor polynomial is also reported

3.1 Interpolation

As seen in Problems 3.1, 3.2 and 3.3, in several applications it may happen that a function is known only through its values at some given points. We are therefore facing a (general) case where $n+1$ couples $\{x_i, y_i\}$, $i = 0, \ldots, n$, are given; the points x_i are all distinct and are called *nodes*.

For instance in the case of Table 3.1, n is equal to 12, the nodes x_i are the values of the latitude reported in the first column, while the y_i are the corresponding values (of the temperature) in the remaining columns.

In such a situation it seems natural to require the approximate function $\tilde{f}$ to satisfy the set of relations

$$\boxed{\tilde{f}(x_i) = y_i, \; i = 0, 1, \ldots, n} \qquad (3.1)$$

Such an $\tilde{f}$ is called *interpolant* of the set of data $\{y_i\}$ and equations (3.1) are the interpolation conditions.

Several kinds of interpolants could be envisaged, such as:

- *polynomial interpolant*:

$$\tilde{f}(x) = a_0 + a_1 x + a_2 x^2 + \ldots + a_n x^n;$$

- *trigonometric interpolant*:

$$\tilde{f}(x) = a_{-M} e^{-iMx} + \ldots + a_0 + \ldots + a_M e^{iMx}$$

where M is an integer equal to $n/2$ if n is even, $(n-1)/2$ if n is odd, and i is the imaginary unit;
- *rational interpolant*:

$$\tilde{f}(x) = \frac{a_0 + a_1 x + \ldots + a_k x^k}{a_{k+1} + a_{k+2} x + \ldots + a_{k+n+1} x^n}.$$

For simplicity we only consider those interpolants which depend linearly on the unknown coefficients a_i. Both polynomial and trigonometric interpolation fall into this category, whereas the rational interpolant does not.

3.1.1 Lagrangian polynomial interpolation

Let us focus on the polynomial interpolation. The following result holds:

> **Proposition 3.1** *For any set of couples* $\{x_i, y_i\}$, $i = 0, \ldots, n$, *with distinct nodes* x_i, *there exists a unique polynomial of degree less than or equal to* n, *which we indicate by* Π_n *and call interpolating polynomial of the values* y_i *at the nodes* x_i, *such that*
>
> $$\Pi_n(x_i) = y_i, \ i = 0, \ldots, n \qquad (3.2)$$
>
> *In the case where the* $\{y_i, i = 0, \ldots, n\}$ *represent the values of a continuous function* f, Π_n *is called interpolating polynomial of* f *(in short, interpolant of* f) *and will be denoted by* $\Pi_n f$.

To verify uniqueness we proceed by contradiction and suppose that there exist two distinct polynomials of degree n, Π_n and Π_n^*, both satisfying the nodal relation (3.2). Their difference, $\Pi_n - \Pi_n^*$, would be a polynomial of degree n which vanishes at $n+1$ distinct points. Owing to a well known theorem of Algebra, such a polynomial should vanish identically, and then Π_n^* must coincide with Π_n.

In order to obtain an expression for Π_n, we start from a very special case where y_i vanishes for all i apart from $i = k$ (for a fixed k) for which $y_k = 1$. Then setting $\varphi_k(x) = \Pi_n(x)$, we must have (see Figure 3.4)

$$\varphi_k \in \mathbb{P}_n, \ \varphi_k(x_j) = \delta_{jk} = \begin{cases} 1 & \text{if } j = k, \\ 0 & \text{otherwise}, \end{cases}$$

where δ_{jk} is the Kronecker symbol.

The functions φ_k have the following expression:

$$\varphi_k(x) = \prod_{\substack{j=0 \\ j \neq k}}^{n} \frac{x - x_j}{x_k - x_j}, \qquad k = 0, \ldots, n. \qquad (3.3)$$

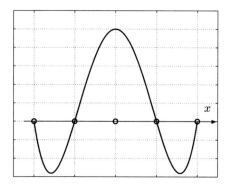

Fig. 3.4. The polynomial $\varphi_2 \in \mathbb{P}_4$ associated with a set of 5 equispaced nodes

We move now to the general case where $\{y_i, i = 0, \ldots, n\}$ is a set of arbitrary values. Using an obvious superposition principle we can obtain the following expression for Π_n

$$\Pi_n(x) = \sum_{k=0}^{n} y_k \varphi_k(x) \qquad (3.4)$$

Indeed, this polynomial satisfies the interpolation conditions (3.2), since

$$\Pi_n(x_i) = \sum_{k=0}^{n} y_k \varphi_k(x_i) = \sum_{k=0}^{n} y_k \delta_{ik} = y_i, \quad i = 0, \ldots, n.$$

Due to their special role, the functions φ_k are called *Lagrange characteristic polynomials*, and (3.4) is the *Lagrange form* of the interpolant. In MATLAB we can store the n+1 couples $\{(x_i, y_i)\}$ in the vectors x

polyfit and y, and then the instruction c=polyfit(x,y,n) will provide the coefficients of the interpolating polynomial. Precisely, c(1) will contain the coefficient of x^n, c(2) that of x^{n-1}, ... and c(n+1) the value of $\Pi_n(0)$. (More on this command can be found in Section 3.4.) As already seen in Chapter 1, we can then use the instruction p=polyval(c,z) to compute the value p(j) attained by the interpolating polynomial at z(j), j=1,...,m, the latter being a set of m arbitrary points.

In the case when the explicit form of the function f is available, we can use the instruction y=eval(f) in order to obtain the vector y of values of f at some specific nodes (which should be stored in a vector x).

Example 3.1 (Climatology) To obtain the interpolating polynomial for the data of Problem 3.1 relating to the value $K = 0.67$ (first column of Table 3.1), using only the values of the temperature for the latitudes 65, 35, 5, -25, -55, we can use the following MATLAB instructions:

3.1 Interpolation   77

```
x=[-55 -25 5 35 65]; y=[-3.25 -3.2 -3.02 -3.32 -3.1];
format short e; c=polyfit(x,y,4)

c =
   8.2819e-08  -4.5267e-07  -3.4684e-04   3.7757e-04  -3.0132e+00
```

The graph of the interpolating polynomial can be obtained as follows:
```
z=linspace(x(1),x(end),100);
p=polyval(c,z);
plot(z,p);hold on;plot(x,y,'o');grid on;
```
In order to get a smooth curve we have evaluated our polynomial at 101 equispaced points in the interval $[-55, 65]$ (as a matter of fact, MATLAB plots are always constructed on piecewise linear interpolation between neighboring points). Note that the instruction x(end) picks up directly the last component of the vector x, without specifying the length of the vector. In Figure 3.5 the filled circles correspond to those values which have been used to construct the interpolating polynomial, whereas the empty circles correspond to values that have not been used. We can appreciate the qualitative agreement between the curve and the data distribution. ∎

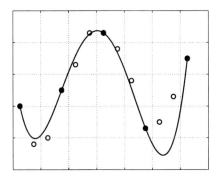

Fig. 3.5. The interpolating polynomial of degree 4 introduced in Example 3.1

Using the following result we can evaluate the error obtained by replacing f with its interpolating polynomial $\Pi_n f$:

Proposition 3.2 *Let I be a bounded interval, and consider $n+1$ distinct interpolation nodes $\{x_i, i = 0, \ldots, n\}$ in I. Let f be continuously differentiable up to order $n+1$ in I.*

Then $\forall x \in I \; \exists \xi \in I$ *such that*

$$E_n f(x) = f(x) - \Pi_n f(x) = \frac{f^{(n+1)}(\xi)}{(n+1)!} \prod_{i=0}^{n}(x - x_i) \qquad (3.5)$$

Obviously, $E_n f(x_i) = 0$, $i = 0, \ldots, n$.

Result (3.5) can be better specified in the case of a uniform distribution of nodes, that is when $x_i = x_{i-1} + h$ for $i = 1, \ldots, n$, for a given $h > 0$ and a given x_0. As stated in Exercise 3.1, $\forall x \in (x_0, x_n)$ one can verify that

$$\left|\prod_{i=0}^{n}(x - x_i)\right| \leq n! \frac{h^{n+1}}{4}, \qquad (3.6)$$

and therefore

$$\max_{x \in I}|E_n f(x)| \leq \frac{\max_{x \in I}|f^{(n+1)}(x)|}{4(n+1)} h^{n+1}. \qquad (3.7)$$

Unfortunately, we cannot deduce from (3.7) that the error tends to 0 when $n \to \infty$, in spite of the fact that $h^{n+1}/[4(n+1)]$ tends to 0. In fact, as shown in Example 3.2, there exist functions f for which the limit can even be infinite, that is

$$\lim_{n \to \infty} \max_{x \in I} |E_n f(x)| = \infty.$$

This striking result indicates that by increasing the degree n of the interpolating polynomial we do not necessarily obtain a better reconstruction of f. For instance, should we use all data of the second column of Table 3.1, we would obtain the interpolating polynomial $\Pi_{12} f$ represented in Figure 3.6, whose behavior in the vicinity of the left-hand of the interval is far less satisfactory than that obtained in Figure 3.5 using a much smaller number of nodes. An even worse result may arise for a special class of functions, as we report in the next example.

Example 3.2 (Runge) If the function $f(x) = 1/(1 + x^2)$ is interpolated at equispaced nodes in the interval $I = (-5, 5)$, the error $\max_{x \in I} |E_n f(x)|$ tends to infinity when $n \to \infty$. This is due to the fact that if $n \to \infty$ the order of magnitude of $\max_{x \in I} |f^{(n+1)}(x)|$ outweighs the infinitesimal order of $h^{n+1}/[4(n+1)]$. This conclusion can be verified by computing the maximum of f and its derivatives up to the order 21 by means of the following MATLAB instructions:

```
syms x; n=20; f=1/(1+x^2); df=diff(f,1);
cdf = char(df);
for i = 1:n+1, df = diff(df,1); cdfn = char(df);
  x = fzero(cdfn,0); M(i) = abs(eval(cdf)); cdf = cdfn;
end
```

3.1 Interpolation

The maximum of the absolute values of the functions $f^{(n)}$, $n = 1, \ldots, 21$, are stored in the vector M. Notice that the command `char` converts the symbolic expression df into a string that can be evaluated by the function `fzero`. In particular, the absolute values of $f^{(n)}$ for $n = 3, 9, 15, 21$ are:

```
>> M([3,9,15,21]) =
ans =
    4.6686e+00   3.2426e+05   1.2160e+12   4.8421e+19
```

while the corresponding values of the maximum of $\prod_{i=0}^{n}(x - x_i)/(n + 1)!$ are

```
z = linspace(-5,5,10000);
for n=0:20; h=10/(n+1); x=[-5:h:5];
    c=poly(x);
    r(n+1)=max(polyval(c,z));
    r(n+1)=r(n+1)/prod([1:n+2]);
end
r([3,9,15,21])

ans =
    2.8935e+00   5.1813e-03   8.5854e-07   2.1461e-11
```

`c=poly(x)` is a vector whose components are the coefficients of that polynomial whose roots are the elements of the vector x. It follows that $\max_{x \in I} |E_n f(x)|$ attains the following values:

```
>> format short e;
    1.3509e+01   1.6801e+03   1.0442e+06   1.0399e+09
```

for $n = 3, 9, 15, 21$, respectively.

The lack of convergence is also indicated by the presence of severe oscillations in the graph of the interpolating polynomial with respect to the graph of f, especially near the endpoints of the interval (see Figure 3.6, right). This behavior is known as *Runge's phenomenon*. ∎

Besides (3.7), the following inequality can also be proved:

$$\max_{x \in I} |f'(x) - (\Pi_n f)'(x)| \leq Ch^n \max_{x \in I} |f^{(n+1)}(x)|,$$

where C is a constant independent of h. Therefore, if we approximate the first derivative of f by the first derivative of $\Pi_n f$, we loose an order of convergence with respect to h.

In MATLAB, $(\Pi_n f)'$ can be computed using the instruction `[d]= polyder(c)`, where c is the input vector in which we store the coefficients of the interpolating polynomial, while d is the output vector where we store the coefficients of its first derivative (see Section 1.4.2).

Octave 3.1 The analogous command in Octave is `[d]=polyderiv (c)`. ∎

See the Exercises 3.1-3.4.

80   3 Approximation of functions and data

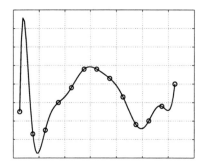 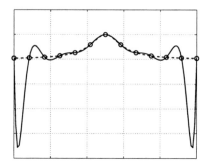

Fig. 3.6. Two examples of Runge's phenomenon: to the left, Π_{12} computed for the data of Table 3.1, column $K = 0.67$; to the right, $\Pi_{12}f$ (*solid line*) computed on 13 equispaced nodes for the function $f(x) = 1/(1+x^2)$ (*dashed line*)

3.1.2 Chebyshev interpolation

Runge's phenomenon can be avoided if a suitable distribution of nodes is used. In particular, in an arbitrary interval $[a, b]$, we can consider the so called *Chebyshev nodes* (see Figure 3.7, right):

$$x_i = \frac{a+b}{2} + \frac{b-a}{2}\widehat{x}_i, \text{ where } \widehat{x}_i = -\cos(\pi i/n), i = 0, \ldots, n \quad (3.8)$$

Obviously, $x_i = \widehat{x}_i$, $i = 0, \ldots, n$, when $[a, b] = [-1, 1]$.

Indeed, for this special distribution of nodes it is possible to prove that, if f is a continuous and differentiable function in $[a, b]$, $\Pi_n f$ converges to f as $n \to \infty$ for all $x \in [a, b]$.

The Chebyshev nodes, which are the abscissas of equispaced nodes on the unit semi-circumference, lie inside $[a, b]$ and are clustered near the endpoints of this interval (see Figure 3.7).

Another non-uniform distribution of nodes in the interval (a, b), sharing the same convergence properties of Chebyshev nodes, is provided by:

$$x_i = \frac{a+b}{2} - \frac{b-a}{2}\cos\left(\frac{2i+1}{n+1}\frac{\pi}{2}\right), i = 0, \ldots, n \quad (3.9)$$

Example 3.3 We consider anew the function f of Runge's example and compute its interpolating polynomial 1at Chebyshev nodes. The latter can be obtained through the following MATLAB instructions:

```
xc = -cos(pi*[0:n]/n);  x = (a+b)*0.5+(b-a)*xc*0.5;
```

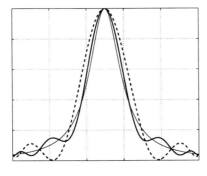
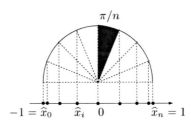

Fig. 3.7. The left side picture shows the comparison between the function $f(x) = 1/(1+x^2)$ (*thin solid line*) and its Chebyshev interpolating polynomials of degree 8 (*dashed line*) and 12 (*solid line*). Note that the amplitude of spurious oscillations decreases as the degree increases. The right side picture shows the distribution of Chebyshev nodes in the interval $[-1, 1]$

where n+1 is the number of nodes, while a and b are the endpoints of the interpolation interval (in the sequel we choose a=-5 and b=5). Then we compute the interpolating polynomial by the following instructions:

```
f= '1./(1+x.^2)'; y = eval(f); c = polyfit(x,y,n);
```

Now let us compute the absolute values of the differences between f and its Chebyshev interpolant at as many as 1001 equispaced points in the interval $[-5, 5]$ and take the maximum error values:

```
x = linspace(-5,5,1000); p=polyval(c,x);
fx = eval(f); err = max(abs(p-fx));
```

As we see in Table 3.3, the maximum of the error decreases when n increases. ∎

n	5	10	20	40
E_n	0.6386	0.1322	0.0177	0.0003

Table 3.3. The Chebyshev interpolation error for Runge's function $f(x) = 1/(1+x^2)$

3.1.3 Trigonometric interpolation and FFT

We want to approximate a periodic function $f : [0, 2\pi] \to \mathbb{C}$, i.e. one satisfying $f(0) = f(2\pi)$, by a trigonometric polynomial $\tilde{f}$ which interpolates f at the $n+1$ nodes $x_j = 2\pi j/(n+1)$, $j = 0, \ldots, n$, i.e.

$$\tilde{f}(x_j) = f(x_j), \text{ for } j = 0, \ldots, n. \tag{3.10}$$

The *trigonometric interpolant* $\tilde{f}$ is obtained by a linear combination of sines and cosines.

In particular, if n is even, $\tilde{f}$ will have the form

$$\tilde{f}(x) = \frac{a_0}{2} + \sum_{k=1}^{M}[a_k \cos(kx) + b_k \sin(kx)], \qquad (3.11)$$

where $M = n/2$ while, if n is odd,

$$\tilde{f}(x) = \frac{a_0}{2} + \sum_{k=1}^{M}[a_k \cos(kx) + b_k \sin(kx)] + a_{M+1}\cos((M+1)x), \qquad (3.12)$$

where $M = (n-1)/2$. We can rewrite (3.11) as

$$\tilde{f}(x) = \sum_{k=-M}^{M} c_k e^{ikx}, \qquad (3.13)$$

i being the imaginary unit. The complex coefficients c_k are related to the coefficients a_k and b_k (complex too) as follows:

$$a_k = c_k + c_{-k}, \quad b_k = i(c_k - c_{-k}), \quad k = 0,\ldots,M. \qquad (3.14)$$

Indeed, from (1.5) it follows that $e^{ikx} = \cos(kx) + i\sin(kx)$ and

$$\sum_{k=-M}^{M} c_k e^{ikx} = \sum_{k=-M}^{M} c_k (\cos(kx) + i\sin(kx))$$
$$= \sum_{k=1}^{M} [c_k(\cos(kx) + i\sin(kx)) + c_{-k}(\cos(kx) - i\sin(kx))] + c_0.$$

Therefore we derive (3.11), thanks to the relations (3.14).

Analogously, when n is odd, (3.12) becomes

$$\tilde{f}(x) = \sum_{k=-(M+1)}^{M+1} c_k e^{ikx}, \qquad (3.15)$$

where the coefficients c_k for $k = 0,\ldots,M$ are the same as before, while $c_{M+1} = c_{-(M+1)} = a_{M+1}/2$. In both cases, we could write

$$\tilde{f}(x) = \sum_{k=-(M+\mu)}^{M+\mu} c_k e^{ikx}, \qquad (3.16)$$

with $\mu = 0$ if n is even and $\mu = 1$ if n is odd. Should f be real valued, its coefficients c_k satisfy $c_{-k} = \bar{c}_k$; from (3.14) it follows that the coefficients a_k and b_k are all real.

3.1 Interpolation   83

Because of its analogy with Fourier series, $\tilde{f}$ is called a *discrete Fourier series*. Imposing the interpolation condition at the nodes $x_j = jh$, with $h = 2\pi/(n+1)$, we find that

$$\sum_{k=-(M+\mu)}^{M+\mu} c_k e^{ikjh} = f(x_j), \qquad j = 0, \ldots, n. \tag{3.17}$$

For the computation of the coefficients $\{c_k\}$ let us multiply equations (3.17) by $e^{-imx_j} = e^{-imjh}$, where m is an integer between 0 and n, and then sum with respect to j:

$$\sum_{j=0}^{n} \sum_{k=-(M+\mu)}^{M+\mu} c_k e^{ikjh} e^{-imjh} = \sum_{j=0}^{n} f(x_j) e^{-imjh}. \tag{3.18}$$

We now require the following identity:

$$\sum_{j=0}^{n} e^{ijh(k-m)} = (n+1)\delta_{km}.$$

This identity is obviously true if $k = m$. When $k \neq m$, we have

$$\sum_{j=0}^{n} e^{ijh(k-m)} = \frac{1 - (e^{i(k-m)h})^{n+1}}{1 - e^{i(k-m)h}}.$$

The numerator on the right hand side is null, since

$$1 - e^{i(k-m)h(n+1)} = 1 - e^{i(k-m)2\pi}$$

$$= 1 - \cos((k-m)2\pi) - i\sin((k-m)2\pi).$$

Therefore, from (3.18) we get the following explicit expression for the coefficients of $\tilde{f}$:

$$\boxed{c_k = \frac{1}{n+1} \sum_{j=0}^{n} f(x_j) e^{-ikjh}, \qquad k = -(M+\mu), \ldots, M+\mu} \tag{3.19}$$

The computation of all the coefficients $\{c_k\}$ can be accomplished with an order $n \log_2 n$ operations by using the *fast Fourier transform* (FFT), which is implemented in the MATLAB program fft (see Example 3.4). Similar conclusions hold for the inverse transform through which we obtain the values $\{f(x_j)\}$ from the coefficients $\{c_k\}$. The inverse fast Fourier transform is implemented in the MATLAB program ifft.

fft
ifft

Example 3.4 Consider the function $f(x) = x(x - 2\pi)e^{-x}$ for $x \in [0, 2\pi]$. To use the MATLAB program fft we first compute the values of f at the nodes $x_j = j\pi/5$ for $j = 0, \ldots, 9$ by the following instructions (recall that .* is the component-by-component vector product):

84   3 Approximation of functions and data

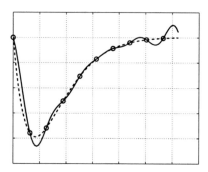

Fig. 3.8. The function $f(x) = x(x - 2\pi)e^{-x}$ (*dashed line*) and the corresponding trigonometric interpolant (*continuous line*) relative to 10 equispaced nodes

```
x=pi/5*[0:9]; y=x.*(x-2*pi).*exp(-x);
```

Now by the FFT we compute the vector of the Fourier coefficients, $Y = (n+1)[c_0, \ldots, c_{M+\mu}, c_{-M}, \ldots, c_{-1}]$, by the following instructions:
```
Y=fft(y);
```

```
Y =
Columns 1 and 2:
  -6.52032 + 0.00000i   -0.46728 + 4.20012i
Columns 3 and 4:
   1.26805 + 1.62110i    1.09849 + 0.60080i
Columns 5 and 6:
   0.92585 + 0.21398i    0.87010 + 0.00000i
Columns 7 and 8:
   0.92585 - 0.21398i    1.09849 - 0.60080i
Columns 9 and 10:
   1.26805 - 1.62110i   -0.46728 - 4.20012i
```

Note that the program `ifft` achieves the maximum efficiency when n is a power of 2, even though it works for any value of n.   ∎

interpft   The command `interpft` provides the trigonometric interpolant of a set of data. It requires in input an integer m and a vector of values which represent the values taken by a function (periodic with period p) at the set of points $x_j = jp/(n+1)$, $j = 0, \ldots, n$. `interpft` returns the m values of the trigonometric interpolant, obtained by the Fourier transform, at the nodes $t_i = ip/m$, $i = 0, \ldots, m-1$. For instance, let us reconsider the function of Example 3.4 in $[0, 2\pi]$ and take its values at 10 equispaced nodes $x_j = j\pi/5$, $j = 0, \ldots, 9$. The values of the trigonometric interpolant at, say, the 100 equispaced nodes $t_i = i\pi/100$, $i = 0, \ldots, 99$ can be obtained as follows (see Figure 3.8)
```
x=pi/5*[0:9]; y=x.*(x-2*pi).*exp(-x); z=interpft(y,100);
```

In some cases the accuracy of trigonometric interpolation can dramatically downgrade, as shown in the following example.

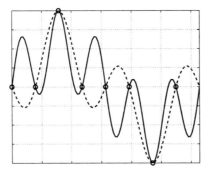

Fig. 3.9. The effects of aliasing: comparison between the function $f(x) = \sin(x)+\sin(5x)$ (*solid line*) and its trigonometric interpolant (3.11) with $M = 3$ (*dashed line*)

Example 3.5 Let us approximate the function $f(x) = f_1(x) + f_2(x)$, with $f_1(x) = \sin(x)$ and $f_2(x) = \sin(5x)$, using nine equispaced nodes in the interval $[0, 2\pi]$. The result is shown in Figure 3.9. Note that in some intervals the trigonometric approximant shows even a phase inversion with respect to the function f. ∎

This lack of accuracy can be explained as follows. At the nodes considered, the function f_2 is indistinguishable from $f_3(x) = -\sin(3x)$ which has a lower frequency (see Figure 3.10). The function that is actually approximated is therefore $F(x) = f_1(x) + f_3(x)$ and not $f(x)$ (in fact, the dashed line of Figure 3.9 does coincide with F).

This phenomenon is known as *aliasing* and may occur when the function to be approximated is the sum of several components having different frequencies. As soon as the number of nodes is not enough to resolve the highest frequencies, the latter may interfere with the low frequencies, giving rise to inaccurate interpolants. To get a better approximation for functions with higher frequencies, one has to increase the number of interpolation nodes.

A real life example of aliasing is provided by the apparent inversion of the sense of rotation of spoked wheels. Once a certain critical velocity is reached the human brain is no longer able to accurately sample the moving image and, consequently, produces distorted images.

Let us summarize

1. Approximating a set of data or a function f in $[a, b]$ consists of finding a suitable function $\tilde{f}$ that represents them with enough accuracy;
2. the interpolation process consists of determining a function $\tilde{f}$ such that $\tilde{f}(x_i) = y_i$, where the $\{x_i\}$ are given nodes and $\{y_i\}$ are either the values $\{f(x_i)\}$ or a set of prescribed values;

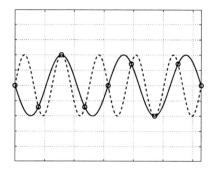

Fig. 3.10. The phenomenon of aliasing: the functions $\sin(5x)$ (*dashed line*) and $-\sin(3x)$ (*dotted line*) take the same values at the interpolation nodes. This circumstance explains the severe loss of accuracy shown in Figure 3.9

3. if the $n+1$ nodes $\{x_i\}$ are distinct, there exists a unique polynomial of degree less than or equal to n interpolating a set of prescribed values $\{y_i\}$ at the nodes $\{x_i\}$;
4. for an equispaced distribution of nodes in $[a,b]$ the interpolation error at any point of $[a,b]$ does not necessarily tend to 0 as n tends to infinity. However, there exist special distributions of nodes, for instance the Chebyshev nodes, for which this convergence property holds true for all continuous functions;
5. trigonometric interpolation is well suited to approximate periodic functions, and is based on choosing $\tilde{f}$ as a linear combination of sine and cosine functions. The FFT is a very efficient algorithm which allows the computation of the Fourier coefficients of a trigonometric interpolant from its node values and admits an equally fast inverse, the IFFT.

3.2 Piecewise linear interpolation

The Chebyshev interpolant provides an accurate approximation of smooth functions f whose expression is known. In the case when f is nonsmooth or when f is only known by its values at a set of given points (which do not coincide with the Chebyshev nodes), one can resort to a different interpolation method which is called linear composite interpolation.

More precisely, given a distribution (not necessarily uniform) of nodes $x_0 < x_1 < \ldots < x_n$, we denote by I_i the interval $[x_i, x_{i+1}]$. We approximate f by a continuous function which, on each interval, is given by the segment joining the two points $(x_i, f(x_i))$ and $(x_{i+1}, f(x_{i+1}))$ (see Figure 3.11). This function, denoted by $\Pi_1^H f$, is called *piecewise linear interpolation polynomial* of f and its expression is:

$$\Pi_1^H f(x) = f(x_i) + \frac{f(x_{i+1}) - f(x_i)}{x_{i+1} - x_i}(x - x_i) \quad \text{for } x \in I_i.$$

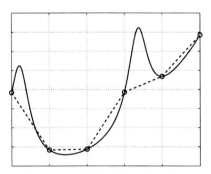

Fig. 3.11. The function $f(x) = x^2 + 10/(\sin(x) + 1.2)$ (*solid line*) and its piecewise linear interpolation polynomial $\Pi_1^H f$ (*dashed line*)

The upper-index H denotes the maximum length of the intervals I_i.

The following result can be inferred from (3.7) setting $n = 1$ and $h = H$:

Proposition 3.3 *If $f \in C^2(I)$, where $I = [x_0, x_n]$, then*

$$\max_{x \in I} |f(x) - \Pi_1^H f(x)| \le \frac{H^2}{8} \max_{x \in I} |f''(x)|.$$

Consequently, for all x in the interpolation interval, $\Pi_1^H f(x)$ tends to $f(x)$ when $H \to 0$, provided that f is sufficiently smooth.

Through the instruction s1=interp1(x,y,z) one can compute the values at arbitrary points, which are stored in the vector z, of the piecewise linear polynomial that interpolates the values y(i) at the nodes x(i), for i = 1,...,n+1. Note that z can have arbitrary dimension. If the nodes are in increasing order (i.e. x(i+1) > x(i), for i=1,...,n) then we can use the quicker version interp1q (q stands for quickly). Notice that interp1q is quicker than interp1 on non-uniformly spaced data because it does not make any input checking.

interp1

interp1q

It is worth mentioning that the command fplot, which is used to display the graph of a function f on a given interval $[a, b]$, does indeed replace the function by its piecewise linear interpolant. The set of interpolating nodes is generated automatically from the function, following the criterion of clustering these nodes around points where f shows strong variations. A procedure of this type is called *adaptive*.

Octave 3.2 `interp1q` is not available in Octave. ∎

3.3 Approximation by spline functions

As done for piecewise linear interpolation, piecewise polynomial interpolation of degree $n \geq 2$ can be defined as well. For instance, the piecewise quadratic interpolation $\Pi_2^H f$ is a continuous function that on each interval I_i replaces f by its quadratic interpolation polynomial at the endpoints of I_i and at its midpoint. If $f \in C^3(I)$, the error $f - \Pi_2^H f$ in the maximum norm decays as H^3 if H tends to zero.

The main drawback of this piecewise interpolation is that $\Pi_k^H f$ with $k \geq 1$, is nothing more than a global continuous function. As a matter of fact, in several applications, e.g. in computer graphics, it is desirable to get approximation by smooth functions which have at least a continuous derivative.

With this aim, we can construct a function s_3 with the following properties:

1. on each interval $I_i = [x_i, x_{i+1}]$, for $i = 0, \ldots, n-1$, s_3 is a polynomial of degree 3 which interpolates the pairs of values $(x_j, f(x_j))$ for $j = i, i+1$;
2. s_3 has continuous first and second derivatives in the nodes x_i, $i = 1, \ldots, n-1$.

For its complete determination, we need four conditions on each interval, therefore a total of $4n$ equations, which we can provide as follows:

- $n+1$ conditions arise from the interpolation requirement at the nodes x_i, $i = 0, \ldots, n$;
- $n-1$ further equations follow from the requirement of continuity of the polynomial at the internal nodes $x_1, \ldots, x_{n-1}$;
- $2(n-1)$ new equations are obtained by requiring that both first and second derivatives be continuous at the internal nodes.

We still lack two further equations, which we can e.g. choose as

$$s_3''(x_0) = 0, \quad s_3''(x_n) = 0. \tag{3.20}$$

The function s_3 which we obtain in this way, is called a *natural interpolating cubic spline*.

By choosing suitably the unknowns (see [QSS06, Section 8.6.1]) to represent s_3 we arrive at a $(n+1) \times (n+1)$ system with a tridiagonal matrix whose solution can be accomplished by a number of operations proportional to n (see Section 5.4) whose solutions are the values $s''(x_i)$ for $i = 0, \ldots, n$.

3.3 Approximation by spline functions

Using Program 3.1, this solution can be obtained with a number of operations equal to the dimension of the system itself (see Section 5.4). The input parameters are the vectors x and y of the nodes and the data to interpolate, plus the vector zi of the abscissae where we want the spline s_3 to be evaluated.

Other conditions can be chosen in place of (3.20) in order to close the system of equations; for instance we could prescribe the value of the first derivative of s_3 at both endpoints x_0 and x_n.

Unless otherwise specified, Program 3.1 computes the natural interpolation cubic spline. The optimal parameters type and der (a vector with two components) serve the purpose of selecting other types of splines. With type=0 Program 3.1 computes the interpolating cubic spline whose first derivative is given by der(1) at x_0 and der(2) at x_n. With type=1 we obtain the interpolating cubic spline whose values of the second derivative at the endpoints is given by der(1) at x_0 and der(2) at x_n.

Program 3.1. cubicspline: interpolating cubic spline

```
function s=cubicspline(x,y,zi,type,der)
%CUBICSPLINE compute a cubic spline
% S=CUBICSPLINE(X,Y,ZI) computes the value at the
% abscissae ZI of the natural interpolating cubic
% spline that interpolates the values Y at the nodes X.
% S=CUBICSPLINE(X,Y,ZI,TYPE,DER) if TYPE=0 computes the
% values at the abscissae ZI of the cubic spline
% interpolating the values Y with first derivative at
% the endpoints equal to the values DER(1) and DER(2).
% If TYPE=1 the values DER(1) and DER(2) are those of
% the second derivative at the endpoints.
[n,m]=size(x);
if n == 1
    x = x';   y = y';   n = m;
end
if nargin == 3
    der0 = 0; dern = 0; type = 1;
else
    der0 = der(1); dern = der(2);
end
h = x(2:end)-x(1:end-1);
e = 2*[h(1); h(1:end-1)+h(2:end); h(end)];
A = spdiags([[h; 0] e [0; h]],-1:1,n,n);
d = (y(2:end)-y(1:end-1))./h;
rhs = 3*(d(2:end)-d(1:end-1));
if type == 0
    A(1,1) = 2*h(1);    A(1,2) = h(1);
    A(n,n) = 2*h(end);  A(end,end-1) = h(end);
    rhs = [3*(d(1)-der0); rhs; 3*(dern-d(end))];
else
    A(1,:) = 0; A(1,1) = 1;
    A(n,:) = 0; A(n,n) = 1;
    rhs = [der0; rhs; dern];
end
S = zeros(n,4);
S(:,3) = A\rhs;
```

90   3 Approximation of functions and data

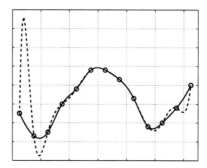

Fig. 3.12. Comparison between the interpolating cubic spline and the Lagrange interpolant for the case considered in Example 3.6

```
for m = 1:n-1
    S(m,4) = (S(m+1,3)-S(m,3))/3/h(m);
    S(m,2) = d(m) - h(m)/3*(S(m + 1,3)+2*S(m,3));
    S(m,1) = y(m);
end
S = S(1:n-1, 4:-1:1);   pp = mkpp(x,S);   s = ppval(pp,zi);
return
```

spline   The MATLAB command spline (see also the toolbox splines) enforces the third derivative of s_3 to be continuous at x_1 and x_{n-1}. To this condition is given the curious name of *not-a-knot condition*. The input parameters are the vectors x and y and the vector zi (same meaning as
mkpp   before). The commands mkpp and ppval that are used in Program 3.1
ppval   are useful to build up and evaluate a composite polynomial.

Example 3.6 Let us reconsider the data of Table 3.1 corresponding to the column $K = 0.67$ and compute the associated interpolating cubic spline s_3. The different values of the latitude provide the nodes x_i, $i = 0, \ldots, 12$. If we are interested in computing the values $s_3(z_i)$, where $z_i = -55 + i$, $i = 0, \ldots, 120$, we can proceed as follows:

```
x = [-55:10:65];
y = [-3.25 -3.37 -3.35 -3.2 -3.12 -3.02 -3.02 ...
     -3.07 -3.17 -3.32 -3.3 -3.22 -3.1];
z = [-55:1:65];
s = spline(x,y,z);
```

The graph of s_3, which is reported in Figure 3.12, looks more plausible than that of the Lagrange interpolant at the same nodes. ∎

Example 3.7 (Robotics) To find the trajectory of the robot satisfying the given constraints, we split the time interval $[0, 5]$ in the two subintervals $[0, 2]$ and $[2, 5]$. Then in each subinterval we look for two splines, $x = x(t)$ and $y = y(t)$, that interpolate the given values and have null derivative at the endpoints. Using Program 3.1 we obtain the desired result by the following instructions:

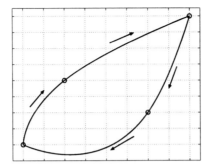

Fig. 3.13. The trajectory in the xy plane of the robot described in Problem 3.4. Circles represent the position of the control points through which the robot should pass during its motion

```
x1   = [0 1 4];  y1  = [0 2 4];
t1   = [0 1 2];  ti1 = [0:0.01:2];
x2   = [0 3 4];  y2  = [0 1 4];
t2   = [0 2 3];  ti2 = [0:0.01:3];  d=[0,0];
six1 = cubicspline(t1,x1,ti1,0,d);
siy1 = cubicspline(t1,y1,ti1,0,d);
six2 = cubicspline(t2,x2,ti2,0,d);
siy2 = cubicspline(t2,y2,ti2,0,d);
```

The trajectory obtained is drawn in Figure 3.13. ∎

The error that we obtain in approximating a function f (continuously differentiable up to its fourth derivative) by the natural interpolating cubic spline satisfies the following inequalities:

$$\max_{x \in I} |f^{(r)}(x) - s_3^{(r)}(x)| \leq C_r H^{4-r} \max_{x \in I} |f^{(4)}(x)|, \ r = 0, 1, 2, 3,$$

where $I = [x_0, x_n]$ and $H = \max_{i=0,\ldots,n-1}(x_{i+1} - x_i)$, while C_r is a suitable constant depending on r, but independent of H. It is then clear that not only f, but also its first, second and third derivatives are well approximated by s_3 when H tends to 0.

Remark 3.1 In general cubic splines do not preserve monotonicity between neighbouring nodes. For instance, by approximating the unitary circumference in the first quarter using the points ($x_k = \sin(k\pi/6), y_k = \cos(k\pi/6)$), for $k = 0, \ldots, 3$, we would obtain an oscillatory spline (see Figure 3.14). In these cases, other approximation techniques can be better suited. For instance, the MATLAB command pchip provides the Hermite piecewise cubic interpolant which is locally monotone and interpolates the function as well as its first derivative at the nodes $\{x_i, i = 1, \ldots, n-1\}$ (see Figure 3.14). The Hermite interpolant can be obtained by using the following instructions: pchip

```
t = linspace(0,pi/2,4)
x = cos(t); y = sin(t);
```

```
xx = linspace(0,1,40);
plot(x,y,'o',xx,[pchip(x,y,xx);spline(x,y,xx)])
```

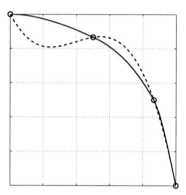

Fig. 3.14. Approximation of the first quarter of the circumference of the unitary circle using only 4 nodes. The dashed line is the cubic spline, while the continuous line is the piecewise cubic Hermite interpolant

See the Exercises 3.5-3.8.

3.4 The least-squares method

As already noticed, a Lagrange interpolation does not guarantee a better approximation of a given function when the polynomial degree gets large. This problem can be overcome by composite interpolation (such as piecewise linear polynomials or splines). However, neither are suitable to extrapolate information from the available data, that is, to generate new values at points lying outside the interval where interpolation nodes are given.

Example 3.8 (Finance) On the basis of the data reported in Figure 3.1, we would like to predict whether the stock price will increase or diminish in the coming days. The Lagrange polynomial interpolation is impractical, as it would require a (tremendously oscillatory) polynomial of degree 719 which will provide a completely erroneous prediction. On the other hand, piecewise linear interpolation, whose graph is reported in Figure 3.1, provides extrapolated results by exploiting only the values of the last two days, thus completely neglecting the previous history. To get a better result we should avoid the interpolation requirement, by invoking least-squares approximation as indicated below. ∎

3.4 The least-squares method

Assume that the data $\{(x_i, y_i), i = 0, \ldots, n\}$ are available, where now y_i could represent the values $f(x_i)$ attained by a given function f at the nodes x_i. For a given integer $m \geq 1$ (usually, $m \ll n$) we look for a polynomial $\tilde{f} \in \mathbb{P}_m$ which satisfies the inequality

$$\sum_{i=0}^{n}[y_i - \tilde{f}(x_i)]^2 \leq \sum_{i=0}^{n}[y_i - p_m(x_i)]^2 \qquad (3.21)$$

for every polynomial $p_m \in \mathbb{P}_m$. Should it exist, $\tilde{f}$ will be called the *least-squares approximation* in $\mathbb{P}_m$ of the set of data $\{(x_i, y_i), i = 0, \ldots, n\}$. Unless $m \geq n$, in general it will not be possible to guarantee that $\tilde{f}(x_i) = y_i$ for all $i = 0, \ldots, n$.

Setting

$$\tilde{f}(x) = a_0 + a_1 x + \ldots + a_m x^m, \qquad (3.22)$$

where the coefficients $a_0, \ldots, a_m$ are unknown, the problem (3.21) can be restated as follows: find $a_0, a_1, \ldots, a_m$ such that

$$\Phi(a_0, a_1, \ldots, a_m) = \min_{\{b_i,\ i=0,\ldots,m\}} \Phi(b_0, b_1, \ldots, b_m)$$

where

$$\Phi(b_0, b_1, \ldots, b_m) = \sum_{i=0}^{n}[y_i - (b_0 + b_1 x_i + \ldots + b_m x_i^m)]^2.$$

We solve this problem in the special case when $m = 1$. Since

$$\Phi(b_0, b_1) = \sum_{i=0}^{n}\left[y_i^2 + b_0^2 + b_1^2 x_i^2 + 2b_0 b_1 x_i - 2b_0 y_i - 2b_1 x_i y_i^2\right],$$

the graph of Φ is a convex paraboloid. The point (a_0, a_1) at which Φ attains its minimum satisfies the conditions

$$\frac{\partial \Phi}{\partial b_0}(a_0, a_1) = 0, \qquad \frac{\partial \Phi}{\partial b_1}(a_0, a_1) = 0,$$

where the symbol $\partial \Phi / \partial b_j$ denotes the partial derivative (that is, the rate of variation) of Φ with respect to b_j, after having frozen the remaining variable (see the definition 8.3).

By explicitly computing the two partial derivatives we obtain

$$\sum_{i=0}^{n}[a_0 + a_1 x_i - y_i] = 0, \ \sum_{i=0}^{n}[a_0 x_i + a_1 x_i^2 - x_i y_i] = 0,$$

which is a system of two equations for the two unknowns a_0 and a_1:

$$a_0(n+1) + a_1 \sum_{i=0}^{n} x_i = \sum_{i=0}^{n} y_i,$$
$$a_0 \sum_{i=0}^{n} x_i + a_1 \sum_{i=0}^{n} x_i^2 = \sum_{i=0}^{n} y_i x_i.$$
(3.23)

Setting $D = (n+1) \sum_{i=0}^{n} x_i^2 - (\sum_{i=0}^{n} x_i)^2$, the solution reads:

$$a_0 = \frac{1}{D}(\sum_{i=0}^{n} y_i \sum_{j=0}^{n} x_j^2 - \sum_{j=0}^{n} x_j \sum_{i=0}^{n} x_i y_i),$$
$$a_1 = \frac{1}{D}((n+1) \sum_{i=0}^{n} x_i y_i - \sum_{j=0}^{n} x_j \sum_{i=0}^{n} y_i).$$
(3.24)

The corresponding polynomial $\tilde{f}(x) = a_0 + a_1 x$ is known as the *least-squares straight line*, or *regression line*.

The previous approach can be generalized in several ways. The first generalization is to the case of an arbitrary m. The associated $(m+1) \times (m+1)$ linear system, which is symmetric, will have the form:

$$a_0(n+1) + a_1 \sum_{i=0}^{n} x_i + \ldots + a_m \sum_{i=0}^{n} x_i^m = \sum_{i=0}^{n} y_i,$$
$$a_0 \sum_{i=0}^{n} x_i + a_1 \sum_{i=0}^{n} x_i^2 + \ldots + a_m \sum_{i=0}^{n} x_i^{m+1} = \sum_{i=0}^{n} x_i y_i,$$
$$\vdots \qquad \vdots \qquad \vdots \qquad \vdots$$
$$a_0 \sum_{i=0}^{n} x_i^m + a_1 \sum_{i=0}^{n} x_i^{m+1} + \ldots + a_m \sum_{i=0}^{n} x_i^{2m} = \sum_{i=0}^{n} x_i^m y_i.$$

When $m = n$, the least-squares polynomial must coincide with the Lagrange interpolating polynomial Π_n (see Exercise 3.9).

The MATLAB command c=polyfit(x,y,m) computes by default the coefficients of the polynomial of degree m which approximates n+1 pairs of data (x(i),y(i)) in the least-squares sense. As already noticed in Section 3.1.1, when m is equal to n it returns the interpolating polynomial.

Example 3.9 (Finance) In Figure 3.15 we draw the graphs of the least-squares polynomials of degree 1, 2 and 4 that approximate in the least-squares sense the data of Figure 3.1. The polynomial of degree 4 reproduces quite reasonably the behavior of the stock price in the considered time interval and suggests that in the near future the quotation will increase. ∎

Example 3.10 (Biomechanics) Using the least-squares method we can answer the question in Problem 3.3 and discover that the line which better approximates the given data has equation $\epsilon(\sigma) = 0.3471\sigma + 0.0654$ (see Figure

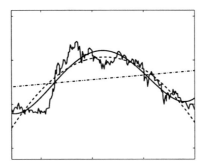

Fig. 3.15. Least-squares approximation of the data of Problem 3.2 of degree 1 (*dashed-dotted line*), degree 2 (*dashed line*) and degree 4 (*thick solid line*). The exact data are represented by the *thin solid line*

3.16); when $\sigma = 0.9$ it provides the estimate $\epsilon = 0.2915$ for the deformation. ∎

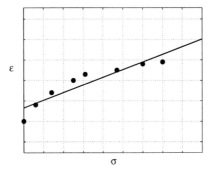

Fig. 3.16. Linear least-squares approximation of the data of Problem 3.3

A further generalization of the least-squares approximation consists of using in (3.21) $\tilde{f}$ and p_m that are no-longer polynomials but functions of a space V_m obtained by linearly combining $m+1$ independent functions $\{\psi_j, j = 0, \ldots, m\}$. Special instances are provided, e.g., by the trigonometric functions $\psi_j(x) = \cos(\gamma j x)$ (for a given parameter $\gamma \neq 0$), by the exponential functions $\psi_j(x) = e^{\delta j x}$ (for some $\delta > 0$), or by a suitable set of spline functions.

The choice of the functions $\{\psi_j\}$ is actually dictated by the conjectured behavior of the law underlying the given data distribution. For instance, in Figure 3.17 we draw the graph of the least-squares approximation of the data of the Example 3.1 computed using the trigonometric functions $\psi_j(x) = \cos(jt(x))$, $j = 0, \ldots, 4$, with $t(x) = 120(\pi/2)(x+55)$. We assume that the data are periodic with period $120(\pi/2)$.

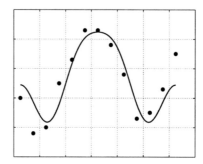

Fig. 3.17. The least-squares approximation of the data of the Problem 3.1 using a cosine basis. The exact data are represented by the small circles

The reader can verify that the unknown coefficients of

$$\tilde{f}(x) = \sum_{j=0}^{m} a_j \psi_j(x),$$

can be obtained by solving the following system (of *normal equations*)

$$\boxed{B^T B a = B^T y} \tag{3.25}$$

where B is the rectangular matrix $(n+1)\times(m+1)$ of entries $b_{ij} = \psi_j(x_i)$, a is the vector of the unknown coefficients, while **y** is the vector of the data.

Let us summarize

1. The composite piecewise linear interpolant of a function f is a piecewise continuous linear function $\tilde{f}$, which interpolates f at a given set of nodes $\{x_i\}$. With this approximation we avoid Runge's type phenomena when the number of nodes increases;
2. interpolation by cubic splines allows the approximation of f by a piecewise cubic function $\tilde{f}$ which is continuous together with its first and second derivatives;
3. in least-squares approximation we look for an approximant $\tilde{f}$ which is a polynomial of degree m (typically, $m \ll n$) that minimizes the mean-square error $\sum_{i=0}^{n}[y_i - \tilde{f}(x_i)]^2$. The same minimization criterium can be applied for a class of functions that are not polynomials.

See the Exercises 3.9-3.14.

3.5 What we haven't told you

For a more general introduction to the theory of interpolation and approximation the reader is referred to, e.g., [Dav63], [Mei67] and [Gau97].

Polynomial interpolation can also be used to approximate data and functions in several dimensions. In particular, composite interpolation, based on piecewise linear or spline functions, is well suited when the region Ω at hand is partitioned into polygons in 2D (triangles or quadrilaterals) and polyhedra in 3D (tetrahedra or prisms).

A special situation occurs when Ω is a rectangle or a parallelepiped in which case the MATLAB commands interp2, and interp3, respectively, can be used. In both cases it is assumed that we want to represent on a regular, fine lattice (or grid) a function whose values are available on a regular, coarser lattice.   interp2 interp3

Consider for instance the values of $f(x,y) = \sin(2\pi x)\cos(2\pi y)$ on a (coarse) 6×6 lattice of equispaced nodes on the square $[0,1]^2$; these values can be obtained using the commands:

```
[x,y]=meshgrid(0:0.2:1,0:0.2:1);
z=sin(2*pi*x).*cos(2*pi*y);
```

By the command interp2 a cubic spline is first computed on this coarse grid, then evaluated at the nodal points of a finer grid of 21×21 equispaced nodes:

```
xi = [0:0.05:1]; yi=[0:0.05:1];
[xf,yf]=meshgrid(xi,yi);
pi3=interp2(x,y,z,xf,yf);
```

The command meshgrid transforms the set of the couples (xi(k),yi(j)) into two matrices xf and yf that can be used to evaluate functions of two variables and to plot three dimensional surfaces. The rows of xf are copies of the vector xi, the columns of yf are copies of yi. Alternatively to the above procedure we can use the command griddata, available also for three-dimensional data (griddata3) and for the approximation of n-dimensional surfaces (griddatan).   meshgrid griddata

The commands described below are for MATLAB only.

When Ω is a two-dimensional domain of arbitrary shape, it can be partitioned into triangles using the graphical interface pdetool.   pdetool

For a general presentation of spline functions see, e.g., [Die93] and [PBP02]. The MATLAB toolbox splines allows one to explore several applications of spline functions. In particular, the spdemos command gives the user the possibility to investigate the properties of the most important type of spline functions. Rational splines, i.e. functions which are the ratio of two splines functions, are accessible through the commands rpmak and rsmak. Special instances are the so-called NURBS splines, which are commonly used in CAGD (*Computer Assisted Geometric Design*).   spdemos rpmak rsmak

98   3 Approximation of functions and data

In the same context of Fourier approximation, we mention the approximation based on *wavelets*. This type of approximation is largely used for image reconstruction and compression and in signal analysis (for an introduction, see [DL92], [Urb02]). A rich family of wavelets (and their applications) can be found in the MATLAB toolbox `wavelet`.

`wavelet`

3.6 Exercises

Exercise 3.1 Prove inequality (3.6).

Exercise 3.2 Provide an upper bound of the Lagrange interpolation error for the following functions:

$$f_1(x) = \cosh(x), \quad f_2(x) = \sinh(x), \quad x_k = -1 + 0.5k, \quad k = 0, \ldots, 4,$$
$$f_3(x) = \cos(x) + \sin(x), \qquad\qquad x_k = -\pi/2 + \pi k/4, \quad k = 0, \ldots, 4.$$

Exercise 3.3 The following data are related to the life expectation of citizens of two European regions:

	1975	1980	1985	1990
Western Europe	72.8	74.2	75.2	76.4
Eastern Europe	70.2	70.2	70.3	71.2

Use the interpolating polynomial of degree 3 to estimate the life expectation in 1970, 1983 and 1988. Then extrapolate a value for the year 1995. It is known that the life expectation in 1970 was 71.8 years for the citizens of the West Europe, and 69.6 for those of the East Europe. Recalling these data, is it possible to estimate the accuracy of life expectation predicted in the 1995?

Exercise 3.4 The price (in euros) of a magazine has changed as follows:

Nov.87	Dec.88	Nov.90	Jan.93	Jan.95	Jan.96	Nov.96	Nov.00
4.5	5.0	6.0	6.5	7.0	7.5	8.0	8.0

Estimate the price in November 2002 by extrapolating these data.

Exercise 3.5 Repeat the computations carried out in Exercise 3.3, using now the cubic interpolating spline computed by the function `spline`. Then compare the results obtained with the two approaches.

Exercise 3.6 In the table below we report the values of the sea water density ρ (in Kg/m^3) corresponding to different values of the temperature T (in degrees Celsius):

T	4°	8°	12°	16°	20°
ρ	1000.7794	1000.6427	1000.2805	999.7165	998.9700

Compute the associated cubic interpolating spline on 4 subintervals of the temperature interval $[4, 20]$. Then compare the results provided by the spline interpolant with the following ones (which correspond to further values of T):

T	6°	10°	14°	18°
ρ	1000.74088	1000.4882	1000.0224	999.3650

Exercise 3.7 The Italian production of citrus fruit has changed as follows:

year	1965	1970	1980	1985	1990	1991
production ($\times 10^5$ Kg)	17769	24001	25961	34336	29036	33417

Use interpolating cubic splines of different kinds to estimate the production in 1962, 1977 and 1992. Compare these results with the real values: 12380, 27403 and 32059, respectively. Compare the results with those that would be obtained using the Lagrange interpolating polynomial.

Exercise 3.8 Evaluate the function $f(x) = \sin(2\pi x)$ at 21 equispaced nodes in the interval $[-1, 1]$. Compute the Lagrange interpolating polynomial and the cubic interpolating spline. Compare the graphs of these two functions with that of f on the given interval. Repeat the same calculation using the following perturbed set of data: $f(x_i) = \sin(2^*\pi^*x_i) + (-1)^{i+1}10^{-4}$, and observe that the Lagrange interpolating polynomial is more sensitive to small perturbations than the cubic spline.

Exercise 3.9 Verify that if $m = n$ the least-squares polynomial of a function f at the nodes $x_0, \ldots, x_n$ coincides with the interpolating polynomial $\Pi_n f$ at the same nodes.

Exercise 3.10 Compute the least-squares polynomial of degree 4 that approximates the values of K reported in the different columns of Table 3.1.

Exercise 3.11 Repeat the computations carried out in Exercise 3.7 using now a least-squares approximation of degree 3.

Exercise 3.12 Express the coefficients of system (3.23) in terms of the average $M = \frac{1}{(n+1)} \sum_{i=0}^{n} x_i$ and the variance $v = \frac{1}{(n+1)} \sum_{i=0}^{n} (x_i - M)^2$ of the set of data $\{x_i, i = 0, \ldots, n\}$.

Exercise 3.13 Verify that the regression line passes through the point whose abscissa is the average of $\{x_i\}$ and ordinate is the average of $\{f(x_i)\}$.

Exercise 3.14 The following values

flow rate	0	35	0.125	5	0	5	1	0.5	0.125	0

represent the measured values of the blood flow-rate in a cross-section of the carotid artery during a heart beat. The frequency of acquisition of the data is constant and is equal to $10/T$, where $T = 1$ s is the beat period. Represent these data by a continuous function of period equal to T.

4
Numerical differentiation and integration

In this chapter we propose methods for the numerical approximation of derivatives and integrals of functions. Concerning integration, quite often for a generic function it is not possible to find a primitive in an explicit form. Even when a primitive is known, its use might not be easy. This is, e.g., the case of the function $f(x) = \cos(4x)\cos(3\sin(x))$, for which we have

$$\int_0^\pi f(x)dx = \pi \left(\frac{3}{2}\right)^4 \sum_{k=0}^\infty \frac{(-9/4)^k}{k!(k+4)!};$$

the task of computing an integral is transformed into the equally troublesome one of summing a series. In other circumstances the function that we want to integrate or differentiate could only be known on a set of nodes (for instance, when the latter represent the results of an experimental measurement), exactly as happens in the case of function approximation, which was discussed in Chapter 3.

In all these situations it is necessary to consider numerical methods in order to obtain an approximate value of the quantity of interest, independently of how difficult is the function to integrate or differentiate.

Problem 4.1 (Hydraulics) The height $q(t)$ reached at time t by a fluid in a straight cylinder of radius $R = 1$ m with a circular hole of radius $r = 0.1$ m on the bottom, has been measured every 5 seconds yielding the following values

t	0	5	10	15	20
$q(t)$	0.6350	0.5336	0.4410	0.3572	0.2822

We want to compute an approximation of the emptying velocity $q'(t)$ of the cylinder, then compare it with the one predicted by Torricelli's law: $q'(t) = -\gamma(r/R)^2\sqrt{2gq(t)}$, where g is the gravity acceleration and $\gamma = 0.6$ is a correction factor. For the solution of this problem, see Example 4.1. ∎

Problem 4.2 (Optics) In order to plan a room for infrared beams we are interested in calculating the energy emitted by a black body (that is, an object capable of irradiating in all the spectrum to the ambient temperature) in the (infrared) spectrum comprised between $3\mu m$ and $14\mu m$ wavelength. The solution of this problem is obtained by computing the integral

$$E(T) = 2.39 \cdot 10^{-11} \int_{3\cdot 10^{-4}}^{14\cdot 10^{-4}} \frac{dx}{x^5(e^{1.432/(Tx)} - 1)}, \quad (4.1)$$

which is the Planck equation for the energy $E(T)$, where x is the wavelength (in cm) and T the temperature (in Kelvin) of the black body. For its computation see Exercise 4.17. ∎

Problem 4.3 (Electromagnetism) Consider an electric wire sphere of arbitrary radius r and conductivity σ. We want to compute the density distribution of the current $\mathbf{j}$ as a function of r and t (the time), knowing the initial distribution of the current density $\rho(r)$. The problem can be solved using the relations between the current density, the electric field and the charge density and observing that, for the symmetry of the problem, $\mathbf{j}(r,t) = j(r,t)\mathbf{r}/|\mathbf{r}|$, where $j = |\mathbf{j}|$. We obtain

$$j(r,t) = \gamma(r)e^{-\sigma t/\varepsilon_0}, \quad \gamma(r) = \frac{\sigma}{\varepsilon_0 r^2}\int_0^r \rho(\xi)\xi^2\, d\xi, \quad (4.2)$$

where $\varepsilon_0 = 8.859 \cdot 10^{-12}$ farad/m is the dielectric constant of the void. For the computation of this integral, see Exercise 4.16. ∎

Problem 4.4 (Demography) We consider a population of a very large number M of individuals. The distribution $N(h)$ of their height can be represented by a "bell" function characterized by the mean value $\bar{h}$ of the height and the standard deviation σ

$$N(h) = \frac{M}{\sigma\sqrt{2\pi}}e^{-(h-\bar{h})^2/(2\sigma^2)}.$$

Then

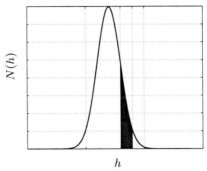

Fig. 4.1. Height distribution of a population of $M = 200$ individuals

$$N = \int_h^{h+\Delta h} N(h)\, dh \tag{4.3}$$

represents the number of individuals whose height is between h and $h + \Delta h$ (for a positive Δh). An instance is provided in Figure 4.1, which corresponds to the case $M = 200$, $\bar{h} = 1.7$ m, $\sigma = 0.1$ m, and the area of the shadowed region gives the number of individuals whose height is in the range 1.8÷1.9 m. For the solution of this problem see Example 4.2. ∎

4.1 Approximation of function derivatives

Consider a function $f : [a, b] \to \mathbb{R}$ continuously differentiable in $[a, b]$. We seek an approximation of the first derivative of f at a generic point $\bar{x}$ in (a, b).

In view of the definition (1.10), for h sufficiently small and positive, we can assume that the quantity

$$\boxed{(\delta_+ f)(\bar{x}) = \frac{f(\bar{x} + h) - f(\bar{x})}{h}} \tag{4.4}$$

is an approximation of $f'(\bar{x})$ which is called the *forward finite difference*. To estimate the error, it suffices to expand f in a Taylor series; if $f \in C^2(a, b)$, we have

$$f(\bar{x} + h) = f(\bar{x}) + h f'(\bar{x}) + \frac{h^2}{2} f''(\xi), \tag{4.5}$$

where ξ is a suitable point in the interval $(\bar{x}, \bar{x} + h)$. Therefore

$$(\delta_+ f)(\bar{x}) = f'(\bar{x}) + \frac{h}{2} f''(\xi), \tag{4.6}$$

and thus $(\delta_+ f)(\bar{x})$ provides a first-order approximation to $f'(\bar{x})$ with respect to h. Still assuming $f \in C^2(a,b)$, with a similar procedure we can derive from the Taylor expansion

$$f(\bar{x} - h) = f(\bar{x}) - hf'(\bar{x}) + \frac{h^2}{2} f''(\eta) \tag{4.7}$$

with $\eta \in (\bar{x} - h, \bar{x})$, the *backward finite difference*

$$\boxed{(\delta_- f)(\bar{x}) = \frac{f(\bar{x}) - f(\bar{x} - h)}{h}} \tag{4.8}$$

which is also first-order accurate. Note that formulae (4.4) and (4.8) can also be obtained by differentiating the linear polynomial interpolating f at the points $\{\bar{x}, \bar{x}+h\}$ and $\{\bar{x}-h, \bar{x}\}$, respectively. In fact, these schemes amount to approximating $f'(\bar{x})$ by the slope of the straight line passing through the two points $(\bar{x}, f(\bar{x}))$ and $(\bar{x}+h, f(\bar{x}+h))$, or $(\bar{x}-h, f(\bar{x}-h))$ and $(\bar{x}, f(\bar{x}))$, respectively (see Figure 4.2).

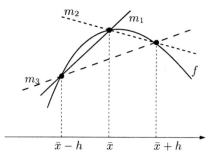

Fig. 4.2. Finite difference approximation of $f'(\bar{x})$: backward (*solid line*), forward (*dotted line*) and centered (*dashed line*). $m_1 = (\delta_- f)(\bar{x})$, $m_2 = (\delta_+ f)(\bar{x})$ and $m_3 = (\delta f)(\bar{x})$ denote the slopes of the three straight lines

Finally, we introduce the *centered finite difference* formula

$$\boxed{(\delta f)(\bar{x}) = \frac{f(\bar{x} + h) - f(\bar{x} - h)}{2h}} \tag{4.9}$$

If $f \in C^3(a,b)$, this formula provides a second-order approximation to $f'(\bar{x})$ with respect to h. Indeed, by expanding $f(\bar{x} + h)$ and $f(\bar{x} - h)$ at the third order around $\bar{x}$ and summing up the two expressions, we obtain

4.2 Numerical integration

$$f'(\bar{x}) - (\delta f)(\bar{x}) = \frac{h^2}{12}[f'''(\xi) + f'''(\eta)], \quad (4.10)$$

where η and ξ are suitable points in the intervals $(\bar{x}-h, \bar{x})$ and $(\bar{x}, \bar{x}+h)$, respectively (see Exercise 4.2).

By (4.9) $f'(\bar{x})$ is approximated by the slope of the straight line passing through the points $(\bar{x} - h, f(\bar{x} - h))$ and $(\bar{x} + h, f(\bar{x} + h))$.

Example 4.1 (Hydraulics) Let us solve Problem 4.1, using formulae (4.4), (4.8) and (4.9), with $h = 5$, to approximate $q'(t)$ at five different points. We obtain:

t	0	5	10	15	20
$q'(t)$	-0.0212	-0.0194	-0.0176	-0.0159	-0.0141
$\delta_+ q$	-0.0203	-0.0185	-0.0168	-0.0150	$--$
$\delta_- q$	$--$	-0.0203	-0.0185	-0.0168	-0.0150
δq	$--$	-0.0194	-0.0176	-0.0159	$--$

The agreement between the exact derivative and the one computed from the finite difference formulae is more satisfactory when using formula (4.9) rather than (4.8) or (4.4). ∎

In general, we can assume that the values of f are available at $n+1$ equispaced points $x_i = x_0 + ih$, $i = 0, \ldots, n$, with $h > 0$. In this case in the numerical derivation $f'(x_i)$ can be approximated by taking one of the previous formulae (4.4), (4.8) or (4.9) with $\bar{x} = x_i$.

Note that the centered formula (4.9) cannot be used at the extrema x_0 and x_n. For these nodes we could use the values

$$\frac{1}{2h}[-3f(x_0) + 4f(x_1) - f(x_2)] \quad \text{at } x_0,$$

$$\frac{1}{2h}[3f(x_n) - 4f(x_{n-1}) + f(x_{n-2})] \text{ at } x_n, \quad (4.11)$$

which are also second-order accurate with respect to h. They are obtained by computing at the point x_0 (respectively, x_n) the first derivative of the polynomial of degree 2 interpolating f at the nodes x_0, x_1, x_2 (respectively, x_{n-2}, x_{n-1}, x_n).

See Exercises 4.1-4.4.

4.2 Numerical integration

In this section we introduce numerical methods suitable for approximating the integral

$$I(f) = \int_a^b f(x)dx,$$

where f is an arbitrary continuous function in $[a,b]$. We start by introducing some simple formulae, which are indeed special instances of the family of Newton-Cotes formulae. Then we will introduce the so-called Gaussian formulae, that feature the highest possible degree of exactness for a given number of evaluations of the function f.

4.2.1 Midpoint formula

A simple procedure to approximate $I(f)$ can be devised by partitioning the interval $[a,b]$ into subintervals $I_k = [x_{k-1}, x_k]$, $k = 1, \ldots, M$, with $x_k = a + kH$, $k = 0, \ldots, M$ and $H = (b-a)/M$. Since

$$I(f) = \sum_{k=1}^{M} \int_{I_k} f(x)dx, \qquad (4.12)$$

on each sub-interval I_k we can approximate the exact integral of f by that of a polynomial $\tilde{f}$ approximating f on I_k. The simplest solution consists in choosing $\tilde{f}$ as the constant polynomial interpolating f at the middle point of I_k:

$$\bar{x}_k = \frac{x_{k-1} + x_k}{2}.$$

In such a way we obtain the *composite midpoint quadrature formula*

$$\boxed{I_{mp}^c(f) = H \sum_{k=1}^{M} f(\bar{x}_k)} \qquad (4.13)$$

The symbol mp stands for midpoint, while c stands for composite. This formula is second-order accurate with respect to H. More precisely, if f is continuously differentiable up to its second derivative in $[a,b]$, we have

$$I(f) - I_{mp}^c(f) = \frac{b-a}{24} H^2 f''(\xi), \qquad (4.14)$$

where ξ is a suitable point in $[a,b]$ (see Exercise 4.6). Formula (4.13) is also called the *composite rectangle quadrature formula* because of its geometrical interpretation, which is evident from Figure 4.3. The classical *midpoint formula* (or *rectangle formula*) is obtained by taking $M = 1$ in (4.13), i.e. using the midpoint rule directly on the interval (a,b):

$$\boxed{I_{mp}(f) = (b-a)f[(a+b)/2]} \qquad (4.15)$$

The error is now given by

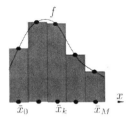 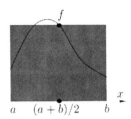

Fig. 4.3. The composite midpoint formula (*left*); the midpoint formula (*right*)

$$I(f) - I_{mp}(f) = \frac{(b-a)^3}{24}f''(\xi), \qquad (4.16)$$

where ξ is a suitable point in $[a,b]$. Relation (4.16) follows as a special case of (4.14), but it can also be proved directly. Indeed, setting $\bar{x} = (a+b)/2$, we have

$$I(f) - I_{mp}(f) = \int_a^b [f(x) - f(\bar{x})]dx$$

$$= \int_a^b f'(\bar{x})(x-\bar{x})dx + \frac{1}{2}\int_a^b f''(\eta(x))(x-\bar{x})^2 dx,$$

where $\eta(x)$ is a suitable point in the interval whose endpoints are x and $\bar{x}$. Then (4.16) follows because $\int_a^b (x-\bar{x})dx = 0$ and, by the mean value theorem for integrals, there exists $\xi \in [a,b]$ such that

$$\frac{1}{2}\int_a^b f''(\eta(x))(x-\bar{x})^2 dx = \frac{1}{2}f''(\xi)\int_a^b (x-\bar{x})^2 dx = \frac{(b-a)^3}{24}f''(\xi).$$

The *degree of exactness* of a quadrature formula is the maximum integer $r \geq 0$ for which the approximate integral (produced by the quadrature formula) of any polynomial of degree r is equal to the exact integral. We can deduce from (4.14) and (4.16) that the midpoint formula has degree of exactness 1, since it integrates exactly all polynomials of degree less than or equal to 1 (but not all those of degree 2).

The midpoint composite quadrature formula is implemented in Program 4.1. Input parameters are the endpoints of the integration interval a and b, the number of subintervals M and the MATLAB function f to define the function f.

Program 4.1. midpointc: composite midpoint quadrature formula

```
function Imp=midpointc(a,b,M,f,varargin)
%MIDPOINTC Composite midpoint numerical integration.
%   IMP = MIDPOINTC(A,B,M,FUN) computes an approximation
```

```
% of the integral of the function FUN via the midpoint
% method (with M equispaced intervals). FUN accepts a
% real vector input x and returns a real vector value.
% FUN can also be an inline object.
% IMP=MIDPOINT(A,B,M,FUN,P1,P2,...) calls the function
% FUN passing the optional parameters P1,P2,... as
% FUN(X,P1,P2,...).
H=(b-a)/M;
x = linspace(a+H/2,b-H/2,M);
fmp=feval(f,x,varargin{:}).*ones(1,M);
Imp=H*sum(fmp);
return
```

See the Exercises 4.5-4.8.

4.2.2 Trapezoidal formula

Another formula can be obtained by replacing f on I_k by the linear polynomial interpolating f at the nodes x_{k-1} and x_k (equivalently, replacing f by $\Pi_1^H f$, see Section 3.2, on the whole interval (a,b)). This yields

$$I_t^c(f) = \frac{H}{2} \sum_{k=1}^{M} [f(x_k) + f(x_{k-1})]$$
$$= \frac{H}{2}[f(a) + f(b)] + H \sum_{k=1}^{M-1} f(x_k) \qquad (4.17)$$

This formula is called the *composite trapezoidal formula*, and is second-order accurate with respect to H. In fact, one can obtain the expression

$$I(f) - I_t^c(f) = -\frac{b-a}{12} H^2 f''(\xi) \qquad (4.18)$$

for the quadrature error for a suitable point $\xi \in [a,b]$, provided that $f \in C^2([a,b])$. When (4.17) is used with $M = 1$, we obtain

$$I_t(f) = \frac{b-a}{2}[f(a) + f(b)] \qquad (4.19)$$

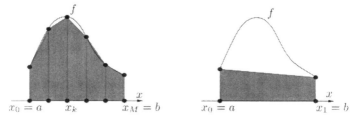

Fig. 4.4. Composite trapezoidal formula (*left*); trapezoidal formula (*right*)

4.2 Numerical integration

which is called the *trapezoidal formula* because of its geometrical interpretation. The error induced is given by

$$I(f) - I_t(f) = -\frac{(b-a)^3}{12} f''(\xi), \qquad (4.20)$$

where ξ is a suitable point in $[a, b]$. We can deduce that (4.19) has degree of exactness equal to 1, as is the case of the midpoint rule.

The composite trapezoidal formula (4.17) is implemented in the MATLAB programs `trapz` and `cumtrapz`. If x is a vector whose components are the abscissae x_k, $k = 0, \ldots, M$ (with $x_0 = a$ and $x_M = b$), and y that of the values $f(x_k)$, $k = 0, \ldots, M$, z=cumtrapz(x,y) returns the vector z whose components are $z_k \simeq \int_a^{x_k} f(x)dx$, the integral being approximated by the composite trapezoidal rule. Thus z(M+1) is an approximation of the integral of f on (a, b).

`trapz`
`cumtrapz`

See the Exercises 4.9-4.11.

4.2.3 Simpson formula

The Simpson formula can be obtained by replacing the integral of f over each I_k by that of its interpolating polynomial of degree 2 at the nodes $x_{k-1}, \bar{x}_k = (x_{k-1} + x_k)/2$ and x_k,

$$\Pi_2 f(x) = \frac{2(x - \bar{x}_k)(x - x_k)}{H^2} f(x_{k-1})$$
$$+ \frac{4(x_{k-1} - x)(x - x_k)}{H^2} f(\bar{x}_k) + \frac{2(x - \bar{x}_k)(x - x_{k-1})}{H^2} f(x_k).$$

The resulting formula is called the *composite Simpson quadrature formula*, and reads

$$I_s^c(f) = \frac{H}{6} \sum_{k=1}^{M} [f(x_{k-1}) + 4f(\bar{x}_k) + f(x_k)] \qquad (4.21)$$

One can prove that it induces the error

$$I(f) - I_s^c(f) = -\frac{b-a}{180} \frac{H^4}{16} f^{(4)}(\xi), \qquad (4.22)$$

where ξ is a suitable point in $[a, b]$, provided that $f \in C^4([a, b])$. It is therefore fourth-order accurate with respect to H. When (4.21) is applied to only one interval, say (a, b), we obtain the so-called *Simpson quadrature formula*

110   4 Numerical differentiation and integration

$$\boxed{I_s(f) = \frac{b-a}{6}\left[f(a) + 4f((a+b)/2) + f(b)\right]} \qquad (4.23)$$

The error is now given by

$$I(f) - I_s(f) = -\frac{1}{16}\frac{(b-a)^5}{180}f^{(4)}(\xi), \qquad (4.24)$$

for a suitable $\xi \in [a, b]$. Its degree of exactness is therefore equal to 3.

The composite Simpson rule is implemented in Program 4.2.

Program 4.2. simpsonc: composite Simpson quadrature formula

```
function [Isic]=simpsonc(a,b,M,f,varargin)
%SIMPSONC Composite Simpson numerical integration.
%  ISIC = SIMPSONC(A,B,M,FUN) computes an approximation
%  of the integral of the function FUN via the Simpson
%  method (using M equispaced intervals). FUN accepts
%  real vector input x and returns a real vector value.
%  FUN can also be an inline object.
%  ISIC = SIMPSONC(A,B,M,FUN,P1,P2,...) calls the
%  function FUN passing the optional parameters
%  P1,P2,... as FUN(X,P1,P2,...).
H=(b-a)/M;
x=linspace(a,b,M+1);
fpm=feval(f,x,varargin{:}).*ones(1,M+1);
fpm(2:end-1) = 2*fpm(2:end-1);
Isic=H*sum(fpm)/6;
x=linspace(a+H/2,b-H/2,M);
fpm=feval(f,x,varargin{:}).*ones(1,M);
Isic = Isic+2*H*sum(fpm)/3;
return
```

Example 4.2 (Demography) Let us consider Problem 4.4. To compute the number of individuals whose height is between 1.8 and 1.9 m, we need to solve the integral (4.3) for $h = 1.8$ and $\Delta h = 0.1$. For that we use the composite Simpson formula with 100 sub-intervals

```
N = inline(['M/(sigma*sqrt(2*pi))*exp(-(h-hbar).^2'...
            './(2*sigma^2))'], 'h', 'M', 'hbar', 'sigma')

N =
  Inline function:
  N(h,M,hbar,sigma) = M/(sigma * sqrt(2*pi)) * exp(-(h -
  hbar).^2./(2*sigma^2))

M = 200; hbar = 1.7; sigma = 0.1;
int = simpsonc(1.8, 1.9, 100, N, M, hbar, sigma)

int =
   27.1810
```

We therefore estimate that the number of individuals in this range of height is 27.1810, corresponding to the 15.39 % of all individuals. ∎

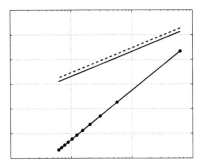

Fig. 4.5. Logarithmic representation of the errors versus H for Simpson (*solid line with circles*), midpoint (*solid line*) and trapezoidal (*dashed line*) composite quadrature formulae

Example 4.3 We want to compare the approximations of the integral $I(f) = \int_0^{2\pi} xe^{-x}\cos(2x)dx = -1/25(10\pi - 3 + 3e^{2\pi})/e^{2\pi} \simeq -0.122122604618968$ obtained by using the composite midpoint, trapezoidal and Simpson formulae. In Figure 4.5 we plot on the logarithmic scale the errors versus H. As pointed out in Section 1.5, in this type of plot the greater the slope of the curve, the higher the order of convergence of the corresponding formula. As expected from the theoretical results, the midpoint and trapezoidal formulae are second-order accurate, whereas the Simpson formula is fourth-order accurate. ∎

4.3 Interpolatory quadratures

All (non-composite) quadrature formulae introduced in the previous sections are remarkable instances of a more general quadrature formula of the form:

$$I_{appr}(f) = \sum_{j=0}^{n} \alpha_j f(y_j) \qquad (4.25)$$

The real numbers $\{\alpha_j\}$ are the *quadrature weights*, while the points $\{y_j\}$ are the *quadrature nodes*. In general, one requires that (4.25) integrates exactly at least a constant function: this property is ensured if $\sum_{j=0}^{n} \alpha_j = b - a$. We can get a degree of exactness equal to (at least) n taking

$$I_{appr}(f) = \int_a^b \Pi_n f(x)dx,$$

where $\Pi_n f \in \mathbb{P}_n$ is the Lagrange interpolating polynomial of the function f at the nodes $y_i, i = 0, \ldots, n$, given by (3.4). This yields the following expression for the weights

$$\alpha_i = \int_a^b \varphi_i(x)dx, \qquad i = 0, \ldots, n,$$

where $\varphi_i \in \mathbb{P}_n$ is the i-th characteristic Lagrange polynomial such that $\varphi_i(y_j) = \delta_{ij}$, for $i, j = 0, \ldots, n$, that was introduced in (3.3).

Example 4.4 For the trapezoidal formula (4.19) we have $n = 1$, $y_0 = a$, $y_1 = b$ and

$$\alpha_0 = \int_a^b \varphi_0(x)dx = \int_a^b \frac{x-b}{a-b}dx = \frac{b-a}{2},$$

$$\alpha_1 = \int_a^b \varphi_1(x)dx = \int_a^b \frac{x-a}{b-a}dx = \frac{b-a}{2}.$$

∎

The question that arises is whether suitable choices of the nodes exist such that the degree of exactness is greater than n, more precisely, equal to $r = n + m$ for some $m > 0$. We can simplify our discussion by restricting ourselves to a reference interval, say $(-1, 1)$. Indeed, once a set of quadrature nodes $\{\bar{y}_j\}$ and weights $\{\bar{\alpha}_j\}$ are available on $[-1, 1]$, then owing to the change of variable (3.8) we can immediately obtain the corresponding nodes and weights,

$$y_j = \frac{a+b}{2} + \frac{b-a}{2}\bar{y}_j, \qquad \alpha_j = \frac{b-a}{2}\bar{\alpha}_j$$

on an arbitrary integration interval $[a, b]$.

The answer to the previous question is furnished by the following result (see, [QSS06, Chapter 10]):

Proposition 4.1 *For a given $m > 0$, the quadrature formula $\sum_{j=0}^n \bar{\alpha}_j f(\bar{y}_j)$ has degree of exactness $n + m$ iff it is of interpolatory type and the nodal polynomial $\omega_{n+1} = \Pi_{i=0}^n(x - \bar{y}_i)$ associated with the nodes $\{\bar{y}_i\}$ is such that*

$$\int_{-1}^1 \omega_{n+1}(x)p(x)dx = 0, \qquad \forall p \in \mathbb{P}_{m-1}. \tag{4.26}$$

The maximum value that m can take is $n + 1$ and is achieved provided ω_{n+1} is proportional to the so-called Legendre polynomial of degree $n+1$, $L_{n+1}(x)$. The Legendre polynomials can be computed recursively, through the following three-term relation

4.3 Interpolatory quadratures

n	$\{\bar{y}_j\}$	$\{\bar{\alpha}_j\}$
1	$\{\pm 1/\sqrt{3}\}$	$\{1\}$
2	$\{\pm\sqrt{15}/5, 0\}$	$\{5/9, 8/9\}$
3	$\{\pm(1/35)\sqrt{525 - 70\sqrt{30}},$ $\pm(1/35)\sqrt{525 + 70\sqrt{30}}\}$	$\{(1/36)(18 + \sqrt{30}),$ $(1/36)(18 - \sqrt{30})\}$
4	$\{0, \pm(1/21)\sqrt{245 - 14\sqrt{70}}$ $\pm(1/21)\sqrt{245 + 14\sqrt{70}}\}$	$\{128/225, (1/900)(322 + 13\sqrt{70})$ $(1/900)(322 - 13\sqrt{70})\}$

Table 4.1. Nodes and weights for some quadrature formulae of Gauss-Legendre on the interval $(-1, 1)$. Weights corresponding to symmetric couples of nodes are reported only once

$$L_0(x) = 1, \quad L_1(x) = x,$$
$$L_{k+1}(x) = \frac{2k+1}{k+1} x L_k(x) - \frac{k}{k+1} L_{k-1}(x), \quad k = 1, 2, \ldots.$$

For every $n = 0, 1, \ldots$, every polynomial in $\mathbb{P}_n$ can be obtained by a linear combination of the polynomials $L_0, L_1, \ldots, L_n$. Moreover, L_{n+1} is orthogonal to all the polynomials of degree less than or equal to n, i.e., $\int_{-1}^{1} L_{n+1}(x) L_j(x) dx = 0$ for all $j = 0, \ldots, n$. This explains why (4.26) is true with m less than or equal to $n + 1$.

The maximum degree of exactness is therefore equal to $2n + 1$, and is obtained for the so-called *Gauss-Legendre formula* (I_{GL} in short), whose nodes and weights are given by:

$$\begin{cases} \bar{y}_j = \text{zeros of } L_{n+1}(x), \\ \bar{\alpha}_j = \frac{2}{(1 - \bar{y}_j^2)[L'_{n+1}(\bar{y}_j)]^2}, \end{cases} \quad j = 0, \ldots, n. \quad (4.27)$$

The weights $\bar{\alpha}_j$ are all positive and the nodes are internal to the interval $(-1, 1)$. In Table 4.1 we report nodes and weights for the Gauss-Legendre quadrature formulae with $n = 1, 2, 3, 4$. If $f \in C^{(2n+2)}([-1, 1])$, the corresponding error is

$$I(f) - I_{GL}(f) = \frac{2^{2n+3}((n+1)!)^4}{(2n+3)((2n+2)!)^3} f^{(2n+2)}(\xi),$$

where ξ is a suitable point in $(-1, 1)$.

It is often useful to include also the endpoints of the interval among the quadrature nodes. By doing so, the Gauss formula with the highest degree of exactness $(2n - 1)$ is the one that employs the so-called *Gauss-Legendre-Lobatto* nodes (briefly, GLL): for $n \geq 1$

$$\bar{y}_0 = -1, \bar{y}_n = 1, \bar{y}_j = \text{zeros of } L'_n(x), \quad j = 1, \ldots, n - 1, \quad (4.28)$$

n	$\{\bar{y}_j\}$	$\{\bar{\alpha}_j\}$
1	$\{\pm 1\}$	$\{1\}$
2	$\{\pm 1, 0\}$	$\{1/3, 4/3\}$
3	$\{\pm 1, \pm\sqrt{5}/5\}$	$\{1/6, 5/6\}$
4	$\{\pm 1, \pm\sqrt{21}/7, 0\}$	$\{1/10, 49/90, 32/45\}$

Table 4.2. Nodes and weights for some quadrature formulae of Gauss-Legendre-Lobatto on the interval $(-1, 1)$. Weights corresponding to symmetric couples of nodes are reported only once

$$\bar{\alpha}_j = \frac{2}{n(n+1)} \frac{1}{[L_n(\bar{y}_j)]^2}, \qquad j = 0, \ldots, n.$$

If $f \in C^{(2n)}([-1, 1])$, the corresponding error is given by

$$I(f) - I_{GLL}(f) = -\frac{(n+1)n^3 2^{2n+1}((n-1)!)^4}{(2n+1)((2n)!)^3} f^{(2n)}(\xi),$$

for a suitable $\xi \in (-1, 1)$. In Table 4.2 we give a table of nodes and weights on the reference interval $(-1, 1)$ for $n = 1, 2, 3, 4$. (For $n = 1$ we recover the trapezoidal rule.)

quadl   Using the MATLAB instruction quadl(fun,a,b) it is possible to compute an integral with a composite Gauss-Legendre-Lobatto quadrature formula. The function fun can be an inline object. For instance, to integrate $f(x) = 1/x$ over $[1, 2]$, we must first define the function

fun=inline('1./x','x');

then call quadl(fun,1,2). Note that in the definition of function f we have used an element by element operation (indeed MATLAB will evaluate this expression component by component on the vector of quadrature nodes).

The specification of the number of subintervals is not requested as it is automatically computed in order to ensure that the quadrature error is below the default tolerance of 10^{-3}. A different tolerance can be provided by the user through the extended command quadl(fun,a,b,tol). In Section 4.4 we will introduce a method to estimate the quadrature error and, consequently, to change H adaptively.

Let us summarize

1. A quadrature formula is a formula to approximate the integral of continuous functions on an interval $[a, b]$;
2. it is generally expressed as a linear combination of the values of the function at specific points (called *nodes*) with coefficients which are called *weights*;

3. the *degree of exactness* of a quadrature formula is the highest degree of the polynomials which are integrated exactly by the formula. It is one for the midpoint and trapezoidal rules, three for the Simpson rule, $2n+1$ for the Gauss-Legendre formula using $n+1$ quadrature nodes, and $2n-1$ for the Gauss-Legendre-Lobatto formula using $n+1$ nodes;
4. the *order of accuracy* of a composite quadrature formula is its order with respect to the size H of the subintervals. The order of accuracy is two for composite midpoint and trapezoidal formulae, four for composite Simpson formula.

See the Exercises 4.12-4.18.

4.4 Simpson adaptive formula

The integration step-length H of a quadrature composite formula can be chosen in order to ensure that the quadrature error is less than a prescribed tolerance $\varepsilon > 0$. For instance, when using the Simpson composite formula, thanks to (4.22) this goal can be achieved if

$$\frac{b-a}{180}\frac{H^4}{16}\max_{x\in[a,b]}|f^{(4)}(x)| < \varepsilon, \qquad (4.29)$$

where $f^{(4)}$ denotes the fourth-order derivative of f. Unfortunately, when the absolute value of $f^{(4)}$ is large only in a small part of the integration interval, the maximum H for which (4.29) holds true can be too small. The goal of the adaptive Simpson quadrature formula is to yield an approximation of $I(f)$ within a fixed tolerance ε by a *nonuniform* distribution of the integration step-sizes in the interval $[a,b]$. In such a way we retain the same accuracy of the composite Simpson rule, but with a lower number of quadrature nodes and, consequently, a reduced number of evaluations of f.

To this end, we must find an error estimator and an automatic procedure to modify the integration step-length H, according to the achievement of the prescribed tolerance. We start by analyzing this procedure, which is independent of the specific quadrature formula that one wants to apply.

In the first step of the adaptive procedure, we compute an approximation $I_s(f)$ of $I(f) = \int_a^b f(x)dx$. We set $H = b-a$ and we try to estimate the quadrature error. If the error is less than the prescribed tolerance, the adaptive procedure is stopped; otherwise the step-size H is halved until the integral $\int_a^{a+H} f(x)dx$ is computed with the prescribed accuracy. When the test is passed, we consider the interval $(a+H, b)$

116   4 Numerical differentiation and integration

and we repeat the previous procedure, choosing as the first step-size the length $b - (a + H)$ of that interval.

We use the following notations:

1. A: the *active* integration interval, i.e. the interval where the integral is being computed;
2. S: the integration interval already examined, for which the error is less than the prescribed tolerance;
3. N: the integration interval yet to be examined.

At the beginning of the integration process we have $N = [a, b]$, $A = N$ and $S = \emptyset$, while the situation at the generic step of the algorithm is depicted in Figure 4.6. Let $J_S(f)$ indicate the computed approximation of $\int_a^\alpha f(x)dx$, with $J_S(f) = 0$ at the beginning of the process; if the algorithm successfully terminates, $J_S(f)$ yields the desired approximation of $I(f)$. We also denote by $J_{(\alpha,\beta)}(f)$ the approximate integral of f over the active interval $[\alpha, \beta]$. This interval is drawn in gray in Figure 4.6. The generic step of the adaptive integration method is organized as follows:

1. if the estimation of the error ensures that the prescribed tolerance is satisfied, then:
   (i) $J_S(f)$ is increased by $J_{(\alpha,\beta)}(f)$, that is $J_S(f) \leftarrow J_S(f) + J_{(\alpha,\beta)}(f)$;
   (ii) we let $S \leftarrow S \cup A$, $A = N$ (corresponding to the path (I) in Figure 4.6) and $\alpha \leftarrow \beta$ and $\beta \leftarrow b$;
2. if the estimation of the error fails the prescribed tolerance, then:
   (j) A is halved, and the new active interval is set to $A = [\alpha, \alpha']$ with $\alpha' = (\alpha + \beta)/2$ (corresponding to the path (II) in Figure 4.6);
   (jj) we let $N \leftarrow N \cup [\alpha', \beta]$, $\beta \leftarrow \alpha'$;
   (jjj) a new error estimate is provided.

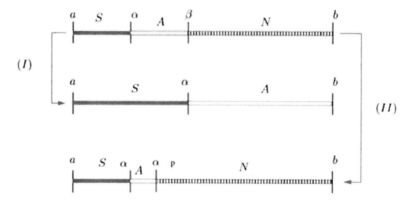

Fig. 4.6. Distribution of the integration intervals at the generic step of the adaptive algorithm and updating of the integration grid

4.4 Simpson adaptive formula

Of course, in order to prevent the algorithm from generating too small step-sizes, it is convenient to monitor the width of A and warn the user, in case of an excessive reduction of the step-length, about the presence of a possible singularity in the integrand function.

The problem now is to find a suitable estimator of the error. To this end, it is convenient to restrict our attention to a generic subinterval $[\alpha, \beta]$ in which we compute $I_s(f)$: of course, if on this interval the error is less than $\varepsilon(\beta - \alpha)/(b - a)$, then the error on the interval $[a, b]$ will be less than the prescribed tolerance ε. Since from (4.24) we get

$$E_s(f; \alpha, \beta) = \int_\alpha^\beta f(x)dx - I_s(f) = -\frac{(\beta - \alpha)^5}{2880} f^{(4)}(\xi),$$

to ensure the achievement of the tolerance, it will be sufficient to verify that $E_s(f; \alpha, \beta) < \varepsilon(\beta - \alpha)/(b - a)$. In practical computation, this procedure is not feasible since the point $\xi \in [\alpha, \beta]$ is unknown.

To estimate the error $E_s(f; \alpha, \beta)$ without using explicitly the value $f^{(4)}(\xi)$, we employ again the composite Simpson formula to compute $\int_\alpha^\beta f(x)dx$, but with a step-length $(\beta - \alpha)/2$. From (4.22) with $a = \alpha$ and $b = \beta$, we deduce that

$$\int_\alpha^\beta f(x)dx - I_s^c(f) = -\frac{(\beta - \alpha)^5}{46080} f^{(4)}(\eta), \qquad (4.30)$$

where η is a suitable point different from ξ. Subtracting the last two equations, we get

$$\Delta I = I_s^c(f) - I_s(f) = -\frac{(\beta - \alpha)^5}{2880} f^{(4)}(\xi) + \frac{(\beta - \alpha)^5}{46080} f^{(4)}(\eta). \quad (4.31)$$

Let us now make the assumption that $f^{(4)}(x)$ is approximately a constant on the interval $[\alpha, \beta]$. In this case $f^{(4)}(\xi) \simeq f^{(4)}(\eta)$. We can compute $f^{(4)}(\eta)$ from (4.31) and, putting this value in the equation (4.30), we obtain the following estimation of the error:

$$\int_\alpha^\beta f(x)dx - I_s^c(f) \simeq \frac{1}{15} \Delta I.$$

The step-length $(\beta - \alpha)/2$ (that is the step-length employed to compute $I_s^c(f)$) will be accepted if $|\Delta I|/15 < \varepsilon(\beta-\alpha)/[2(b-a)]$. The quadrature formula that uses this criterion in the adaptive procedure described previously, is called *adaptive Simpson formula*. It is implemented in Program 4.3. Among the input parameters, f is the string in which the function f is defined, a and b are the endpoints of the integration interval, tol is the prescribed tolerance on the error and hmin is the minimum admissible value for the integration step-length (in order to ensure that the adaptation procedure always terminates).

Program 4.3. simpadpt: adaptive Simpson formula

```
function [JSf,nodes]=simpadpt(f,a,b,tol,hmin,varargin)
%SIMPADPT Numerically evaluate integral, adaptive
% Simpson quadrature.
%
% JSF = SIMPADPT(FUN,A,B,TOL,HMIN) tries to approximate
% the integral of function FUN from A to B to within an
% error of TOL using recursive adaptive Simpson
% quadrature. The inline function Y = FUN(V) should
% accept a vector argument V and return a vector result
% Y, the integrand evaluated at each element of X.
% JSF = SIMPADPT(FUN,A,B,TOL,HMIN,P1,P2,...) calls the
% function FUN passing the optional parameters
% P1,P2,... as FUN(X,P1,P2,...).
% [JSF,NODES] = SIMPADPT(...) returns the distribution
% of nodes used in the quadrature process.
A=[a,b]; N=[]; S=[]; JSf = 0; ba = b - a; nodes=[];
while ~isempty(A),
  [deltaI,ISc]=caldeltai(A,f,varargin{:});
  if abs(deltaI) <= 15*tol*(A(2)-A(1))/ba;
      JSf = JSf + ISc;    S = union(S,A);
      nodes = [nodes, A(1) (A(1)+A(2))*0.5 A(2)];
      S = [S(1), S(end)]; A = N; N = [];
  elseif A(2)-A(1) < hmin
      JSf=JSf+ISc;        S = union(S,A);
      S = [S(1), S(end)]; A=N; N=[];
      warning('Too small integration-step');
  else
      Am = (A(1)+A(2))*0.5;
      A = [A(1) Am];
      N = [Am, b];
  end
end
nodes=unique(nodes);
return

function [deltaI,ISc]=caldeltai(A,f,varargin)
L=A(2)-A(1);
t=[0; 0.25; 0.5; 0.5; 0.75; 1];
x=L*t+A(1);
L=L/6;
w=[1; 4; 1];
fx=feval(f,x,varargin{:}).*ones(6,1);
IS=L*sum(fx([1 3 6]).*w);
ISc=0.5*L*sum(fx.*[w;w]);
deltaI=IS-ISc;
return
```

Example 4.5 Let us compute the integral $I(f) = \int_{-1}^{1} e^{-10(x-1)^2} dx$ by using the adaptive Simpson formula. Using Program 4.3 with

```
>> fun=inline('exp(-10*(x-1).^2)'); tol = 1.e-04; hmin = 1.e-03;
```

we find the approximate value 0.28024765884708, instead of the exact value 0.28024956081990. The error is less than the prescribed tolerance `tol=10`$^{-5}$.

To obtain this result it was sufficient to use only 10 nonuniform subintervals. Note that the corresponding composite formula with uniform step-size would have required 22 subintervals to ensure the same accuracy. ∎

4.5 What we haven't told you

The midpoint, trapezoidal and Simpson formulae are particular cases of a larger family of quadrature rules known as *Newton-Cotes formulae*. For an introduction, see [QSS06, Chapter 10]. Similarly, the Gauss-Legendre and the Gauss-Legendre-Lobatto formulae that we have introduced in Section 4.3 are special cases of a more general family of Gaussian quadrature formulae. These are *optimal* in the sense that they maximize the degree of exactness for a prescribed number of quadrature nodes. For an introduction to Gaussian formulae, see [QSS06, Chapter 10] or [RR85]. Further developments on numerical integration can be found, e.g., in [DR75] and [PdDKÜK83].

Numerical integration can also be used to compute integrals on unbounded intervals. For instance, to approximate $\int_0^\infty f(x)dx$, a first possibility is to find a point α such that the value of $\int_\alpha^\infty f(x)dx$ can be neglected with respect to that of $\int_0^\alpha f(x)dx$. Then we compute by a quadrature formula this latter integral on a bounded interval. A second possibility is to resort to Gaussian quadrature formulae for unbounded intervals (see [QSS06, Chapter 10]).

Finally, numerical integration can also be used to compute multidimensional integrals. In particular, we mention the MATLAB instruction `dblquad('f',xmin,xmax,ymin,ymax)` by which it is possible to compute the integral of a function contained in the MATLAB file `f.m` over the rectangular domain `[xmin,xmax]` × `[ymin,ymax]`. Note that the function `f` must have at least two input parameters corresponding to the variables `x` and `y` with respect to which the integral is computed.

Octave 4.1 In Octave, `dblquad` is not available; however there are some Octave functions featuring the same functionalities:

1. `quad2dg` for two-dimensional integration, which uses a Gaussian quadrature integration scheme;
2. `quad2dc` for two-dimensional integration, which uses a Gaussian-Chebyshev quadrature integration scheme.

∎

4.6 Exercises

Exercise 4.1 Verify that, if $f \in C^3$ in a neighborhood I_0 of x_0 (respectively, I_n of x_n) the error of formula (4.11) is equal to $-\frac{1}{3}f'''(\xi_0)h^2$ (respectively, $-\frac{1}{3}f'''(\xi_n)h^2$), where ξ_0 and ξ_n are two suitable points belonging to I_0 and I_n, respectively.

Exercise 4.2 Verify that if $f \in C^3$ in a neighborhood of $\bar{x}$ the error of the formula (4.9) is equal to (4.10).

Exercise 4.3 Compute the order of accuracy with respect to h of the following formulae for the numerical approximation of $f'(x_i)$:

a. $\dfrac{-11f(x_i) + 18f(x_{i+1}) - 9f(x_{i+2}) + 2f(x_{i+3})}{6h}$,

b. $\dfrac{f(x_{i-2}) - 6f(x_{i-1}) + 3f(x_i) + 2f(x_{i+1})}{6h}$,

c. $\dfrac{-f(x_{i-2}) - 12f(x_i) + 16f(x_{i+1}) - 3f(x_{i+2})}{12h}$.

Exercise 4.4 (Demography) The following values represent the time evolution of the number $n(t)$ of individuals of a given population whose birth rate is constant ($b = 2$) and mortality rate is $d(t) = 0.01n(t)$:

t (months)	0	0.5	1	1.5	2	2.5	3
n	100	147	178	192	197	199	200

Use this data to approximate as accurately as possible the rate of variation of this population. Then compare the obtained results with the exact rate $n'(t) = 2n(t) - 0.01n^2(t)$.

Exercise 4.5 Find the minimum number M of subintervals to approximate with an absolute error less than 10^{-4} the integrals of the following functions:

$$f_1(x) = \frac{1}{1 + (x - \pi)^2} \quad \text{in } [0, 5],$$

$$f_2(x) = e^x \cos(x) \quad \text{in } [0, \pi],$$

$$f_3(x) = \sqrt{x(1-x)} \quad \text{in } [0, 1],$$

using the composite midpoint formula. Verify the results obtained using the Program 4.1.

Exercise 4.6 Prove (4.14) starting from (4.16).

Exercise 4.7 Why does the midpoint formula lose one order of convergence when used in its composite mode?

4.6 Exercises

Exercise 4.8 Verify that, if f is a polynomial of degree less than or equal 1, then $I_{mp}(f) = I(f)$ i.e. the midpoint formula has degree of exactness equal to 1.

Exercise 4.9 For the function f_1 in Exercise 4.5, compute (numerically) the values of M which ensure that the quadrature error is less than 10^{-4} when the integral is approximated by the composite trapezoidal and Gauss quadrature formulae.

Exercise 4.10 Let I_1 and I_2 be two values obtained by the composite trapezoidal formula applied with two different step-lengths, H_1 and H_2, for the approximation of $I(f) = \int_a^b f(x)dx$. Verify that, if $f^{(2)}$ has a mild variation on (a, b), the value

$$I_R = I_1 + (I_1 - I_2)/(H_2^2/H_1^2 - 1) \qquad (4.32)$$

is a better approximation of $I(f)$ than I_1 and I_2. This strategy is called the *Richardson extrapolation method*. Derive (4.32) from (4.18).

Exercise 4.11 Verify that, among all formulae of the form $I_{appx}(f) = \alpha f(\bar{x}) + \beta f(\bar{z})$ where $\bar{x}, \bar{z} \in [a, b]$ are two unknown nodes and α and β two undetermined weights, the Gauss formula with $n = 1$ of Table 4.1 features the maximum degree of exactness.

Exercise 4.12 For the first two functions of Exercise 4.5, compute the minimum number of intervals such that the quadrature error of the composite Simpson quadrature formula is less than 10^{-4}.

Exercise 4.13 Compute $\int_0^2 e^{-x^2/2}dx$ using the Simpson formula (4.23) and the Gauss-Legendre formula of Table 4.1 for $n = 1$, then compare the obtained results.

Exercise 4.14 To compute the integrals $I_k = \int_0^1 x^k e^{x-1}dx$ for $k = 1, 2, \ldots$, one can use the following recursive formula: $I_k = 1 - kI_{k-1}$, with $I_1 = 1/e$. Compute I_{20} using the composite Simpson formula in order to ensure that the quadrature error is less than 10^{-3}. Compare the Simpson approximation with the result obtained using the above recursive formula.

Exercise 4.15 Apply the Richardson extrapolation formula (4.32) for the approximation of the integral $I(f) = \int_0^2 e^{-x^2/2}dx$, with $H_1 = 1$ and $H_2 = 0.5$ using first the Simpson formula (4.23), then the Gauss-Legendre formula for $n = 1$ of Table 4.1. Verify that in both cases I_R is more accurate than I_1 and I_2.

Exercise 4.16 (Electromagnetism) Compute using the composite Simpson formula the function $j(r)$ defined in (4.2) for $r = k/10$ m with $k = 1, \ldots, 10$, with $\rho(\xi) = e^\xi$ and $\sigma = 0.36$ W/(mK). Ensure that the quadrature error is less than 10^{-10}. (Recall that: m=meters, W=watts, K=degrees Kelvin.)

Exercise 4.17 (Optics) By using the composite Simpson and Gauss-Legendre with $n = 1$ formulae compute the function $E(T)$, defined in (4.1), for T equal to 213 K, up to at least 10 exact significant digits.

Exercise 4.18 Develop a strategy to compute $I(f) = \int_0^1 |x^2 - 0.25| dx$ by the composite Simpson formula such that the quadrature error is less than 10^{-2}.

5
Linear systems

In applied sciences, one is quite often led to face a linear system of the form

$$\mathbf{A}\mathbf{x} = \mathbf{b}, \qquad (5.1)$$

where A is a square matrix of dimension $n \times n$ whose elements a_{ij} are either real or complex, while $\mathbf{x}$ and $\mathbf{b}$ are column vectors of dimension n with $\mathbf{x}$ representing the unknown solution and $\mathbf{b}$ a given vector. Component-wise, (5.1) can be written as

$$a_{11}x_1 + a_{12}x_2 + \ldots + a_{1n}x_n = b_1,$$
$$a_{21}x_1 + a_{22}x_2 + \ldots + a_{2n}x_n = b_2,$$
$$\vdots \qquad \vdots \qquad \vdots$$
$$a_{n1}x_1 + a_{n2}x_2 + \ldots + a_{nn}x_n = b_n.$$

We present three different problems that give rise to linear systems.

Problem 5.1 (Hydraulic network) Let us consider the hydraulic network made of the 10 pipelines in Figure 5.1, which is fed by a reservoir of water at constant pressure $p_r = 10$ bar. In this problem, pressure values refer to the difference between the real pressure and the atmospheric one. For the j-th pipeline, the following relationship holds between the flow-rate Q_j (in m^3/s) and the pressure gap Δp_j at pipe-ends:

$$Q_j = kL\Delta p_j, \qquad (5.2)$$

where k is the hydraulic resistance (in m^2 /(bar s)) and L is the length (in m) of the pipeline. We assume that water flows from the outlets (indicated by a black dot) at atmospheric pressure, which is set to 0 bar for coherence with the previous convention.

124   5 Linear systems

A typical problem consists in determining the pressure values at each internal node 1, 2, 3, 4. With this aim, for each $j = 1, 2, 3, 4$ we can supplement the relationship (5.2) with the statement that the algebraic sum of the flow-rates of the pipelines which meet at node j must be null (a negative value would indicate the presence of a seepage).

Denoting by $\mathbf{p} = (p_1, p_2, p_3, p_4)^T$ the pressure vector at the internal nodes, we get a 4×4 system of the form $\mathbf{Ap} = \mathbf{b}$.

In the following table we report the relevant characteristics of the different pipelines:

pipeline	k	L	pipeline	k	L	pipeline	k	L
1	0.01	20	2	0.005	10	3	0.005	14
4	0.005	10	5	0.005	10	6	0.002	8
7	0.002	8	8	0.002	8	9	0.005	10
10	0.002	8						

Correspondingly, A and b take the following values (only the first 4 significant digits are provided):

$$\mathbf{A} = \begin{bmatrix} -0.370 & 0.050 & 0.050 & 0.070 \\ 0.050 & -0.116 & 0 & 0.050 \\ 0.050 & 0 & -0.116 & 0.050 \\ 0.070 & 0.050 & 0.050 & -0.202 \end{bmatrix}, \quad \mathbf{b} = \begin{bmatrix} -2 \\ 0 \\ 0 \\ 0 \end{bmatrix}.$$

The solution of this system is postponed to Example 5.5. ∎

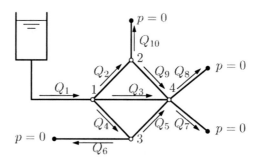

Fig. 5.1. The pipeline network of Problem 5.1

Problem 5.2 (Spectrometry) Let us consider a gas mixture of n non-reactive unknown components. Using a mass spectrometer the compound is bombarded by low-energy electrons: the resulting mixture of ions is analyzed by a galvanometer which shows peaks corresponding to specific ratios mass/charge. We only consider the n most relevant peaks. One may conjecture that the height h_i of the i-th peak is a linear combination of $\{p_j, j = 1, \ldots, n\}$, p_j being the partial pressure of the j-th component (that is the pressure exerted by a single gas when it is part of a mixture), yielding

$$\sum_{j=1}^{n} s_{ij} p_j = h_i, \qquad i = 1, \ldots, n, \tag{5.3}$$

where the s_{ij} are the so-called sensitivity coefficients. The determination of the partial pressures demands therefore the solution of a linear system. For its solution, see Example 5.3. ∎

Problem 5.3 (Economy: input-output analysis) We want to determine the situation of equilibrium between demand and offer of certain goods. In particular, let us consider a production model in which $m \geq n$ factories (or production lines) produce n different products. They must face the internal demand of goods (the input) necessary to the factories for their own production, as well as the external demand (the output) from the consumers. The main assumption of the Leontief model (1930)[1] is that the production model is linear, that is, the amount of a certain output is proportional to the quantity of input used. Under this assumption the activity of the factories is completely described by two matrices, the input matrix $C = (c_{ij}) \in \mathbb{R}^{n \times m}$ and the output matrix $P = (p_{ij}) \in \mathbb{R}^{n \times m}$. ("C" stands for *consumables* and "P" for *products*.) The coefficient c_{ij} (respectively, p_{ij}) represent the quantity of the i-th good absorbed (respectively, produced) by the j-th factory for a fixed period of time. The matrix $A = P - C$ is called *input-output matrix*: a_{ij} positive (respectively, negative) denotes the quantity of the i-th good produced (respectively, absorbed) by the j-th factory. Finally, it is reasonable to assume that the production system satisfies the demand of goods from the market, that can be represented by a vector $\mathbf{b} = (b_i) \in \mathbb{R}^n$ (the vector of the *final demand*). The component b_i represents the quantity of the i-th good absorbed by the market. The equilibrium is reached when the vector $\mathbf{x} = (x_i) \in \mathbb{R}^m$ of the total production equals the total demand, that is,

$$A\mathbf{x} = \mathbf{b}, \qquad \text{where } A = P - C. \tag{5.4}$$

For the solution of this linear system see Exercise 5.17. ∎

The solution of system (5.1) exists iff A is nonsingular. In principle, the solution might be computed using the so-called *Cramer rule*:

$$x_i = \frac{\det(A_i)}{\det(A)}, \qquad i = 1, \ldots, n,$$

---

[1] On 1973 Wassily Leontief was arwarded the Nobel prize in economy for his studies.

126   5 Linear systems

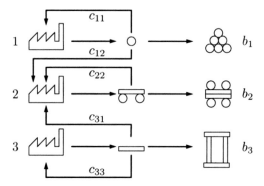

Fig. 5.2. The interaction scheme between three factories and the market

where A_i is the matrix obtained from A by replacing the i-th column by **b** and det(A) denotes the determinant of A. If the $n+1$ determinants are computed by the Laplace expansion (see Exercise 5.1), a total number of approximately $2(n+1)!$ operations is required. As usual, by operation we mean a sum, a subtraction, a product or a division. For instance, a computer capable of carrying out 10^9 *flops* (i.e. 1 giga *flops*), would require about 12 hours to solve a system of dimension $n = 15$, 3240 years if $n = 20$ and 10^{143} years if $n = 100$. The computational cost can be drastically reduced to the order of about $n^{3.8}$ operations if the $n+1$ determinants are computed by the algorithm quoted in Example 1.3. Yet, this cost is still too high for large values of n, which often arise in practical applications.

Two alternative approaches will be pursued: they are called *direct methods* if they yield the solution of the system in a finite number of steps, *iterative methods* if they require (in principle) an infinite number of steps. Iterative methods will be addressed in Section 5.7. We warn the reader that the choice between direct and iterative methods may depend on several factors: primarily, the predicted theoretical efficiency of the scheme, but also the particular type of matrix, the memory storage requirements and, finally, the computer architecture (see, Section 5.11 for more details).

Finally, we note that a system with full matrix cannot be solved by less than n^2 operations. Indeed, if the equations are fully coupled, we should expect that every one of the n^2 matrix coefficients would be involved in an algebraic operation at least once.

5.1 The LU factorization method

Let A be a square matrix of order n. Assume that there exist two suitable matrices L and U, lower triangular and upper triangular, respectively,

5.1 The LU factorization method

such that
$$A = LU \tag{5.5}$$

We call (5.5) an LU-*factorization* (or decomposition) of A. If A is non-singular, so are both L and U, and thus their diagonal elements are nonnull (as observed in Section 1.3).

In such a case, solving $A\mathbf{x} = \mathbf{b}$ leads to the solution of the two triangular systems
$$L\mathbf{y} = \mathbf{b}, \; U\mathbf{x} = \mathbf{y} \tag{5.6}$$

Both systems are easy to solve. Indeed, L being lower triangular, the first row of the system $L\mathbf{y} = \mathbf{b}$ takes the form:
$$l_{11} y_1 = b_1,$$
which provides the value of y_1 since $l_{11} \neq 0$. By substituting this value of y_1 in the subsequent $n-1$ equations we obtain a new system whose unknowns are $y_2, \ldots, y_n$, on which we can proceed in a similar manner. Proceeding forward, equation by equation, we can compute all unknowns with the following *forward substitutions algorithm*:

$$\begin{aligned} y_1 &= \frac{1}{l_{11}} b_1, \\ y_i &= \frac{1}{l_{ii}} \left(b_i - \sum_{j=1}^{i-1} l_{ij} y_j \right), \; i = 2, \ldots, n \end{aligned} \tag{5.7}$$

Let us count the number of operations required by (5.7). Since $i-1$ sums, $i-1$ products and 1 division are needed to compute the unknown y_i, the total number of operations required is
$$\sum_{i=1}^{n} 1 + 2 \sum_{i=1}^{n} (i-1) = 2 \sum_{i=1}^{n} i - n = n^2.$$

The system $U\mathbf{x} = \mathbf{y}$ can be solved by proceeding in a similar manner. This time, the first unknown to be computed is x_n, then, by proceeding backward, we can compute the remaining unknowns x_i, for $i = n-1$ to $i = 1$:

$$\begin{aligned} x_n &= \frac{1}{u_{nn}} y_n, \\ x_i &= \frac{1}{u_{ii}} \left(y_i - \sum_{j=i+1}^{n} u_{ij} x_j \right), \; i = n-1, \ldots, 1 \end{aligned} \tag{5.8}$$

128   5 Linear systems

This is called *backward substitutions algorithm* and requires n^2 operations too. At this stage we need an algorithm that allows an effective computation of the factors L and U of the matrix A. We illustrate a general procedure starting from a couple of examples.

Example 5.1 Let us write the relation (5.5) for a generic matrix $A \in \mathbb{R}^{2\times 2}$

$$\begin{bmatrix} l_{11} & 0 \\ l_{21} & l_{22} \end{bmatrix} \begin{bmatrix} u_{11} & u_{12} \\ 0 & u_{22} \end{bmatrix} = \begin{bmatrix} a_{11} & a_{12} \\ a_{21} & a_{22} \end{bmatrix}.$$

The 6 unknown elements of L and U must satisfy the following (nonlinear) equations:

$$\begin{array}{l} (e_1)\ l_{11}u_{11} = a_{11},\ (e_2)\ l_{11}u_{12} = a_{12}, \\ (e_3)\ l_{21}u_{11} = a_{21},\ (e_4)\ l_{21}u_{12} + l_{22}u_{22} = a_{22}. \end{array} \quad (5.9)$$

System (5.9) is *underdetermined* as it features less equations than unknowns. We can complete it by assigning *arbitrarily* the diagonal elements of L, for instance setting $l_{11} = 1$ and $l_{22} = 1$. Now system (5.9) can be solved by proceeding as follows: we determine the elements u_{11} and u_{12} of the first row of U using (e_1) and (e_2). If u_{11} is nonnull then from (e_3) we deduce l_{21} (that is the first column of L, since l_{11} is already available). Now we can obtain from (e_4) the only nonzero element u_{22} of the second row of U. ∎

Example 5.2 Let us repeat the same computations in the case of a 3×3 matrix. For the 12 unknown coefficients of L and U we have the following 9 equations:

(e_1) $l_{11}u_{11} = a_{11}$, (e_2) $l_{11}u_{12} = a_{12}$,   (e_3) $l_{11}u_{13} = a_{13}$,
(e_4) $l_{21}u_{11} = a_{21}$, (e_5) $l_{21}u_{12} + l_{22}u_{22} = a_{22}$, ($e_6$) $l_{21}u_{13} + l_{22}u_{23} = a_{23}$,
(e_7) $l_{31}u_{11} = a_{31}$, (e_8) $l_{31}u_{12} + l_{32}u_{22} = a_{32}$, ($e_9$) $l_{31}u_{13} + l_{32}u_{23} + l_{33}u_{33} = a_{33}$.

Let us complete this system by setting $l_{ii} = 1$ for $i = 1, 2, 3$. Now, the coefficients of the first row of U can be obtained by using (e_1), (e_2) and (e_3). Next, using (e_4) and (e_7), we can determine the coefficients l_{21} and l_{31} of the first column of L. Using (e_5) and (e_6) we can now compute the coefficients u_{22} and u_{23} of the second row of U. Then, using (e_8), we obtain the coefficient l_{32} of the second column of L. Finally, the last row of U (which consists of the only element u_{33}) can be determined by solving (e_9). ∎

On a matrix of arbitrary dimension n we can proceed as follows:

1. the elements of L and U satisfy the system of nonlinear equations

$$\sum_{r=1}^{\min(i,j)} l_{ir} u_{rj} = a_{ij},\ i, j = 1, \ldots, n; \quad (5.10)$$

2. system (5.10) is underdetermined; indeed there are n^2 equations and $n^2 + n$ unknowns, thus the factorization LU cannot be unique;

5.1 The LU factorization method

3. By forcing the n diagonal elements of L to be equal to 1, (5.10) turns into a determined system which can be solved by the following *Gauss algorithm*: set $A^{(1)} = A$ i.e. $a_{ij}^{(1)} = a_{ij}$ for $i, j = 1, \ldots, n$;

$$
\begin{aligned}
&\text{for } k = 1, \ldots, n-1 \\
&\quad \text{for } i = k+1, \ldots, n \\
&\quad\quad l_{ik} = \frac{a_{ik}^{(k)}}{a_{kk}^{(k)}}, \\
&\quad\quad \text{for } j = k+1, \ldots, n \\
&\quad\quad\quad a_{ij}^{(k+1)} = a_{ij}^{(k)} - l_{ik} a_{kj}^{(k)}
\end{aligned}
\tag{5.11}
$$

The elements $a_{kk}^{(k)}$ must all be different from zero and are called *pivot elements*. For every $k = 1, \ldots, n-1$ the matrix $A^{(k+1)} = (a_{ij}^{(k+1)})$ has $n - k$ rows and columns.

At the end of this procedure the elements of the upper triangular matrix U are given by $u_{ij} = a_{ij}^{(i)}$ for $i = 1, \ldots, n$ and $j = i, \ldots, n$, whereas those of L are given by the coefficients l_{ij} generated by this algorithm. In (5.11) there is no computation of the diagonal elements of L, as we already know that their value is equal to 1.

This factorization is called the *Gauss factorization*; determining the elements of the factors L and U requires about $2n^3/3$ operations (see Exercise 5.4).

Example 5.3 (Spectrometry) For the Problem 5.2 we consider a gas mixture that, after a spectroscopic inspection, presents the following seven most relevant peaks: $h_1 = 17.1$, $h_2 = 65.1$, $h_3 = 186.0$, $h_4 = 82.7$, $h_5 = 84.2$, $h_6 = 63.7$ and $h_7 = 119.7$. We want to compare the measured total pressure, equal to 38.78 μm of Hg (which accounts also for those components that we might have neglected in our simplified model) with that obtained using relations (5.3) with $n = 7$, where the sensitivity coefficients are given in Table 5.1 (taken from [CLW69, p.331]). The partial pressures can be computed solving the system (5.3) for $n = 7$ using the LU factorization. We obtain

```
partpress=
    0.6525
    2.2038
    0.3348
    6.4344
    2.9975
    0.5505
   25.6317
```

Using these values we compute an approximate total pressure (given by sum(partpress)) of the gas mixture which differs from the measured value by 0.0252 μm of Hg. ∎

	Components and indices						
Peak index	Hydrogen 1	Methane 2	Etilene 3	Ethane 4	Propylene 5	Propane 6	n-Pentane 7
1	16.87	0.1650	0.2019	0.3170	0.2340	0.1820	0.1100
2	0.0	27.70	0.8620	0.0620	0.0730	0.1310	0.1200
3	0.0	0.0	22.35	13.05	4.420	6.001	3.043
4	0.0	0.0	0.0	11.28	0.0	1.110	0.3710
5	0.0	0.0	0.0	0.0	9.850	1.1684	2.108
6	0.0	0.0	0.0	0.0	0.2990	15.98	2.107
7	0.0	0.0	0.0	0.0	0.0	0.0	4.670

Table 5.1. The sensitivity coefficients for a gas mixture

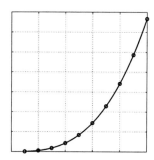

Fig. 5.3. The number of floating-point operations necessary to generate the Gauss factorization LU of the Vandermonde matrix, as a function of the matrix dimension n. This function is a cubic polynomial obtained by approximating in the least-squares sense the values (represented by circles) corresponding to $n = 10, 20, \ldots, 100$

Example 5.4 Consider the Vandermonde matrix

$$A = (a_{ij}) \text{ with } a_{ij} = x_i^{n-j}, \ i, j = 1, \ldots, n, \qquad (5.12)$$

where the x_i are n distinct abscissae. It can be constructed using the MATLAB command `vander`. In Figure 5.3 we report the number of floating-point operations required to compute the Gauss factorization of A, versus n. Several values of n (precisely, $n = 10, 20, \ldots, 100$) are considered and the corresponding number of operations are indicated with circles. The curve reported in the picture is a polynomial in n of third degree representing the least-squares approximation of the above data. The computation of the number of operations was made using a MATLAB command (`flops`) that was present in MATLAB version 5.3.1 and earlier. ∎

Storing the matrices $A^{(k)}$ in the algorithm (5.11) is not necessary; actually we can overlap the $(n-k) \times (n-k)$ elements of $A^{(k+1)}$ on the corresponding last $(n-k) \times (n-k)$ elements of the original matrix A.

5.1 The LU factorization method

Moreover, since at step k, the subdiagonal elements of the k-th column don't have any effect on the final U, they can be replaced by the entries of the k-th column of L, as done in Program 5.1. Then, at step k of the process the elements stored at location of the original entries of A are

$$\begin{bmatrix} a_{11}^{(1)} & a_{12}^{(1)} & \cdots & & \cdots & a_{1n}^{(1)} \\ l_{21} & a_{22}^{(2)} & & & & a_{2n}^{(2)} \\ \vdots & \ddots & \ddots & & & \vdots \\ l_{k1} & \cdots & l_{k,k-1} & a_{kk}^{(k)} & \cdots & a_{kn}^{(k)} \\ \vdots & & \vdots & \vdots & & \vdots \\ l_{n1} & \cdots & l_{n,k-1} & a_{nk}^{(k)} & \cdots & a_{nn}^{(k)} \end{bmatrix},$$

where the boxed submatrix is $A^{(k)}$. The Gauss factorization is the basis of several MATLAB commands:

- [L,U]=lu(A) whose mode of use will be discussed in Section 5.2;   lu
- inv that allows the computation of the inverse of a matrix;   inv
- \ by which it is possible to solve a linear system with matrix A and \
  right hand side b by simply writing A\b (see Section 5.6).

Remark 5.1 (Computing a determinant) By means of the LU factorization one can compute the determinant of A with a computational cost of $\mathcal{O}(n^3)$ operations, noting that (see Sect.1.3)

$$\det(A) = \det(L) \det(U) = \prod_{k=1}^{n} u_{kk}.$$

As a matter of fact, this procedure is also at the basis of the MATLAB command det.   • det

In Program 5.1 we implement the algorithm (5.11). The factor L is stored in the (strictly) lower triangular part of A and U in the upper triangular part of A (for the sake of storage saving). After the program execution, the two factors can be recovered by simply writing: L = eye(n) + tril(A,-1) and U = triu(A), where n is the size of A.

Program 5.1. lugauss: Gauss factorization

```
function A=lugauss(A)
%LUGAUSS LU factorization without pivoting.
% A = LUGAUSS(A) stores an upper triangular matrix in
% the upper triangular part of A and a lower triangular
% matrix in the strictly lower part of A (the diagonal
% elements of L are 1).
[n,m]=size(A);
if n ~= m; error('A is not a square matrix'); else
 for k = 1:n-1
  for i = k+1:n
```

```
    A(i,k) = A(i,k)/A(k,k);
    if A(k,k) == 0, error('Null diagonal element'); end
    j = [k+1:n]; A(i,j) = A(i,j) - A(i,k)*A(k,j);
   end
  end
end
return
```

Example 5.5 Let us compute the solution of the system encountered in Problem 5.1 by using the LU factorization, then applying the backward and forward substitution algorithms. We need to compute the matrix A and the right-hand side b and execute the following instructions:

```
A=lugauss(A);
y(1)=b(1);
for i=2:4; y=[y; b(i)-A(i,1:i-1)*y(1:i-1)]; end
x(4)=y(4)/A(4,4);
for i=3:-1:1;x(i)=(y(i)-A(i,i+1:4)*x(i+1:4)')/A(i,i);end
```

The result is $\mathbf{p} = (8.1172, 5.9893, 5.9893, 5.7779)^T$. ■

Example 5.6 Suppose that we solve $A\mathbf{x} = \mathbf{b}$ with

$$A = \begin{bmatrix} 1 & 1-\varepsilon & 3 \\ 2 & 2 & 2 \\ 3 & 6 & 4 \end{bmatrix}, \mathbf{b} = \begin{bmatrix} 5-\varepsilon \\ 6 \\ 13 \end{bmatrix}, \varepsilon \in \mathbb{R}, \tag{5.13}$$

whose solution is $\mathbf{x} = (1, 1, 1)^T$ (independently of the value of ε).

Let us set $\varepsilon = 1$. The Gauss factorization of A obtained by the Program 5.1 yields

$$L = \begin{bmatrix} 1 & 0 & 0 \\ 2 & 1 & 0 \\ 3 & 3 & 1 \end{bmatrix}, U = \begin{bmatrix} 1 & 0 & 3 \\ 0 & 2 & -4 \\ 0 & 0 & 7 \end{bmatrix}.$$

If we set $\varepsilon = 0$, despite the fact that A is non singular, the Gauss factorization cannot be carried out since the algorithm (5.11) would involve divisions by 0. ■

The previous example shows that, unfortunately, the Gauss factorization A=LU does not necessarily exist for every nonsingular matrix A. In this respect, the following result can be proven:

Proposition 5.1 *For a given matrix* $A \in \mathbb{R}^{n \times n}$, *its Gauss factorization exists and is unique iff the principal submatrices* A_i *of A of order* $i = 1, \ldots, n-1$ *(that is those obtained by restricting A to its first i rows and columns) are nonsingular.*

Going back to Example 5.6, we can notice that when $\varepsilon = 0$ the second principal submatrix A_2 of the matrix A is singular.

We can identify special classes of matrices for which the hypotheses of Proposition 5.1 are fulfilled. In particular, we mention:

1. symmetric and positive definite matrices. A matrix $A \in \mathbb{R}^{n \times n}$ is *positive definite* if

$$\forall \mathbf{x} \in \mathbb{R}^n \text{ with } \mathbf{x} \neq \mathbf{0}, \qquad \mathbf{x}^T A \mathbf{x} > 0;$$

2. diagonally dominant matrices. A matrix is *diagonally dominant by row* if

$$|a_{ii}| \geq \sum_{\substack{j=1 \\ j \neq i}}^{n} |a_{ij}|, \quad i = 1, \ldots, n,$$

by column if

$$|a_{ii}| \geq \sum_{\substack{j=1 \\ j \neq i}}^{n} |a_{ji}|, \quad i = 1, \ldots, n.$$

A special case occurs when in the previous inequalities we can replace $\geq$ by $>$. Then the matrix A is called *strictly* diagonally dominant (by row or by column, respectively).

If A is symmetric and positive definite, it is moreover possible to construct a special factorization:

$$\boxed{A = HH^T} \tag{5.14}$$

where H is a lower triangular matrix with positive diagonal elements. This is the so-called *Cholesky factorization* and requires about $n^3/3$ operations (half of those required by the Gauss LU factorization). Further, let us note that, due to the symmetry, only the lower part of A is stored, and H can be stored in the same area.

The elements of H can be computed by the following algorithm: we set $h_{11} = \sqrt{a_{11}}$ and for $i = 2, \ldots, n$,

$$\boxed{\begin{aligned} h_{ij} &= \frac{1}{h_{jj}} \left(a_{ij} - \sum_{k=1}^{j-1} h_{ik} h_{jk} \right), \, j = 1, \ldots, i-1, \\ h_{ii} &= \sqrt{a_{ii} - \sum_{k=1}^{i-1} h_{ik}^2} \end{aligned}} \tag{5.15}$$

Cholesky factorization is available in MATLAB by setting `R=chol(A)`, `chol` where R is the triangular *upper* factor H^T.

See Exercises 5.1-5.5.

5.2 The pivoting technique

We are going to introduce a special technique that allows us to achieve the LU factorization for every nonsingular matrix, even if the hypotheses of Proposition 5.1 are not fulfilled.

Let us go back to the case described in Example 5.6 and take $\varepsilon = 0$. Setting $A^{(1)} = A$ after carrying out the first step ($k = 1$) of the procedure, the new entries of A are

$$\begin{bmatrix} 1 & 1 & 3 \\ 2 & 0 & -4 \\ 3 & 3 & -5 \end{bmatrix}. \tag{5.16}$$

Since the *pivot* a_{22} is equal to zero, this procedure cannot be continued further. On the other hand, should we interchange the second and third rows beforehand, we would obtain the matrix

$$\begin{bmatrix} 1 & 1 & 3 \\ 3 & 3 & -5 \\ 2 & 0 & -4 \end{bmatrix}$$

and thus the factorization could be accomplished without involving a division by 0.

We can state that *permutation* in a suitable manner of the rows of the original matrix A would make the entire factorization procedure feasible even if the hypotheses of Proposition 5.1 are not verified, provided that $\det(A) \neq 0$. Unfortunately, we cannot know *a priori* which rows should be permuted. However, this decision can be made at every step k at which a null diagonal element $a_{kk}^{(k)}$ is generated.

Let us return to the matrix in (5.16): since the coefficient in position $(2, 2)$ is null, let us interchange the third and second row of this matrix and check whether the new generated coefficient in position $(2, 2)$ is still null. By executing the second step of the factorization procedure we find the same matrix that we would have generated by an *a priori* permutation of the same two rows of A.

We can therefore perform a row permutation as soon as this becomes necessary, without carrying out any *a priori* transformation on A. Since a row permutation entails changing the *pivot element*, this technique is given the name of *pivoting by row*. The factorization generated in this way returns the original matrix up to a row permutation. Precisely we obtain

$$\boxed{PA = LU} \tag{5.17}$$

P is a suitable *permutation matrix* initially set equal to the identity matrix. If in the course of the procedure the rows r and s of A are permuted, the same permutation must be performed on the homologous rows of P. Correspondingly, we should now solve the following triangular systems

$$\mathbf{Ly} = \mathbf{Pb}, \qquad \mathbf{Ux} = \mathbf{y}. \qquad (5.18)$$

From the second equation of (5.11) we see that not only null pivot elements $a_{kk}^{(k)}$ are troublesome, but so are those which are very small. Indeed, should $a_{kk}^{(k)}$ be near zero, possible roundoff errors affecting the coefficients $a_{kj}^{(k)}$ will be severely amplified.

Example 5.7 Consider the nonsingular matrix

$$A = \begin{bmatrix} 1 & 1 + 0.5 \cdot 10^{-15} & 3 \\ 2 & 2 & 20 \\ 3 & 6 & 4 \end{bmatrix}.$$

During the factorization procedure by Program 5.1 no null pivot elements are obtained. Yet, the factors L and U turn out to be quite inaccurate, as one can realize by computing the residual matrix A − LU (which should be the null matrix if all operations were carried out in exact arithmetic):

$$A - LU = \begin{bmatrix} 0 & 0 & 0 \\ 0 & 0 & 0 \\ 0 & 0 & 4 \end{bmatrix}.$$

∎

It is therefore recommended to carry out the pivoting at every step of the factorization procedure, by searching among all virtual pivot elements $a_{ik}^{(k)}$ with $i = k, \ldots, n$, the one with maximum modulus. The algorithm (5.11) with pivoting by row carried out at each step takes the following form:

$$\begin{aligned} &\text{for } k = 1, \ldots, n \\ &\quad \text{for } i = k+1, \ldots, n \\ &\qquad \text{find } \bar{r} \text{ such that } |a_{\bar{r}k}^{(k)}| = \max_{r=k,\ldots,n} |a_{rk}^{(k)}|, \\ &\qquad \text{exchange row } k \text{ with row } \bar{r}, \\ &\qquad l_{ik} = \frac{a_{ik}^{(k)}}{a_{kk}^{(k)}}, \\ &\qquad \text{for } j = k+1, \ldots, n \\ &\qquad\quad a_{ij}^{(k+1)} = a_{ij}^{(k)} - l_{ik} a_{kj}^{(k)} \end{aligned} \qquad (5.19)$$

The MATLAB program lu that we have mentioned previously computes the Gauss factorization with pivoting by row. Its complete syntax is indeed [L,U,P]=lu(A), P being the permutation matrix. When called in the shorthand mode [L,U]=lu(A), the matrix L is equal to P*M, where M is lower triangular and P is the permutation matrix generated by the pivoting by row. The program lu activates automatically the pivoting by row when a null (or very small) pivot element is computed.

See Exercises 5.6-5.8.

5.3 How accurate is the LU factorization?

We have already noticed in Example 5.7 that, due to roundoff errors, the product LU does not reproduce A exactly. Even though the pivoting strategy damps these errors, yet the result could sometimes be rather unsatisfactory.

Example 5.8 Consider the linear system $A_n x_n = b_n$, where $A_n \in \mathbb{R}^{n \times n}$ is the so-called *Hilbert matrix* whose elements are

$$a_{ij} = 1/(i+j-1), \qquad i,j = 1, \ldots, n,$$

while b_n is chosen in such a way that the exact solution is $x_n = (1, 1, \ldots, 1)^T$. The matrix A_n is clearly symmetric and one can prove that it is also positive definite.

For different values of n we use the MATLAB function lu to get the Gauss factorization of A_n with pivoting by row. Then we solve the associated linear systems (5.18) and denote by $\widehat{x}_n$ the computed solution. In Figure 5.4 we report (in logarithmic scale) the relative errors

$$E_n = \|x_n - \widehat{x}_n\|/\|x_n\|, \qquad (5.20)$$

having denoted by $\|\cdot\|$ the Euclidean norm introduced in the Section 1.3.1. We have $E_n \geq 10$ if $n \geq 13$ (that is a relative error on the solution higher than 1000%!), whereas $R_n = L_n U_n - P_n A_n$ is the null matrix (up to machine accuracy) for any given value of n. ∎

On the ground of the previous remark, we could speculate by saying that, when a linear system $Ax = b$ is solved numerically, one is indeed looking for the *exact* solution $\widehat{x}$ of a *perturbed* system

$$(A + \delta A)\widehat{x} = b + \delta b, \qquad (5.21)$$

where δA and δb are respectively a matrix and a vector which depend on the specific numerical method which is being used. We start by considering the case where $\delta A = 0$ and $\delta b \neq 0$ which is simpler than the most general case. Moreover, for simplicity we will also assume that A is symmetric and positive definite.

By comparing (5.1) and (5.21) we find $x - \widehat{x} = -A^{-1}\delta b$, and thus

$$\|x - \widehat{x}\| = \|A^{-1}\delta b\|. \qquad (5.22)$$

5.3 How accurate is the LU factorization?   137

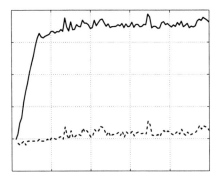

Fig. 5.4. Behavior versus n of E_n (*solid line*) and of $\max_{i,j=1,\ldots,n} |r_{ij}|$ (*dashed line*) in logarithmic scale, for the Hilbert system of Example 5.8. The r_{ij} are the coefficients of the matrix R

In order to find an upper bound for the right-hand side of (5.22), we proceed as follows. Since A is symmetric and positive definite, the set of its eigenvectors $\{\mathbf{v}_i\}_{i=1}^n$ provides an orthonormal basis of $\mathbb{R}^n$ (see [QSS06, Chapter 5]). This means that

$$A\mathbf{v}_i = \lambda_i \mathbf{v}_i, \ i=1,\ldots,n,$$

$$\mathbf{v}_i^T \mathbf{v}_j = \delta_{ij}, \ i,j=1,\ldots,n,$$

where λ_i is the eigenvalue of A associated with $\mathbf{v}_i$ and δ_{ij} is the Kronecker symbol. Consequently, a generic vector $\mathbf{w} \in \mathbb{R}^n$ can be written as

$$\mathbf{w} = \sum_{i=1}^n w_i \mathbf{v}_i,$$

for a suitable (and unique) set of coefficients $w_i \in \mathbb{R}$. We have

$$\begin{aligned}\|A\mathbf{w}\|^2 &= (A\mathbf{w})^T(A\mathbf{w}) \\ &= [w_1(A\mathbf{v}_1)^T + \ldots + w_n(A\mathbf{v}_n)^T][w_1 A\mathbf{v}_1 + \ldots + w_n A\mathbf{v}_n] \\ &= (\lambda_1 w_1 \mathbf{v}_1^T + \ldots + \lambda_n w_n \mathbf{v}_n^T)(\lambda_1 w_1 \mathbf{v}_1 + \ldots + \lambda_n w_n \mathbf{v}_n) \\ &= \sum_{i=1}^n \lambda_i^2 w_i^2.\end{aligned}$$

Denote by λ_{max} the largest eigenvalue of A. Since $\|\mathbf{w}\|^2 = \sum_{i=1}^n w_i^2$, we conclude that

$$\|A\mathbf{w}\| \leq \lambda_{max} \|\mathbf{w}\| \quad \forall \mathbf{w} \in \mathbb{R}^n. \tag{5.23}$$

In a similar manner, we obtain

$$\|A^{-1}\mathbf{w}\| \leq \frac{1}{\lambda_{min}} \|\mathbf{w}\|,$$

upon recalling that the eigenvalues of A^{-1} are the reciprocals of those of A. This inequality enables us to draw from (5.22) that

$$\frac{\|\mathbf{x}-\widehat{\mathbf{x}}\|}{\|\mathbf{x}\|} \leq \frac{1}{\lambda_{min}} \frac{\|\delta \mathbf{b}\|}{\|\mathbf{x}\|}. \tag{5.24}$$

Using (5.23) once more and recalling that $A\mathbf{x} = \mathbf{b}$, we finally obtain

$$\boxed{\frac{\|\mathbf{x}-\widehat{\mathbf{x}}\|}{\|\mathbf{x}\|} \leq \frac{\lambda_{max}}{\lambda_{min}} \frac{\|\delta \mathbf{b}\|}{\|\mathbf{b}\|}} \tag{5.25}$$

We can conclude that the relative error in the solution depends on the relative error in the data through the following constant (≥ 1)

$$\boxed{K(A) = \frac{\lambda_{max}}{\lambda_{min}}} \tag{5.26}$$

cond which is called *spectral condition number of the matrix* A. $K(A)$ can be computed in MATLAB using the command **cond**. Other definitions for the condition number are available for nonsymmetric matrices, see [QSS06, Chapter 3].

Remark 5.2 The MATLAB command **cond(A)** allows the computation of the condition number of any type of matrix A, even those which are not sym-
condest metric and positive definite. A special MATLAB command **condest(A)** is available to compute an approximation of the condition number of a sparse ma-
rcond trix A, and one **rcond(A)** for its reciprocal, with a substantial saving of floating point operations. If the matrix A is ill-conditioned (i.e. $K(A) \gg 1$), the computation of its condition number can be very inaccurate. Consider for instance the tridiagonal matrices $A_n = \text{tridiag}(-1,2,-1)$ for different values of n. A_n is symmetric and positive definite, its eigenvalues are $\lambda_j = 2 - 2\cos(j\theta)$, for $j = 1, \ldots, n$, with $\theta = \pi/(n+1)$, hence $K(A_n)$ can be computed exactly. In Figure 5.5 we report the value of the error $E_K(n) = |K(A_n) - \text{cond}(A_n)|/K(A_n)$. Note that $E_K(n)$ increases when n increases. •

A more involved proof would lead to the following more general result in the case where δA is an arbitrary symmetric and positive definite matrix "small enough" to satisfy $\lambda_{max}(\delta A) < \lambda_{min}(A)$:

$$\boxed{\frac{\|\mathbf{x}-\widehat{\mathbf{x}}\|}{\|\mathbf{x}\|} \leq \frac{K(A)}{1 - \lambda_{max}(\delta A)/\lambda_{min}} \left(\frac{\lambda_{max}(\delta A)}{\lambda_{max}} + \frac{\|\delta \mathbf{b}\|}{\|\mathbf{b}\|} \right)} \tag{5.27}$$

If $K(A)$ is "small", that is of the order of the unity, A is said to be *well conditioned*. In that case, small errors in the data will lead to errors of the same order of magnitude in the solution. This would not occur in the case of *ill conditioned* matrices.

5.3 How accurate is the LU factorization?

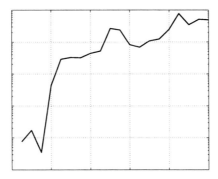

Fig. 5.5. Behavior of $E_K(n)$ as a function of n (in logarithmic scale)

Example 5.9 For the Hilbert matrix introduced in Example 5.8, $K(A_n)$ is a rapidly increasing function of n. One has $K(A_4) > 15000$, while if $n > 13$ the condition number is so high that MATLAB warns that the matrix is "close to singular". Actually, $K(A_n)$ grows at an exponential rate: $K(A_n) \simeq e^{3.5n}$ (see, [Hig02]). This provides an indirect explanation of the bad results obtained in Example 5.8. ∎

Inequality (5.25) can be reformulated by the help of the *residual* **r**:

$$\mathbf{r} = \mathbf{b} - A\widehat{\mathbf{x}}. \tag{5.28}$$

Should $\widehat{\mathbf{x}}$ be the exact solution, the residual would be the null vector. Thus, in general, **r** can be regarded as an *estimator* of the error $\mathbf{x} - \widehat{\mathbf{x}}$. The extent to which the residual is a good error estimator depends on the size of the condition number of A. Indeed, observing that $\delta \mathbf{b} = A(\widehat{\mathbf{x}} - \mathbf{x}) = A\widehat{\mathbf{x}} - \mathbf{b} = -\mathbf{r}$, we deduce from (5.25) that

$$\boxed{\frac{\|\mathbf{x} - \widehat{\mathbf{x}}\|}{\|\mathbf{x}\|} \leq K(A) \frac{\|\mathbf{r}\|}{\|\mathbf{b}\|}} \tag{5.29}$$

Thus if $K(A)$ is "small", we can be sure that the error is small provided that the residual is small, whereas this might not be true when $K(A)$ is "large".

Example 5.10 The residuals associated with the computed solution of the linear systems of Example 5.8 are very small (their norms vary between 10^{-16} and 10^{-11}); however the computed solutions differ remarkably from the exact solution. ∎

See Exercises 5.9-5.10.

5.4 How to solve a tridiagonal system

In many applications (see for instance Chapter 8), we have to solve a system whose matrix has the form

$$A = \begin{bmatrix} a_1 & c_1 & & 0 \\ e_2 & a_2 & \ddots & \\ & \ddots & \ddots & c_{n-1} \\ 0 & & e_n & a_n \end{bmatrix}.$$

This matrix is called *tridiagonal* since the only elements that can be nonnull belong to the main diagonal and to the first super and sub diagonals.

If the Gauss LU factorization of A exists, the factors L and U must be *bidiagonals* (lower and upper, respectively), more precisely:

$$L = \begin{bmatrix} 1 & & & 0 \\ \beta_2 & 1 & & \\ & \ddots & \ddots & \\ 0 & & \beta_n & 1 \end{bmatrix}, U = \begin{bmatrix} \alpha_1 & c_1 & & 0 \\ & \alpha_2 & \ddots & \\ & & \ddots & c_{n-1} \\ 0 & & & \alpha_n \end{bmatrix}.$$

The unknown coefficients α_i and β_i can be determined by requiring that the equality LU = A holds. This yields the following recursive relations for the computation of the L and U factors:

$$\alpha_1 = a_1, \; \beta_i = \frac{e_i}{\alpha_{i-1}}, \; \alpha_i = a_i - \beta_i c_{i-1}, \; i = 2, \ldots, n. \tag{5.30}$$

Using (5.30), we can easily solve the two bidiagonal systems $Ly = b$ and $Ux = y$, to obtain the following formulae:

$$(Ly = b) \quad y_1 = b_1, \; y_i = b_i - \beta_i y_{i-1}, \; i = 2, \ldots, n, \tag{5.31}$$

$$(Ux = y) \quad x_n = \frac{y_n}{\alpha_n}, \; x_i = (y_i - c_i x_{i+1})/\alpha_i, \; i = n-1, \ldots, 1. \tag{5.32}$$

This is known as the *Thomas algorithm* and allows the solution of the original system with a computational cost of the order of n operations.

spdiags　　The MATLAB command spdiags allows the construction of a tridiagonal matrix. For instance, the commands
```
b=ones(10,1); a=2*b; c=3*b;
T=spdiags([b a c],-1:1,10,10);
```

compute the tridiagonal matrix $T \in \mathbb{R}^{10 \times 10}$ with elements equal to 2 on the main diagonal, 1 on the first subdiagonal and 3 on the first superdiagonal.

Note that T is stored in a *sparse mode*, according to which the only elements stored are those different than 0. A matrix $A \in \mathbb{R}^{n \times n}$ is *sparse* if it has a number of nonzero entries of the order of n (and not n^2). We call *pattern* of a sparse matrix the set of its nonzero coefficients.

When a system is solved by invoking the command \, MATLAB is able to recognize the type of matrix (in particular, whether it has been generated in a sparse mode) and select the most appropriate solution algorithm. In particular, when A is a tridiagonal matrix generated in sparse mode, the Thomas algorithm is the selected algorithm.

5.5 Overdetermined systems

A linear system Ax=b with $A \in \mathbb{R}^{m \times n}$ is called *overdetermined* if $m > n$, *underdetermined* if $m < n$.

An overdetermined system generally has no solution unless the right side b is an element of range(A), where

$$\text{range}(A) = \{\mathbf{y} \in \mathbb{R}^m : \mathbf{y} = A\mathbf{x} \text{ for } \mathbf{x} \in \mathbb{R}^n\}. \tag{5.33}$$

In general, for an arbitrary right-hand side b we can search a vector $\mathbf{x}^* \in \mathbb{R}^n$ that minimizes the Euclidean norm of the residual, that is,

$$\Phi(\mathbf{x}^*) = \|A\mathbf{x}^* - \mathbf{b}\|_2^2 \leq \min_{\mathbf{x} \in \mathbb{R}^n} \|A\mathbf{x} - \mathbf{b}\|_2^2 = \min_{\mathbf{x} \in \mathbb{R}^n} \Phi(\mathbf{x}). \tag{5.34}$$

Such a vector $\mathbf{x}^*$ is called *least-squares solution* of the overdetermined system Ax=b.

Similarly to what was done in Section 3.4, the solution of (5.34) can be found by imposing the condition that the gradient of the function Φ must be equal to zero at $\mathbf{x}^*$. With similar calculations we find that $\mathbf{x}^*$ is in fact the solution of the square linear system

$$\boxed{A^T A \mathbf{x}^* = A^T \mathbf{b}} \tag{5.35}$$

which is called the system of *normal equations*. This system is nonsingular if A has *full rank* (that is $\text{rank}(A) = \min(m,n)$, where the *rank* of A, rank(A), is the maximum order of the nonvanishing determinants extracted from A). In such a case $B = A^T A$ is a symmetric and positive definite matrix, then the least-squares solution exists and is unique.

To compute it one could use the Cholesky factorization (5.14). However, due to roundoff errors, the computation of $A^T A$ may be affected by a loss of significant digits, with a consequent loss of the positive definiteness of the matrix itself. Instead, it is more convenient to use the so-called QR factorization. Any full rank matrix $A \in \mathbb{R}^{m \times n}$, with $m \geq n$, admits a unique *QR factorization*, that is, that is there exist a matrix $Q \in \mathbb{R}^{m \times m}$ with the orthogonal property $Q^T Q = I$, and an upper trapezoidal matrix $R \in \mathbb{R}^{m \times n}$ with null rows from the $n+1$-th one on, such that

$$\boxed{A = QR} \qquad (5.36)$$

Then the unique solution of (5.34) is given by

$$\mathbf{x}^* = \tilde{R}^{-1} \tilde{Q}^T \mathbf{b}, \qquad (5.37)$$

where $\tilde{R} \in \mathbb{R}^{n \times n}$ and $\tilde{Q} \in \mathbb{R}^{m \times n}$ are the following matrices

$$\tilde{Q} = Q(1:m, 1:n), \qquad \tilde{R} = R(1:n, 1:n).$$

Notice that $\tilde{R}$ is not singular.

Example 5.11 Consider an alternative approach to the problem of finding the regression line $\epsilon(\sigma) = a_1 \sigma + a_0$ (see Section 3.4) of the data of Problem 3.3. Using the data of Table 3.2 and imposing the interpolating conditions we obtain the overdetermined system $A\mathbf{a} = \mathbf{b}$, where $\mathbf{a} = (a_1, a_0)^T$ and

$$A = \begin{bmatrix} 0 & 1 \\ 0.06 & 1 \\ 0.14 & 1 \\ 0.25 & 1 \\ 0.31 & 1 \\ 0.47 & 1 \\ 0.60 & 1 \\ 0.70 & 1 \end{bmatrix}, \quad \mathbf{b} = \begin{bmatrix} 0 \\ 0.08 \\ 0.14 \\ 0.20 \\ 0.23 \\ 0.25 \\ 0.28 \\ 0.29 \end{bmatrix}.$$

In order to compute its least-squares solution we use the following instructions
```
[Q,R]=qr(A);
Qt=Q(:,1:2); Rt=R(1:2,:);
xstar = Rt \ (Qt'*b)

xstar =
    0.3741
    0.0654
```

These are precisely the same coefficients for the regression line computed in the Example 3.10. Notice that this procedure is directly implemented in the command \: in fact, the instruction xstar = A\b produces the same xstar vector. ∎

5.6 What is hidden behind the command \

It is useful to know that the specific algorithm used by MATLAB when the \ command is invoked depends upon the structure of the matrix A. To determine the structure of A and select the appropriate algorithm, MATLAB follows this precedence (in the case of a real A):

1. if A is sparse and banded, then banded solvers are used (like the Thomas algorithm of Section 5.4). We say that a matrix $A \in \mathbb{R}^{m \times n}$ (or in $\mathbb{C}^{m \times n}$) has *lower band* p if $a_{ij} = 0$ when $i > j + p$ and *upper band* q if $a_{ij} = 0$ when $j > i + q$. The maximum between p and q is called the *bandwidth* of the matrix;
2. if A is an upper or lower triangular matrix (or else a permutation of a triangular matrix), then the system is solved by a backward substitution algorithm for upper triangular matrices, or by a forward substitution algorithm for lower triangular matrices. The check for triangularity is done for full matrices by testing for zero elements and for sparse matrices by accessing the sparse data structure;
3. if A is symmetric and has real positive diagonal elements (which does not imply that A is positive definite), then a Cholesky factorization is attempted (chol). If A is sparse, a preordering algorithm is applied first;
4. if none of previous criteria are fulfilled, then a general triangular factorization is computed by Gaussian elimination with partial pivoting (lu);
5. if A is sparse, then the UMFPACK library is used to compute the solution of the system;
6. if A is not square, proper methods based on the QR factorization for undetermined systems are used (for the overdetermined case, see Section 5.5).

The command \ is available also in Octave. For a system with dense matrix, Octave only uses the LU or the QR factorization. When the matrix is sparse Octave follows this procedure:

1. if the matrix is upper (with column permutations) or lower (with row permutations) triangular, perform a sparse forward or backward substitution;
2. if the matrix is square, symmetric with a positive diagonal, attempt sparse Cholesky factorization;
3. if the sparse Cholesky factorization failed or the matrix is not symmetric with a positive diagonal, factorize using the UMFPACK library;
4. if the matrix is square, banded and if the band density is "small enough" continue, else goto 3;
    a) if the matrix is tridiagonal and the right-hand side is not sparse continue, else goto b);

        i. if the matrix is symmetric, with a positive diagonal, attempt Cholesky factorization;
        ii. if the above failed or the matrix is not symmetric with a positive diagonal use Gaussian elimination with pivoting;
    b) if the matrix is symmetric with a positive diagonal, attempt Cholesky factorization;
    c) if the above failed or the matrix is not symmetric with a positive diagonal use Gaussian elimination with pivoting;
5. if the matrix is not square, or any of the previous solvers flags a singular or near singular matrix, find a solution in the least-squares sense.

Let us summarize

1. The LU factorization of A consists in computing a lower triangular matrix L and an upper triangular matrix U such that A = LU;
2. the LU factorization, provided it exists, is not unique. However, it can be determined unequivocally by providing an additional condition such as, e.g., setting the diagonal elements of L equal to 1. This is called Gauss factorization;
3. the Gauss factorization exists and is unique if and only if the principal submatrices of A of order 1 to $n-1$ are nonsingular (otherwise at least one pivot element is null);
4. if a null pivot element is generated, a new pivot element can be obtained by exchanging in a suitable manner two rows (or columns) of our system. This is the pivoting strategy;
5. the computation of the Gauss factorization requires about $2n^3/3$ operations, and only an order of n operations in the case of tridiagonal systems;
6. for symmetric and positive definite matrices we can use the Cholesky factorization $A = HH^T$, where H is a lower triangular matrix, and the computational cost is of the order of $n^3/3$ operations;
7. the sensitivity of the result to perturbation of data depends on the condition number of the system matrix; more precisely, the accuracy of the computed solution can be low for ill conditioned matrices;
8. the solution of an overdetermined linear system can be intended in the least-squares sense and can be computed using the QR factorization.

5.7 Iterative methods

An iterative method for the solution of the linear system (5.1) consists in setting up a sequence of vectors $\{\mathbf{x}^{(k)}, k \geq 0\}$ of $\mathbb{R}^n$ that *converges* to

5.7 Iterative methods

the exact solution **x**, that is

$$\lim_{k \to \infty} \mathbf{x}^{(k)} = \mathbf{x}, \qquad (5.38)$$

for any given initial vector $\mathbf{x}^{(0)} \in \mathbb{R}^n$. A possible strategy able to realize this process can be based on the following recursive definition

$$\mathbf{x}^{(k+1)} = \mathbf{B}\mathbf{x}^{(k)} + \mathbf{g}, \qquad k \geq 0, \qquad (5.39)$$

where B is a suitable matrix (depending on A) and **g** is a suitable vector (depending on A and **b**), which must satisfy the relation

$$\mathbf{x} = \mathbf{B}\mathbf{x} + \mathbf{g}. \qquad (5.40)$$

Since $\mathbf{x} = \mathbf{A}^{-1}\mathbf{b}$ this yields $\mathbf{g} = (\mathbf{I} - \mathbf{B})\mathbf{A}^{-1}\mathbf{b}$.

Let $\mathbf{e}^{(k)} = \mathbf{x} - \mathbf{x}^{(k)}$ define the error at step k. By subtracting (5.39) from (5.40), we obtain

$$\mathbf{e}^{(k+1)} = \mathbf{B}\mathbf{e}^{(k)}.$$

For this reason B is called the *iteration matrix* associated with (5.39). If B is symmetric and positive definite, by (5.23) we have

$$\|\mathbf{e}^{(k+1)}\| = \|\mathbf{B}\mathbf{e}^{(k)}\| \leq \rho(\mathbf{B})\|\mathbf{e}^{(k)}\|, \qquad \forall k \geq 0.$$

We have denoted by $\rho(\mathbf{B})$ the *spectral radius* of B, that is, the maximum modulus of eigenvalues of B. By iterating the same inequality backward, we obtain

$$\|\mathbf{e}^{(k)}\| \leq [\rho(\mathbf{B})]^k \|\mathbf{e}^{(0)}\|, \quad k \geq 0. \qquad (5.41)$$

Thus $\mathbf{e}^{(k)} \to \mathbf{0}$ as $k \to \infty$ for every possible $\mathbf{e}^{(0)}$ (and henceforth $\mathbf{x}^{(0)}$) provided that $\rho(\mathbf{B}) < 1$. Actually, this property is also necessary for convergence.

Should, by any chance, an approximate value of $\rho(\mathbf{B})$ be available, (5.41) would allow us to deduce the minimum number of iterations k_{min} that are needed to damp the initial error by a factor ε. Indeed, k_{min} would be the lowest positive integer for which $[\rho(\mathbf{B})]^{k_{min}} \leq \varepsilon$.

In conclusion, for a generic matrix the following result holds:

Proposition 5.2 *For an iterative method of the form (5.39) whose iteration matrix satisfies (5.40), convergence for any $\mathbf{x}^{(0)}$ holds iff $\rho(\mathbf{B}) < 1$. Moreover, the smaller $\rho(\mathbf{B})$, the fewer the number of iterations necessary to reduce the initial error by a given factor.*

5.7.1 How to construct an iterative method

A general technique to devise an iterative method is based on a *splitting* of the matrix A, $A = P - (P - A)$, being P a suitable nonsingular matrix (called the *preconditioner* of A). Then

$$Px = (P - A)x + b,$$

has the form (5.40) provided that we set $B = P^{-1}(P - A) = I - P^{-1}A$ and $g = P^{-1}b$. Correspondingly, we can define the following iterative method:

$$P(x^{(k+1)} - x^{(k)}) = r^{(k)}, \qquad k \geq 0,$$

where

$$\boxed{r^{(k)} = b - Ax^{(k)}} \qquad (5.42)$$

denotes the residual vector at iteration k. A generalization of this iterative method is the following

$$\boxed{P(x^{(k+1)} - x^{(k)}) = \alpha_k r^{(k)}, \qquad k \geq 0} \qquad (5.43)$$

where $\alpha_k \neq 0$ is a parameter that may change at every iteration k and which, a priori, will be useful to improve the convergence properties of the sequence $\{x^{(k)}\}$.

The method (5.43) requires to find at each step the so-called *preconditioned residual* $z^{(k)}$ which is the solution of the linear system

$$Pz^{(k)} = r^{(k)}, \qquad (5.44)$$

then the new iterate is defined by $x^{(k+1)} = x^{(k)} + \alpha_k z^{(k)}$. For that reason the matrix P ought to be chosen in such a way that the computational cost for the solution of (5.44) be quite low (e.g., every P either diagonal or triangular or tridiagonal will serve the purpose). Let us now consider some special instance of iterative methods which take the form (5.43).

The Jacobi method

If the diagonal entries of A are nonzero, we can set $P = D = \text{diag}(a_{11}, a_{22}, \ldots, a_{nn})$, where D is the diagonal matrix containing the diagonal entries of A. The Jacobi method corresponds to this choice with the assumption $\alpha_k = 1$ for all k. Then from (5.43) we obtain

$$Dx^{(k+1)} = b - (A - D)x^{(k)}, \qquad k \geq 0,$$

or, componentwise,

5.7 Iterative methods

$$x_i^{(k+1)} = \frac{1}{a_{ii}} \left(b_i - \sum_{j=1,j\neq i}^{n} a_{ij} x_j^{(k)} \right), \ i = 1, \ldots, n \quad (5.45)$$

where $k \geq 0$ and $\mathbf{x}^{(0)} = (x_1^{(0)}, x_2^{(0)}, \ldots, x_n^{(0)})^T$ is the initial vector.

The iteration matrix is therefore

$$\mathrm{B} = \mathrm{D}^{-1}(\mathrm{D} - \mathrm{A}) = \begin{bmatrix} 0 & -a_{12}/a_{11} & \cdots & -a_{1n}/a_{11} \\ -a_{21}/a_{22} & 0 & & -a_{2n}/a_{22} \\ \vdots & & \ddots & \vdots \\ -a_{n1}/a_{nn} & -a_{n2}/a_{nn} & \cdots & 0 \end{bmatrix}. \quad (5.46)$$

The following result allows the verification of Proposition 5.2 without explicitly computing $\rho(\mathrm{B})$:

Proposition 5.3 *If the matrix A is strictly diagonally dominant by row, then the Jacobi method converges.*

As a matter of fact, we can verify that $\rho(\mathrm{B}) < 1$, where B is given in (5.46). To start with, we note that the diagonal elements of A are nonnull owing to the strict diagonal dominance. Let λ be a generic eigenvalue of B and $\mathbf{x}$ an associated eigenvector. Then

$$\sum_{j=1}^{n} b_{ij} x_j = \lambda x_i, \ i = 1, \ldots, n.$$

Assume for simplicity that $\max_{k=1,\ldots,n} |x_k| = 1$ (this is not restrictive since an eigenvector is defined up to a multiplicative constant) and let x_i be the component whose modulus is equal to 1. Then

$$|\lambda| = \left| \sum_{j=1}^{n} b_{ij} x_j \right| = \left| \sum_{j=1,j\neq i}^{n} b_{ij} x_j \right| \leq \sum_{j=1,j\neq i}^{n} \left| \frac{a_{ij}}{a_{ii}} \right|,$$

having noticed that B has only null diagonal elements. Therefore $|\lambda| < 1$ thanks to the assumption made on A.

The Jacobi method is implemented in the Program 5.2 setting in the input parameter P='J'. Input parameters are: the system matrix A, the right hand side b, the initial vector x0 and the maximum number of iterations allotted, nmax. The iterative procedure is terminated as soon as the ratio between the Euclidean norm of the current residual and that of the initial residual is less than a prescribed tolerance tol (for a justification of this stopping criterion, see Section 5.10).

Program 5.2. itermeth: general iterative method

```
function [x,iter]= itermeth(A,b,x0,nmax,tol,P)
%ITERMETH   General iterative method
% X = ITERMETH(A,B,X0,NMAX,TOL,P) attempts to solve the
% system of linear equations A*X=B for X. The N-by-N
% coefficient matrix A must be non-singular and the
% right hand side column vector B must have length
% N. If P='J' the Jacobi method is used, if P='G' the
% Gauss-Seidel method is selected. Otherwise, P is a
% N-by-N matrix that plays the role of a preconditioner
% for the dynamic Richardson method. TOL specifies the
% tolerance of the method. NMAX specifies the maximum
% number of iterations.
[n,n]=size(A);
if nargin == 6
  if ischar(P)==1
    if P=='J'
      L = diag(diag(A));
      U = eye(n);
      beta = 1;
      alpha = 1;
    elseif P == 'G'
      L = tril(A);
      U = eye(n);
      beta = 1;
      alpha = 1;
    end
  else
    [L,U]=lu(P);
    beta = 0;
  end
else
  L = eye(n);
  U = L;
  beta = 0;
end
iter = 0;
r = b - A * x0;
r0 = norm(r);
err = norm (r);
x = x0;
while err > tol & iter < nmax
  iter = iter + 1;
  z = L\r;
  z = U\z;
  if beta == 0
    alpha = z'*r/(z'*A*z);
  end
  x = x + alpha*z;
  r = b - A * x;
  err = norm (r) / r0;
end
```

The Gauss-Seidel method

When applying the Jacobi method, each component of the new vector, say $x_i^{(k+1)}$, is computed independently of the others. This may suggest that a faster convergence could be (hopefully) achieved if the new components already available $x_j^{(k+1)}$, $j = 1, \ldots, i-1$, together with the old ones $x_j^{(k)}$, $j \geq i$, are used for the calculation of $x_i^{(k+1)}$. This would lead to modifying (5.45) as follows: for $k \geq 0$ (still assuming that $a_{ii} \neq 0$ for $i = 1, \ldots, n$)

$$x_i^{(k+1)} = \frac{1}{a_{ii}} \left(b_i - \sum_{j=1}^{i-1} a_{ij} x_j^{(k+1)} - \sum_{j=i+1}^{n} a_{ij} x_j^{(k)} \right), i = 1, .., n \quad (5.47)$$

The updating of the components is made in *sequential* mode, whereas in the original Jacobi method it is made *simultaneously* (or in parallel). The new method, which is called the *Gauss-Seidel method*, corresponds to the choice $P = D - E$ and $\alpha_k = 1$, $k \geq 0$, in (5.43), where E is a lower triangular matrix whose non null entries are $e_{ij} = -a_{ij}$, $i = 2, \ldots, n$, $j = 1, \ldots, i-1$. The corresponding iteration matrix is then

$$B = (D - E)^{-1}(D - E - A).$$

A possible generalization is the so-called *relaxation method* in which $P = \frac{1}{\omega} D - E$, where $\omega \neq 0$ is the relaxation parameter, and $\alpha_k = 1$, $k \geq 0$ (see Exercise 5.13).

Also for the Gauss-Seidel method there exist special matrices A whose associated iteration matrices satisfy the assumptions of Proposition 5.2 (those guaranteeing convergence). Among them let us mention:

1. matrices which are strictly diagonally dominant by row;
2. matrices which are symmetric and positive definite.

The Gauss-Seidel method is implemented in Program 5.2 setting the input parameter P equal to 'G'.

There are no general results stating that the Gauss-Seidel method converges faster than Jacobi's. However, in some special instances this is the case, as stated by the following proposition:

Proposition 5.4 *Let A be a tridiagonal $n \times n$ nonsingular matrix whose diagonal elements are all nonnull. Then the Jacobi method and the Gauss-Seidel method are either both divergent or both convergent. In the latter case, the Gauss-Seidel method is faster than Jacobi's; more precisely the spectral radius of its iteration matrix is equal to the square of that of Jacobi.*

150   5 Linear systems

Example 5.12 Let us consider a linear system $Ax = b$, where b is chosen in such a way that the solution is the unit vector $(1, 1, \ldots, 1)^T$ and A is the 10×10 tridiagonal matrix whose diagonal entries are all equal to 3, the entries of the first lower diagonal are equal to -2 and those of the upper diagonal are all equal to -1. Both Jacobi and Gauss-Seidel methods converge since the spectral radii of their iteration matrices are strictly less than 1. More precisely, by starting from a null initial vector and setting tol $=10^{-12}$, the Jacobi method converges in 277 iterations while only 143 iterations are requested from Gauss-Seidel's. To get this result we have used the following instructions:

```
n=10;
A=3*eye(n)-2*diag(ones(n-1,1),1)-diag(ones(n-1,1),-1);
b=A*ones(n,1);
[x,iter]=itermeth(A,b,zeros(n,1),400,1.e-12,'J'); iter

iter =
    277

[x,iter]=itermeth(A,b,zeros(n,1),400,1.e-12,'G'); iter

iter =
    143
```

■

See Exercises 5.11-5.14.

5.8 Richardson and gradient methods

Let us now consider methods (5.43) for which the acceleration parameters α_k are nonnull. We call *stationary* the case when $\alpha_k = \alpha$ (a given constant) for any $k \geq 0$, *dynamic* the case in which α_k may change along the iterations. In this framework the nonsingular matrix P is still called a *preconditioner* of A.

The crucial issue is the way the parameters are chosen. In this respect, the following result holds (see, e.g., [QV94, Chapter 2], [Axe94]).

5.8 Richardson and gradient methods

Proposition 5.5 *If both* P *and* A *are symmetric and positive definite, the stationary Richardson method converges for every possible choice of* $\mathbf{x}^{(0)}$ *iff* $0 < \alpha < 2/\lambda_{max}$, *where* $\lambda_{max}(>0)$ *is the maximum eigenvalue of* $P^{-1}A$. *Moreover, the spectral radius* $\rho(B_\alpha)$ *of the iteration matrix* $B_\alpha = I - \alpha P^{-1}A$ *is minimal when* $\alpha = \alpha_{opt}$, *where*

$$\alpha_{opt} = \frac{2}{\lambda_{min} + \lambda_{max}} \quad (5.48)$$

λ_{min} *being the minimum eigenvalue of* $P^{-1}A$.
Under the same assumption on P *and* A, *the dynamic Richardson method converges if for instance* α_k *is chosen in the following way:*

$$\alpha_k = \frac{(\mathbf{z}^{(k)})^T \mathbf{r}^{(k)}}{(\mathbf{z}^{(k)})^T A \mathbf{z}^{(k)}} \quad \forall k \geq 0 \quad (5.49)$$

where $\mathbf{z}^{(k)} = P^{-1}\mathbf{r}^{(k)}$ *is the preconditioned residual defined in (5.44). The method (5.43) with this choice of* α_k *is called the preconditioned gradient method, or simply the gradient method when the preconditioner* P *is the identity matrix.*
For both choices, (5.48) and (5.49), the following convergence estimate holds:

$$\|\mathbf{e}^{(k)}\|_A \leq \left(\frac{K(P^{-1}A) - 1}{K(P^{-1}A) + 1}\right)^k \|\mathbf{e}^{(0)}\|_A, \quad k \geq 0, \quad (5.50)$$

where $\|\mathbf{v}\|_A = \sqrt{\mathbf{v}^T A \mathbf{v}}$, $\forall \mathbf{v} \in \mathbb{R}^n$, *is the so-called energy norm associated with the matrix* A.

The dynamic version should therefore be preferred to the stationary one since it does not require the knowledge of the extreme eigenvalues of $P^{-1}A$. Rather, the parameter α_k is determined in terms of quantities which are already available from the previous iteration.

We can rewrite the preconditioned gradient method more efficiently through the following algorithm (derivation is left as an exercise): given $\mathbf{x}^{(0)}$, $\mathbf{r}^{(0)} = \mathbf{b} - A\mathbf{x}^{(0)}$, do

152   5 Linear systems

$$\begin{aligned}
&\text{for } k = 0, 1, \ldots \\
&\mathrm{Pz}^{(k)} = \mathbf{r}^{(k)}, \\
&\alpha_k = \frac{(\mathbf{z}^{(k)})^T \mathbf{r}^{(k)}}{(\mathbf{z}^{(k)})^T \mathrm{A}\mathbf{z}^{(k)}}, \\
&\mathbf{x}^{(k+1)} = \mathbf{x}^{(k)} + \alpha_k \mathbf{z}^{(k)}, \\
&\mathbf{r}^{(k+1)} = \mathbf{r}^{(k)} - \alpha_k \mathrm{A}\mathbf{z}^{(k)}
\end{aligned} \qquad (5.51)$$

The same algorithm can be used to implement the stationary Richardson method by simply replacing α_k with the constant value α.

From (5.50), we deduce that if $P^{-1}A$ is ill conditioned the convergence rate will be very low even for $\alpha = \alpha_{opt}$ (as in that case $\rho(B_{\alpha_{opt}}) \simeq 1$). This circumstance can be avoided provided that a convenient choice of P is made. This is the reason why P is called the preconditioner or the preconditioning matrix.

If A is a generic matrix it may be a difficult task to find a preconditioner which guarantees an optimal trade-off between damping the condition number and keeping the computational cost for the solution of the system (5.44) reasonably low.

The dynamic Richardson method is implemented in Program 5.2 where the input parameter P stands for the preconditioning matrix (when not prescribed, the program implements the unpreconditioned method by setting P=I).

Example 5.13 This example, of theoretical interest only, has the purpose of comparing the convergence behavior of Jacobi, Gauss-Seidel and gradient methods applied to solve the following (mini) linear system:

$$2x_1 + x_2 = 1, \; x_1 + 3x_2 = 0 \qquad (5.52)$$

with initial vector $\mathbf{x}^{(0)} = (1, 1/2)^T$. Note that the system matrix is symmetric and positive definite, and that the exact solution is $\mathbf{x} = (3/5, -1/5)^T$. We report in Figure 5.6 the behavior of the relative residual $E^{(k)} = \|\mathbf{r}^{(k)}\|/\|\mathbf{r}^{(0)}\|$ (versus k) for the three methods above. Iterations are stopped at the first iteration k_{min} for which $E^{(k_{min})} \leq 10^{-14}$. The gradient method appears to converge the fastest. ∎

Example 5.14 Let us consider a system $A\mathbf{x} = \mathbf{b}$, where $A \in \mathbb{R}^{100 \times 100}$ is a pentadiagonal matrix whose main diagonal has all entries equal to 4, while the first and third lower and upper diagonals have all entries equal to -1. As customary, $\mathbf{b}$ is chosen in such a way that $\mathbf{x} = (1, \ldots, 1)^T$ is the exact solution of our system. Let P be the tridiagonal matrix whose diagonal elements are all equal to 2, while the elements on the lower and upper diagonal are all equal to -1. Both A and P are symmetric and positive definite. With such a P as

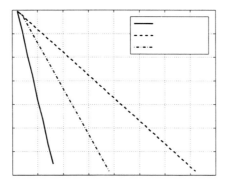

Fig. 5.6. Convergence history for Jacobi, Gauss-Seidel and gradient methods applied to system (5.52)

preconditioner, Program 5.2 can be used to implement the dynamic preconditioner Richardson method. We fix tol=1.e-05, nmax=5000, x0=zeros(100,1). The method converges in 18 iterations. The same Program 5.2, used with P='G', implements the Gauss-Seidel method; this time as many as 2421 iterations are required before satisfying the same stopping criterion. ∎

5.9 The conjugate gradient method

In iterative schemes like (5.51) the new iterate $\mathbf{x}^{(k+1)}$ is obtained by adding to the old iterate $\mathbf{x}^{(k)}$ a vector $\mathbf{z}^{(k)}$ that is either the residual or the preconditioned residual. A natural question is whether it is possible to find instead of $\mathbf{z}^{(k)}$ an optimal sequence of vectors, say $\mathbf{p}^{(k)}$, that ensure the convergence of the method in a minimum number of iterations.

When the matrix A is symmetric and positive definite, the conjugate gradient method (in short, CG) makes use of a sequence of vectors that are *A-orthogonal* (or *A-conjugate*), that is, $\forall k \geq 1$,

$$(A\mathbf{p}^{(j)})^T \mathbf{p}^{(k)} = 0, \qquad j = 0, 1, \ldots, k-1. \tag{5.53}$$

Then, setting $\mathbf{r}^{(0)} = \mathbf{b} - A\mathbf{x}^{(0)}$ and $\mathbf{p}^{(0)} = \mathbf{r}^{(0)}$, the k-th iteration of the conjugate gradient method takes the following form:

for $k = 0, 1, \ldots$

$$\alpha_k = \frac{\mathbf{p}^{(k)T}\mathbf{r}^{(k)}}{\mathbf{p}^{(k)T}\mathbf{A}\mathbf{p}^{(k)}},$$

$$\mathbf{x}^{(k+1)} = \mathbf{x}^{(k)} + \alpha_k \mathbf{p}^{(k)},$$

$$\mathbf{r}^{(k+1)} = \mathbf{r}^{(k)} - \alpha_k \mathbf{A}\mathbf{p}^{(k)}, \qquad (5.54)$$

$$\beta_k = \frac{(\mathbf{A}\mathbf{p}^{(k)})^T \mathbf{r}^{(k+1)}}{(\mathbf{A}\mathbf{p}^{(k)})^T \mathbf{p}^{(k)}},$$

$$\mathbf{p}^{(k+1)} = \mathbf{r}^{(k+1)} - \beta_k \mathbf{p}^{(k)}$$

The constant α_k guarantees that the error is minimized along the descent direction $\mathbf{p}^{(k)}$, while β_k is chosen to ensure that the new direction $\mathbf{p}^{(k+1)}$ is A-conjugate with $\mathbf{p}^{(k)}$. For a complete derivation of the method, see for instance [QSS06, Chapter 4] or [Saa96]. It is possible to prove the following important result:

Proposition 5.6 *Let A be a symmetric and positive definite matrix. The conjugate gradient method for solving (5.1) converges after at most n steps (in exact arithmetic). Moreover, the error $\mathbf{e}^{(k)}$ at the k-th iteration (with $k < n$) is orthogonal to $\mathbf{p}^{(j)}$, for $j = 0, \ldots, k-1$ and*

$$\|\mathbf{e}^{(k)}\|_A \leq \frac{2c^k}{1+c^{2k}} \|\mathbf{e}^{(0)}\|_A, \text{ with } c = \frac{\sqrt{K_2(\mathrm{A})} - 1}{\sqrt{K_2(\mathrm{A})} + 1}. \qquad (5.55)$$

Therefore, in absence of rounding errors, the CG method can be regarded as a direct method, since it terminates after a finite number of steps. However, for matrices of large size, it is usually employed as an iterative scheme, where the iterations are stopped when the error gets below a fixed tolerance. In this respect, the dependence of the error reduction factor on the condition number of the matrix is more favorable than for the gradient method (thanks to the presence of the square root of $K_2(\mathrm{A})$).

Also for the CG method it is possible to consider a preconditioned version (the PCG method), with a preconditioner P symmetric and positive definite, which reads as follows: given $\mathbf{x}^{(0)}$ and setting $\mathbf{r}^{(0)} = \mathbf{b} - \mathbf{A}\mathbf{x}^{(0)}$, $\mathbf{z}^{(0)} = \mathbf{P}^{-1}\mathbf{r}^{(0)}$ and $\mathbf{p}^{(0)} = \mathbf{z}^{(0)}$,

5.9 The conjugate gradient method

for $k = 0, 1, \ldots$

$$\alpha_k = \frac{\mathbf{p}^{(k)^T} \mathbf{r}^{(k)}}{\mathbf{p}^{(k)^T} \mathbf{A} \mathbf{p}^{(k)}},$$

$$\mathbf{x}^{(k+1)} = \mathbf{x}^{(k)} + \alpha_k \mathbf{p}^{(k)},$$

$$\mathbf{r}^{(k+1)} = \mathbf{r}^{(k)} - \alpha_k \mathbf{A} \mathbf{p}^{(k)}, \qquad (5.56)$$

$$\mathbf{P} \mathbf{z}^{(k+1)} = \mathbf{r}^{(k+1)},$$

$$\beta_k = \frac{(\mathbf{A} \mathbf{p}^{(k)})^T \mathbf{z}^{(k+1)}}{(\mathbf{A} \mathbf{p}^{(k)})^T \mathbf{p}^{(k)}},$$

$$\mathbf{p}^{(k+1)} = \mathbf{z}^{(k+1)} - \beta_k \mathbf{p}^{(k)}$$

The PCG method is implemented in the MATLAB function pcg   pcg

Example 5.15 (Factorization vs iterative methods on the Hilbert system)
Let us go back to Example 5.8 on the Hilbert matrix and solve the system (for different values of n) by the preconditioned gradient (PG) and the preconditioned conjugate gradient (PCG) methods, using as preconditioner the diagonal matrix D made of the diagonal entries of the Hilbert matrix. We fix $\mathbf{x}^{(0)}$ to be the null vector and iterate untill the relative residual is less than 10^{-6}. In Table 5.2 we report the absolute errors (with respect to the exact solution) obtained with PG and PCG methods and the errors obtained using the MATLAB command \. In the latter case the error degenerates when n gets large. On the other hand, we can appreciate the beneficial effect that a suitable iterative method such as the PCG scheme can have on the number of iterations. ∎

		\	PG		PCG	
n	$K(\mathbf{A}_n)$	Error	Error	Iter.	Error	Iter.
4	1.55e+04	2.96e-13	1.74-02	995	2.24e-02	3
6	1.50e+07	4.66e-10	8.80e-03	1813	9.50e-03	9
8	1.53e+10	4.38e-07	1.78e-02	1089	2.13e-02	4
10	1.60e+13	3.79e-04	2.52e-03	875	6.98e-03	5
12	1.79e+16	0.24e+00	1.76e-02	1355	1.12e-02	5
14	4.07e+17	0.26e+02	1.46e-02	1379	1.61e-02	5

Table 5.2. Errors obtained using the preconditioned gradient method (PG), the preconditioned conjugate gradient method (PCG) and the direct method implemented in the MATLAB command \ for the solution of the Hilbert system. For the iterative methods we report also the number of iterations

156   5 Linear systems

Remark 5.3 (Non-symmetric systems) The CG method is a special instance of the so-called *Krylov* (or *Lanczos*) *methods* that can be used for the solution of systems which are not necessarily symmetric. Some of them share with the CG method the notable property of finite termination, that is, in exact arithmetic they provide the exact solution in a finite number of iterations also for nonsymmetric systems. A remarkable example is the *GMRES* (Generalized Minimum RESidual) *method*.

gmres   Their description is provided, e.g., in [Axe94], [Saa96] and [vdV03]. They are available in the MATLAB toolbox sparfun under the name of gmres. Another method of this family without the property of finite termination, which however requires a less computational effort than GMRES, is the *conjugate gradient squared* (CGS) *method* and its variant, the Bi-CGStab method, that is characterized by a more regular convergence than CGS. All these methods are available in the MATLAB toolbox sparfun.   •

Octave 5.1 Octave provides only an implementation of the preconditioned conjugate gradient (PCG) method through the command pcg and the preconditioned conjuguate residuals (PCR/Richardson) through the command pcr. Other iterative methods such as GMRES, CGS, Bi-CGStab are not yet implemented.   ■

See Exercises 5.15-5.17.

5.10 When should an iterative method be stopped?

In theory iterative methods require an infinite number of iterations to converge to the exact solution of a linear system. In practice, this is neither reasonable nor necessary. Indeed we do not really need to achieve the exact solution, but rather an approximation $\mathbf{x}^{(k)}$ for which we can guarantee that the error be lower than a desired tolerance ϵ. On the other hand, since the error is itself unknown (as it depends on the exact solution), we need a suitable *a posteriori* error estimator which predicts the error starting from quantities that have already been computed.

The first type of estimator is represented by the residual at the k-th iteration, see (5.42). More precisely, we could stop our iterative method at the first iteration step k_{min} for which

$$\|\mathbf{r}^{(k_{min})}\| \leq \varepsilon \|\mathbf{b}\|.$$

Setting $\widehat{\mathbf{x}} = \mathbf{x}^{(k_{min})}$ and $\mathbf{r} = \mathbf{r}^{(k_{min})}$ in (5.29) we would obtain

$$\frac{\|\mathbf{e}^{(k_{min})}\|}{\|\mathbf{x}\|} \leq \varepsilon K(\mathbf{A}),$$

5.10 When should an iterative method be stopped?

which is an estimate for the relative error. We deduce that the control on the residual is meaningful only for those matrices whose condition number is reasonably small.

Example 5.16 Let us consider the linear system (5.1) where A=A_{20} is the Hilbert matrix of dimension 20 introduced in Example 5.8 and **b** is constructed in such a way that the exact solution is $\mathbf{x} = (1, 1, \ldots, 1)^T$. Since A is symmetric and positive definite the Gauss-Seidel method surely converges. We use Program 5.2 to solve this system taking x0 to be the null initial vector and setting a tolerance on the residual equal to 10^{-5}. The method converges in 472 iterations; however the relative error is very large and equals 0.26. This is due to the fact that A is extremely ill conditioned, having $K(A) \simeq 10^{17}$. In Figure 5.7 we show the behavior of the residual (normalized to the initial one) and that of the error as the number of iterations increases. ∎

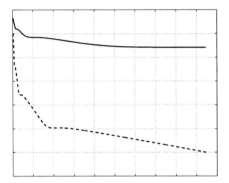

Fig. 5.7. Behavior of the normalized residual $\|\mathbf{r}^{(k)}\|/\|\mathbf{r}^{(0)}\|$ (*dashed line*) and of the error $\|\mathbf{x} - \mathbf{x}^{(k)}\|$ (*solid line*) for Gauss-Seidel iterations applied to the system of Example 5.16

An alternative approach is based on the use of a different error estimator, namely the *increment* $\boldsymbol{\delta}^{(k)} = \mathbf{x}^{(k+1)} - \mathbf{x}^{(k)}$. More precisely, we can stop our iterative method at the first iteration step k_{min} for which

$$\|\boldsymbol{\delta}^{(k_{min})}\| \leq \varepsilon \|\mathbf{b}\|.$$

In the special case where B is symmetric and positive definite, we have

$$\|\mathbf{e}^{(k)}\| = \|\mathbf{e}^{(k+1)} - \boldsymbol{\delta}^{(k)}\| \leq \rho(B)\|\mathbf{e}^{(k)}\| + \|\boldsymbol{\delta}^{(k)}\|.$$

Since $\rho(B)$ should be less than 1 in order for the method to converge, we deduce

$$\boxed{\|\mathbf{e}^{(k)}\| \leq \frac{1}{1 - \rho(B)} \|\boldsymbol{\delta}^{(k)}\|} \tag{5.57}$$

From the last inequality we see that the control on the increment is meaningful only if $\rho(B)$ is much smaller than 1 since in that case the error will be of the same size as the increment.

In fact, the same conclusion holds even if B is not symmetric and positive definite (as it occurs for the Jacobi and Gauss-Seidel methods); however in that case (5.57) is no longer true.

Example 5.17 Let us consider a system whose matrix $A \in \mathbb{R}^{50 \times 50}$ is tridiagonal and symmetric with entries equal to 2.001 on the main diagonal and equal to 1 on the two other diagonals. As usual, the right hand side **b** is chosen in such a way that the unit vector $(1, \ldots, 1)^T$ is the exact solution. Since A is tridiagonal with strict diagonal dominance, the Gauss-Seidel method will converge about twice as fast as the Jacobi method (in view of Proposition 5.4). Let us use Program 5.2 to solve our system in which we replace the stopping criterion based on the residual by that based on the increment. Using a null initial vector and setting the tolerance tol$= 10^{-5}$, after 1604 iterations the program returns a solution whose error 0.0029 is quite large. The reason is that the spectral radius of the iteration matrix is equal to 0.9952, which is very close to 1. Should the diagonal entries be set equal to 3, after only 17 iterations we would have obtained an error equal to 10^{-5}. In fact in that case the spectral radius of the iteration matrix would be equal to 0.428. ■

Let us summarize

1. An iterative method for the solution of a linear system starts from a given initial vector $\mathbf{x}^{(0)}$ and builds up a sequence of vectors $\mathbf{x}^{(k)}$ which we require to converge to the exact solution as $k \to \infty$;
2. an iterative method converges for every possible choice of the initial vector $\mathbf{x}^{(0)}$ iff the spectral radius of the iteration matrix is strictly less than 1;
3. classical iterative methods are those of Jacobi and Gauss-Seidel. A sufficient condition for convergence is that the system matrix be strictly diagonally dominant by row (or symmetric and definite positive in the case of Gauss-Seidel);
4. in the Richardson method convergence is accelerated thanks to the introduction of a parameter and (possibly) a convenient preconditioning matrix;
5. with the conjugate gradient method the exact solution of a symmetric positive definite system can be computed in a finite number of iterations (in exact arithmetic). This method can be generalized to the nonsymmetric case;
6. there are two possible stopping criteria for an iterative method: controlling the residual or controlling the increment. The former is meaningful if the system matrix is well conditioned, the latter if the spectral radius of the iteration matrix is not close to 1.

5.11 To wrap-up: direct or iterative?

In this section we compare direct and iterative methods on several simple test cases. For a linear system of small size, it doesn't really matter since every method will make the job. Instead, for large scale systems, the choice will depend primarily on the matrix properties (such as symmetry, positive definiteness, sparsity pattern, condition number), but also on the kind of available computer resources (memory access, fast processors, etc.). We must admit that in our tests the comparison will not be fully loyal. One direct solver that we will in fact use is the MATLAB built-in function \ which is compiled and optimized, whereas the iterative solvers are not. Our computations were carried out on a processor Intel Pentium M 1.60 GHz with 2048KB cache and 1GByte RAM.

A sparse, banded linear system with small bandwidth

The first test case concerns linear systems arising from the 5-point finite difference discretizations of the Poisson problem on the square $(-1,1)^2$ (see Section 8.1.3). Uniform grids of step $h = 1/N$ in both spatial coordinates are considered, for several values of N. The corresponding finite difference matrices, with N^2 rows and columns, are generated using Program 8.2. On Figure 5.8, left, we plot the matrix structure corresponding to the value $N^2 = 256$: it is sparse, banded, with only 5 nonnull entries per row. Any such matrix is symmetric and positive definite but ill conditioned: its spectral condition number behaves like a constant time h^{-2} for all values of h. To solve the associated linear systems we will use the Cholesky factorization, the preconditioned conjugate gradient method (PCG) with preconditioner given by the incomplete Cholesky factorization (available through the command cholinc) and the MATLAB command \ that, in the current case, is in fact an ad hoc algorithm for pentadiagonal symmetric matrices. The stopping criterion for the PCG method is that the norm of the relative residual be lower than 10^{-14}; the CPU time is also inclusive of the time necessary to construct the preconditioner.

In Figure 5.8, right, we compare the CPU time for the three different methods versus the matrix size. The direct method hidden by the command \ is by far the cheapest: in fact, it is based on a variant of the Gaussian elimination that is particularly effective for sparse banded matrices with small bandwith.

The PCG method, in its turn, is more convenient than the Cholesky factorization, provided a suitable preconditioner is used. For instance, if $N^2 = 4096$ the PCG method requires 19 iterations, whereas the CG method (with no preconditioning) would require 325 iterations, resulting in fact less convenient than the simple Cholesky factorization.

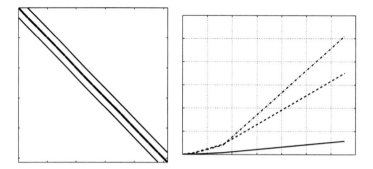

Fig. 5.8. The structure of the matrix for the first test case (*left*), and the CPU time needed for the solution of the associated linear system (*right*): the *solid line* refers to the command \, the *dashed-dotted line* to the use of the Cholesky factorization, the *dashed line* to the PCG iterative method

The case of a wide band

We still consider the same Poisson equation, however this time the discretization is based on spectral methods with quadrature formulae of Gauss-Lobatto-Legendre (see, for instance, [CHQZ06]). Even though the number of grid-nodes is the same as for the finite differences, with spectral methods the derivatives are approximated using many more nodes (in fact, at any given node the x-derivatives are approximated using all the nodes sitting on the same row, whereas all those on the same column are used to compute y-derivatives). The corresponding matrices are still sparse and structured, however the number of non-null entries is definitely higher. This is clear from the example in Figure 5.9, left, where the spectral matrix has still $N^2 = 256$ rows and columns, but the number of nonzero entries is 7936 instead of the 1216 of the finite difference matrix of Figure 5.8.

The CPU time reported in Figure 5.9, right, shows that for this matrix the PCG algorithm, using the incomplete Cholesky factorization as preconditioner, performs much better than the other two methods.

A first conclusion to draw is that for sparse symmetric and positive definite matrices with large bandwidth, PCG is more efficient than the direct method implemented in MATLAB (which does not use the Cholesky factorization since the matrix is stored with the format `sparse`). We point out that a suitable preconditioner is however crucial in order for the PCG method to become competitive.

Finally, we shoud keep in mind that direct methods require more memory storage than iterative methods, a difficulty that could become insurmontable in large scale applications.

5.11 To wrap-up: direct or iterative?

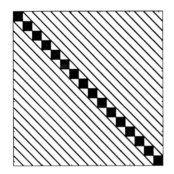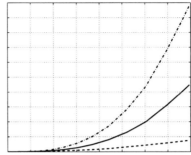

Fig. 5.9. The structure of the matrix used in the second test case (*left*), and the CPU time needed to solve the associated linear system (*right*): the *solid line* refers to the command \, the *dashed-dotted line* to the use of the Cholesky factorization, the *dashed line* to the PCG iterative method

Systems with full matrices

With the MATLAB command gallery we can get access to a collection of matrices featuring different structure and properties. In particular for our third test case, by the command A=gallery('riemann',n) we select the so-called Riemann matrix of dimension n, that is a n × n full, non symmetric matrix whose determinant behaves like det(A) = $\mathcal{O}(n!n^{-1/2+\epsilon})$ for all $\epsilon > 0$. The associated linear system is solved by the iterative GMRES method (see section 5.3) and the iterations will be stopped as soon as the norm of the relative residual is less than 10^{-14}. Alternatively, we will use the MATLAB command \ that, in the case at hand, implements the LU factorization.

For several values of n we will solve the corresponding linear system whose exact solution is the unitary vector: the right-hand side is computed accordingly. The GMRES iterations are obtained without preconditioning and with a special diagonal preconditioner. The latter is obtained by the command luinc(A,1.e0) based on the so-called *incomplete LU factorization*, a matrix that is generated from an algebraic manipulation of the entries of the L and U factors of A, see [QSS06]. In Figure 5.10, right, we report the CPU time for n ranging between 100 and 1000. On the left we report the condition number of A, cond(A). As we can see, the direct factorization method is far less expensive than the un-preconditioned GMRES method, however it becomes more expensive for large n when a suitable preconditioner is used.

gallery

luinc

Octave 5.2 The gallery command is not available in Octave. However a few are available such as the Hilbert, Hankel or Vandermonde matrices, see the commands hankel, hilb, invhilb sylvester_matrix, toeplitz and vander. Moreover if you have access to MATLAB, you can save a matrix

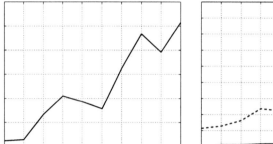
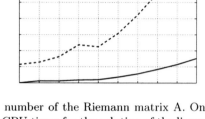

Fig. 5.10. On the left, the condition number of the Riemann matrix A. On the right, the comparison between the CPU times for the solution of the linear system: the solid line refers to the command \, the dashed line refers to the GMRES iterative method with no preconditioning. The values in abscissa refer to the matrix dimension

defined in the gallery using the save command and then load it in Octave using load. Here is an example:
In MATLAB:
```
riemann10=gallery('riemann',10);
save 'riemann10' riemann10
```
In Octave:
```
load 'riemann10' riemann10
```
Note that only Octave version 2.9 can load Mat-files properly from MATLAB version 7. ∎

Systems with sparse, nonsymmetric matrices

We consider linear systems that are generated by the finite element discretization of diffusion-transport-reaction boundary-value problems in two dimensions. These problems are similar to the one reported in (8.17) which refers to a one-dimensional case. Its finite element approximation, that is illustrated at the end of Section 8.17 in the one-dimensional case, makes use of piecewise linear polynomials to represent the solution in each triangular element of a grid that partitions the region where the boundary-value problem is set up. The unknowns of the associated algebraic system is the set of values attained by the solution at the vertices of the internal triangles. We refer to, e.g., [QV94] for a description of this method, as well as for the determination of the entries of the matrix. Let us simply point out that this matrix is sparse, but not banded (its sparsity pattern depends on the way the vertices are numbered) and nonsymmetric, due to the presence of the transport term. The lack of

5.11 To wrap-up: direct or iterative? 163

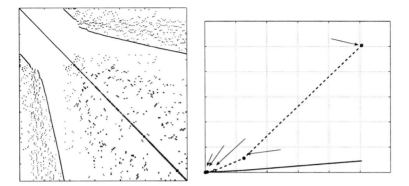

Fig. 5.11. The structure of one of the matrices used for the fourth test case (*left*), and the CPU time needed for the solution of the associated linear system (*right*): the *solid line* refers to the command \, the *dashed line* to the Bi-CGStab iterative method

symmetry, however, is not evident from the representation of its structure in Figure 5.11, left.

The smaller the *diameter h* of the triangles (i.e. the lengths of their longest edge), the higher the matrix size. We have compared the CPU time necessary to solve the linear system corresponding to the case $h = 0.1, 0.05, 0.025, 0.0125$ and 0.0063. We have used the MATLAB command \, that in this case use the UMFPACK library and the (MATLAB implementation of the) iterative method Bi-CGStab which can be regarded as a generalization to nonsymmetric systems of the conjugate gradient method. In abscissae we have reported the number of unknowns that range from 64 (for $h = 0.1$) and 101124 (for $h = 0.0063$). Also in this case, the direct method is less expensive than the iterative one. Should we use as preconditioner for the Bi-CGStab method the incomplete LU factorization, the number of iterations would reduce, however the CPU time would be higher than the one for the unpreconditioned case.

In conclusion

The comparisons that we have carried out, although very limited, outlines a few relevant aspects. In general, direct methods (especially if implemented in their most sophisticated versions, such as in the \ MATLAB command) are more efficient than iterative methods when the latter are used without efficient preconditioners. However, they are more sensitive to the matrix ill conditioning (see for instance the Example 5.15) and may require a substantial amount of storage.

A further aspect that is worth mentioning is that direct methods require the knowledge of the matrix entries, whereas iterative methods don't. In fact, what is nedeed at each iteration is the computation of matrix-vector products for given vectors. This aspect makes iterative methods especially interesting for those problems in which the matrix is not explicitely generated.

5.12 What we haven't told you

Several efficient variants of the Gauss LU factorization are available for sparse systems of large dimension. Among the most advanced, we quote the so-called *multifrontal method* which makes use of a suitable reordering of the system unknowns in order to keep the triangular factors L and U as sparse as possible. The multifrontal method is implemented in the software package UMFPACK. More on this issue is available on [GL96] and [DD99].

Concerning iterative methods, both the conjugate gradient method and the GMRES method are special instances of Krylov methods. For a description of Krylov methods see e.g. [Axe94], [Saa96] and [vdV03].

As it was pointed out, iterative methods converge slowly if the system matrix is severely ill conditioned. Several preconditioning strategies have been developed (see, e.g., [dV89] and [vdV03]). Some of them are purely algebraic, that is, they are based on incomplete (or inexact) factorizations of the given system matrix, and are implemented in the MATLAB functions `luinc` or the already quoted `cholinc`. Other strategies are developed *ad hoc* by exploiting the physical origin and the structure of the problem which has generated the linear system at hand.

Finally it is worthwhile to mention the *multigrid methods* which are based on the sequential use of a hierarchy of systems of variable dimensions that "resemble" the original one, allowing a clever error reduction strategy (see, e.g., [Hac85], [Wes04] and [Hac94]).

Octave 5.3 In Octave, `cholinc` is not yet available. Only `luinc` has been implemented. ∎

5.13 Exercises

Exercise 5.1 For a given matrix $A \in \mathbb{R}^{n \times n}$ find the number of operations (as a function of n) that are needed for computing its determinant by the recursive formula (1.8).

Exercise 5.2 Use the MATLAB command `magic(n)`, n=3, 4, ..., 500, to construct the magic squares of order n, that is, those matrices having entries for which the sum of the elements by rows, columns or diagonals are identical.

Then compute their determinants by the command det introduced in Section 1.3 and the CPU time that is needed for this computation using the cputime command. Finally, approximate this data by the least-squares method and deduce that the CPU time scales approximately as n^3.

Exercise 5.3 Find for which values of ε the matrix defined in (5.13) does not satisfy the hypotheses of Proposition 5.1. For which value of ε does this matrix become singular? Is it possible to compute the LU factorization in that case?

Exercise 5.4 Verify that the number of operations necessary to compute the LU factorization of a square matrix A of dimension n is approximately $2n^3/3$.

Exercise 5.5 Show that the LU factorization of A can be used for computing the inverse matrix A^{-1}. (Observe that the j-th column vector of A^{-1} satisfies the linear system $Ay_j = e_j$, e_j being the vector whose components are all null except the j-th component which is 1.)

Exercise 5.6 Compute the factors L and U of the matrix of Example 5.7 and verify that the LU factorization is inaccurate.

Exercise 5.7 Explain why partial pivoting by row is not convenient for symmetric matrices.

Exercise 5.8 Consider the linear system $Ax = b$ with

$$A = \begin{bmatrix} 2 & -2 & 0 \\ \varepsilon - 2 & 2 & 0 \\ 0 & -1 & 3 \end{bmatrix},$$

and b such that the corresponding solution is $x = (1, 1, 1)^T$ and ε is a positive real number. Compute the Gauss factorization of A and note that $l_{32} \to \infty$ when $\varepsilon \to 0$. In spite of that, verify that the computed solution is accurate.

Exercise 5.9 Consider the linear systems $A_i x_i = b_i$, $i = 1, 2, 3$, with

$$A_1 = \begin{bmatrix} 15 & 6 & 8 & 11 \\ 6 & 6 & 5 & 3 \\ 8 & 5 & 7 & 6 \\ 11 & 3 & 6 & 9 \end{bmatrix}, \ A_i = (A_1)^i, \ i = 2, 3,$$

and b_i such that the solution is always $x_i = (1, 1, 1, 1)^T$. Solve the system by the Gauss factorization using partial pivoting by row, and comment on the obtained results.

Exercise 5.10 Show that for a symmetric and positive definite matrix A we have $K(A^2) = (K(A))^2$.

Exercise 5.11 Analyse the convergence properties of the Jacobi and Gauss-Seidel methods for the solution of a linear system whose matrix is

$$A = \begin{bmatrix} \alpha & 0 & 1 \\ 0 & \alpha & 0 \\ 1 & 0 & \alpha \end{bmatrix}, \quad \alpha \in \mathbb{R}.$$

Exercise 5.12 Provide a sufficient condition on β so that both the Jacobi and Gauss-Seidel methods converge when applied for the solution of a system whose matrix is

$$A = \begin{bmatrix} -10 & 2 \\ \beta & 5 \end{bmatrix}. \tag{5.58}$$

Exercise 5.13 For the solution of the linear system $A\mathbf{x} = \mathbf{b}$ with $A \in \mathbb{R}^{n \times n}$, consider the *relaxation method*: given $\mathbf{x}^{(0)} = (x_1^{(0)}, \ldots, x_n^{(0)})^T$, for $k = 0, 1, \ldots$ compute

$$r_i^{(k)} = b_i - \sum_{j=1}^{i-1} a_{ij} x_j^{(k+1)} - \sum_{j=i+1}^{n} a_{ij} x_j^{(k)}, \quad x_i^{(k+1)} = (1-\omega) x_i^{(k)} + \omega \frac{r_i^{(k)}}{a_{ii}},$$

for $i = 1, \ldots, n$, where ω is a real parameter. Find the explicit form of the corresponding iterative matrix, then verify that the condition $0 < \omega < 2$ is necessary for the convergence of this method. Note that if $\omega = 1$ this method reduces to the Gauss-Seidel method. If $1 < \omega < 2$ the method is known as *SOR (successive over-relaxation)*.

Exercise 5.14 Consider the linear system $A\mathbf{x} = \mathbf{b}$ with $A = \begin{bmatrix} 3 & 2 \\ 2 & 6 \end{bmatrix}$ and say whether the Gauss-Seidel method converges, without explicitly computing the spectral radius of the iteration matrix.

Exercise 5.15 Compute the first iteration of the Jacobi, Gauss-Seidel and preconditioned gradient method (with preconditioner given by the diagonal of A) for the solution of system (5.52) with $\mathbf{x}^{(0)} = (1, 1/2)^T$.

Exercise 5.16 Prove (5.48), then show that

$$\rho(B_{\alpha_{opt}}) = \frac{\lambda_{max} - \lambda_{min}}{\lambda_{max} + \lambda_{min}} = \frac{K(P^{-1}A) - 1}{K(P^{-1}A) + 1}. \tag{5.59}$$

Exercise 5.17 Let us consider a set of $n = 20$ factories which produce 20 different goods. With reference to the Leontief model introduced in Problem 5.3, suppose that the matrix C has the following integer entries: $c_{ij} = i + j - 1$ for $i, j = 1, \ldots, n$, while $b_i = i$, for $i = 1, \ldots, 20$. Is it possible to solve this system by the gradient method? Propose a method based on the gradient method noting that, if A is nonsingular, the matrix $A^T A$ is symmetric and positive definite.

6
Eigenvalues and eigenvectors

Given a square matrix $A \in \mathbb{C}^{n \times n}$, the eigenvalue problem consists in finding a scalar λ (real or complex) and a nonnull vector $\mathbf{x}$ such that

$$\boxed{A\mathbf{x} = \lambda \mathbf{x}} \qquad (6.1)$$

Any such λ is called an *eigenvalue* of A, while $\mathbf{x}$ is the associated *eigenvector*. The latter is not unique; indeed all its multiples $\alpha \mathbf{x}$ with $\alpha \neq 0$, real or complex, are also eigenvectors associated with λ. Should $\mathbf{x}$ be known, λ can be recovered by using the *Rayleigh quotient* $\mathbf{x}^H A \mathbf{x} / \|\mathbf{x}\|^2$, $\mathbf{x}^H$ being the vector whose i-th component is equal to $\bar{x}_i$.

A number λ is an eigenvalue of A if it is a root of the following polynomial of degree n (called the *characteristic polynomial* of A):

$$p_A(\lambda) = \det(A - \lambda I).$$

Consequently, a square matrix of dimension n has exactly n eigenvalues (real or complex), not necessarily distinct. Also, if A has real entries, $p_A(\lambda)$ has real coefficients, and therefore complex eigenvalues of A necessarily occur in complex conjugate pairs.

A matrix $A \in \mathbb{C}^{n \times n}$ is diagonalizable if there exists a nonsingular matrix $U \in \mathbb{C}^{n \times n}$ such that

$$U^{-1} A U = \Lambda = \text{diag}(\lambda_1, \ldots, \lambda_n). \qquad (6.2)$$

The columns of U are the eigenvectors of A and form a basis for $\mathbb{C}^n$.

If $A \in \mathbb{C}^{m \times n}$, there exist two unitary matrices $U \in \mathbb{C}^{m \times m}$ and $V \in \mathbb{C}^{n \times n}$ such that

$$U^* A V = \Sigma = \text{diag}(\sigma_1, \ldots, \sigma_p) \in \mathbb{R}^{m \times n}, \qquad (6.3)$$

where $p = \min(m, n)$ and $\sigma_1 \geq \ldots \geq \sigma_p \geq 0$. (A matrix U is called unitary if $A^H A = A A^H = I$.)

Formula (6.3) is called *singular value decomposition* (SVD) of A and the numbers σ_i (or $\sigma_i(A)$) are called *singular values* of A.

Problem 6.1 (Elastic springs) Consider the system of Figure 6.1 made of two pointwise bodies P_1 and P_2 of mass m, connected by two springs and free to move along the line joining P_1 and P_2. Let $x_i(t)$ denote the position occupied by P_i at time t for $i = 1, 2$. Then from the second law of dynamics we obtain

$$m\ddot{x}_1 = K(x_2 - x_1) - Kx_1, \qquad m\ddot{x}_2 = K(x_1 - x_2),$$

where K is the elasticity coefficient of both springs. We are interested in free oscillations whose corresponding solution is $x_i = a_i \sin(\omega t + \phi)$, $i = 1, 2$, with $a_i \neq 0$. In this case we find that

$$-ma_1\omega^2 = K(a_2 - a_1) - Ka_1, \qquad -ma_2\omega^2 = K(a_1 - a_2). \quad (6.4)$$

This is a 2×2 homogeneous system which has a non-trivial solution a_1, a_2 iff the number $\lambda = m\omega^2/K$ is an eigenvalue of the matrix

$$A = \begin{bmatrix} 2 & -1 \\ -1 & 1 \end{bmatrix}.$$

With this definition of λ, (6.4) becomes $A\mathbf{a} = \lambda \mathbf{a}$. Since $p_A(\lambda) = (2 - \lambda)(1 - \lambda) - 1$, the two eigenvalues are $\lambda_1 \simeq 2.618$ and $\lambda_2 \simeq 0.382$ and correspond to the frequencies of oscillation $\omega_i = \sqrt{K\lambda_i/m}$ which are admitted by our system. ∎

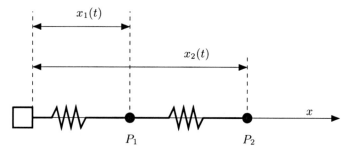

Fig. 6.1. The system of two pointwise bodies of equal mass connected by springs

Problem 6.2 (Population dynamics) Several mathematical models have been proposed in order to predict the evolution of certain species (either human or animal). The simplest population model, which was introduced in 1920 by Lotka and formalized by Leslie 20 years later, is based on the rate of mortality and fecundity for different age intervals, say $i = 0, \ldots, n$. Let $x_i^{(t)}$ denote the number of females (males don't matter in this context) whose age at time t falls in the i-th interval. The values of $x_i^{(0)}$ are given. Moreover, let s_i denote the rate of survival of the females belonging to the i-th interval, and m_i the average number of females generated from a female in the i-th interval.

The model by Lotka and Leslie is described by the set of equations

$$x_{i+1}^{(t+1)} = x_i^{(t)} s_i \quad i = 0, \ldots, n-1,$$

$$x_0^{(t+1)} = \sum_{i=0}^{n} x_i^{(t)} m_i.$$

The n first equations describe the population development, the last its reproduction. In matrix form we have

$$\mathbf{x}^{(t+1)} = A\mathbf{x}^{(t)},$$

where $\mathbf{x}^{(t)} = (x_0^{(t)}, \ldots, x_n^{(t)})^T$ and A is the *Leslie matrix*:

$$A = \begin{bmatrix} m_0 & m_1 & \ldots & \ldots & m_n \\ s_0 & 0 & \ldots & \ldots & 0 \\ 0 & s_1 & \ddots & & \vdots \\ \vdots & \ddots & \ddots & \ddots & \vdots \\ 0 & 0 & 0 & s_{n-1} & 0 \end{bmatrix}.$$

We will see in Section 6.1 that the dynamics of this population is determined by the eigenvalue of maximum modulus of A, say λ_1, whereas the distribution of the individuals in the different age intervals (normalized with respect to the whole population), is obtained as the limit of $\mathbf{x}^{(t)}$ for $t \to \infty$ and satisfies $A\mathbf{x} = \lambda_1 \mathbf{x}$. This problem will be solved in Exercise 6.2. ■

Problem 6.3 (Interurban viability) For n given cities, let A be the matrix whose entry a_{ij} is equal to 1 if the i-th city is directly connected to the j-th city, and 0 otherwise. One can show that the components of the eigenvector $\mathbf{x}$ (of unit length) associated with the maximum eigenvalue provides the accessibility rate (which is a measure of the ease of access) to the various cities. In Example 6.2 we will compute this vector for the case of the railways system of the eleven most important cities in Lombardy (see Figure 6.2). ■

Problem 6.4 (Image compression) The problem of image compression can be faced using the singular-value decomposition of a matrix. Indeed, a black and white image can be represented by a real $m \times n$ rectangular matrix A where m and n represent the number of *pixels* that

170   6 Eigenvalues and eigenvectors

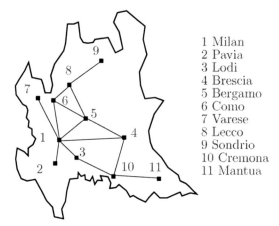

Fig. 6.2. A schematic representation of the railway network between the main cities of Lombardy

are present in the horizontal and vertical direction, respectively, and the coefficient a_{ij} represents the intensity of gray of the (i,j)-th pixel. Considering the singular value decomposition (6.3) of A, and denoting by $\mathbf{u}_i$ and $\mathbf{v}_i$ the i-th column vectors of U and V, respectively, we find

$$A = \sigma_1 \mathbf{u}_1 \mathbf{v}_1^T + \sigma_2 \mathbf{u}_2 \mathbf{v}_2^T + \ldots + \sigma_p \mathbf{u}_p \mathbf{v}_p^T. \tag{6.5}$$

We can approximate A by the matrix A_k which is obtained by truncating the sum (6.5) to the first k terms, for $1 \le k \le p$. If the singular values σ_i are in decreasing order, $\sigma_1 \ge \sigma_2 \ge \ldots \ge \sigma_p$, disregarding the latter $p-k$ should not significantly affect the quality of the image. To transfer the "compressed" image A_k (for instance from one computer to another) we simply need to transfer the vectors $\mathbf{u}_i$, $\mathbf{v}_i$ and the singular values σ_i for $i = 1, \ldots, k$ and not all the entries of A. In Example 6.9 we will see this technique in action. ∎

In the special case where A is either diagonal or triangular, its eigenvalues are nothing but its diagonal entries. However, if A is a general matrix and its dimension n is sufficiently large, seeking the zeros of $p_A(\lambda)$ is not a convenient approach. Ad hoc algorithms are better suited, and one of them is described in the next section.

6.1 The power method

As noticed in Problems 6.2 and 6.3, the knowledge of the whole *spectrum* of A (that is the the set of all its eigenvalues) is not always required. Often, only the *extremal* eigenvalues matter, that is, those having largest and smallest modulus.

Suppose that A is a square matrix of dimension n, with real entries, and assume that its eigenvalues are ordered as follows

$$|\lambda_1| > |\lambda_2| \geq |\lambda_3| \geq \ldots \geq |\lambda_n|. \tag{6.6}$$

Note, in particular, that $|\lambda_1|$ is distinct from the other moduli of the eigenvalues of A. Let us indicate by $\mathbf{x}_1$ the eigenvector (with unit length) associated with λ_1. If the eigenvectors of A are linearly independent, λ_1 and $\mathbf{x}_1$ can be computed by the following iterative procedure, commonly known as the *power method*:

given an arbitrary initial vector $\mathbf{x}^{(0)} \in \mathbb{C}^n$ and setting $\mathbf{y}^{(0)} = \mathbf{x}^{(0)}/\|\mathbf{x}^{(0)}\|$, compute

$$\text{for } k = 1, 2, \ldots$$
$$\mathbf{x}^{(k)} = A\mathbf{y}^{(k-1)}, \quad \mathbf{y}^{(k)} = \frac{\mathbf{x}^{(k)}}{\|\mathbf{x}^{(k)}\|}, \quad \lambda^{(k)} = (\mathbf{y}^{(k)})^H A \mathbf{y}^{(k)} \tag{6.7}$$

Note that, by recursion, one finds $\mathbf{y}^{(k)} = \beta^{(k)} A^k \mathbf{y}^{(0)}$ where $\beta^{(k)} = (\Pi_{i=1}^k \|\mathbf{x}^{(i)}\|)^{-1}$ for $k \geq 1$. The presence of the powers of A justifies the name given to this method.

In the next section we will see that this method generates a sequence of vectors $\{\mathbf{y}^{(k)}\}$ with unit length which, as $k \to \infty$, align themselves along the direction of the eigenvector $\mathbf{x}_1$. The error $\|\lambda^{(k)} - \lambda_1\|$ is proportional to the ratio $|\lambda_2/\lambda_1|^k$ in the case of a generic matrix, and to $|\lambda_2/\lambda_1|^{2k}$ when the matrix A is hermitian. Consequently one obtains that $\lambda^{(k)} \to \lambda_1$ for $k \to \infty$.

An implementation of the power method is given in the Program 6.1. The iterative procedure is stopped at the first iteration k when

$$|\lambda^{(k)} - \lambda^{(k-1)}| < \varepsilon |\lambda^{(k)}|,$$

where ε is a desired tolerance. The input parameters are the real matrix A, the initial vector x0, the tolerance tol for the stopping test and the maximum admissible number of iterations nmax. Output parameters are the maximum modulus eigenvalue lambda, the associated eigenvector and the actual number of iterations which have been carried out.

Program 6.1. eigpower: power method

```
function [lambda,x,iter]=eigpower(A,tol,nmax,x0)
%EIGPOWER Numerically evaluate one eigenvalue of a real
%   matrix.
%   LAMBDA=EIGPOWER(A) computes with the power method the
%   eigenvalue of A of maximum modulus from an initial
%   guess which by default is an all one vector.
%   LAMBDA=EIGPOWER(A,TOL,NMAX,X0) uses an absolute error
%   tolerance TOL (the default is 1.e-6) and a maximum
%   number of iterations NMAX (the default is 100),
```

172   6 Eigenvalues and eigenvectors

```
%    starting from the initial vector X0.
%    [LAMBDA,V,ITER]=EIGPOWER(A,TOL,NMAX,X0) also returns
%    the eigenvector V such that A*V=LAMBDA*V and the
%    iteration number at which V was computed.
[n,m] = size(A);
if n ~= m, error('Only for square matrices'); end
if nargin == 1
   tol = 1.e-06;
   x0 = ones(n,1);
   nmax = 100;
end
x0 = x0/norm(x0);
pro = A*x0;
lambda = x0'*pro;
err = tol*abs(lambda) + 1;
iter = 0;
while err>tol*abs(lambda)&abs(lambda)~=0&iter<=nmax
   x = pro; x = x/norm(x);
   pro = A*x; lambdanew = x'*pro;
   err = abs(lambdanew - lambda);
   lambda = lambdanew;
   iter = iter + 1;
end
return
```

Example 6.1 Consider the family of matrices

$$A(\alpha) = \begin{bmatrix} \alpha & 2 & 3 & 13 \\ 5 & 11 & 10 & 8 \\ 9 & 7 & 6 & 12 \\ 4 & 14 & 15 & 1 \end{bmatrix}, \quad \alpha \in \mathbb{R}.$$

We want to approximate the eigenvalue with largest modulus by the power method. When $\alpha = 30$, the eigenvalues of the matrix are given by $\lambda_1 = 39.396$, $\lambda_2 = 17.8208$, $\lambda_3 = -9.5022$ and $\lambda_4 = 0.2854$ (only the first four significant digits are reported). The method approximates λ_1 in 22 iterations with a tolerance $\varepsilon = 10^{-10}$ and $\mathbf{x}^{(0)} = \mathbf{1}$. However, if $\alpha = -30$ we need as many as 708 iterations. The different behavior can be explained by noting that in the latter case one has $\lambda_1 = -30.643$, $\lambda_2 = 29.7359$, $\lambda_3 = -11.6806$ and $\lambda_4 = 0.5878$. Thus, $|\lambda_2|/|\lambda_1| = 0.9704$, close to unity. ∎

Example 6.2 (Interurban viability) We denote by $A \in \mathbb{R}^{11 \times 11}$ the matrix associated to the railways system of Figure 6.2, i.e. the matrix whose entry a_{ij} is equal to one if there is a direct connection between the i-th and the j-th cities, zero otherwise. Setting `tol=1.e-12` and `x0=ones(11,1)`, after 26 iterations Program 6.1 returns the following approximation of the eigenvector (of unitary length) associated to the eigenvalue of maximum modulus of A:

```
x' =
  Columns 1 through 8
  0.5271  0.1590  0.2165  0.3580  0.4690  0.3861  0.1590  0.2837
  Columns 9 through 11
  0.0856  0.1906  0.0575
```

The most reachable city is Milan, which is the one associated to the first component of **x** (the highest in modulus), the least one is Mantua, which is associated to the last component of **x**, that of minimum modulus. Of course our analysis accounts solely for the existence of connections among the cities but not on how frequent these connections are. ∎

6.1.1 Convergence analysis

Since we have assumed that the eigenvectors $\mathbf{x}_1, \ldots, \mathbf{x}_n$ of A are linearly independent, these eigenvectors form a basis for $\mathbb{C}^n$. Thus the vectors $\mathbf{x}^{(0)}$ and $\mathbf{y}^{(0)}$ can be written as

$$\mathbf{x}^{(0)} = \sum_{i=1}^{n} \alpha_i \mathbf{x}_i, \quad \mathbf{y}^{(0)} = \beta^{(0)} \sum_{i=1}^{n} \alpha_i \mathbf{x}_i, \quad \text{with } \beta^{(0)} = 1/\|\mathbf{x}^{(0)}\| \text{ and } \alpha_i \in \mathbb{C}.$$

At the first step the power method gives

$$\mathbf{x}^{(1)} = A\mathbf{y}^{(0)} = \beta^{(0)} A \sum_{i=1}^{n} \alpha_i \mathbf{x}_i = \beta^{(0)} \sum_{i=1}^{n} \alpha_i \lambda_i \mathbf{x}_i$$

and, similarly,

$$\mathbf{y}^{(1)} = \beta^{(1)} \sum_{i=1}^{n} \alpha_i \lambda_i \mathbf{x}_i, \quad \beta^{(1)} = \frac{1}{\|\mathbf{x}^{(0)}\| \, \|\mathbf{x}^{(1)}\|}.$$

At a given step k we will have

$$\mathbf{y}^{(k)} = \beta^{(k)} \sum_{i=1}^{n} \alpha_i \lambda_i^k \mathbf{x}_i, \quad \beta^{(k)} = \frac{1}{\|\mathbf{x}^{(0)}\| \cdots \|\mathbf{x}^{(k)}\|}$$

and therefore

$$\mathbf{y}^{(k)} = \lambda_1^k \beta^{(k)} \left(\alpha_1 \mathbf{x}_1 + \sum_{i=2}^{n} \alpha_i \frac{\lambda_i^k}{\lambda_1^k} \mathbf{x}_i \right).$$

Since $|\lambda_i/\lambda_1| < 1$ for $i = 2, \ldots, n$, the vector $\mathbf{y}^{(k)}$ tends to align along the same direction as the eigenvector $\mathbf{x}_1$ when k tends to $+\infty$, provided $\alpha_1 \neq 0$. The condition on α_1, which is impossible to ensure in practice since $\mathbf{x}_1$ is unknown, is in fact not restrictive. Actually, the effect of roundoff errors is the appearance of a non-null component along the direction of $\mathbf{x}_1$, even though this was not the case for the initial vector $\mathbf{x}^{(0)}$. (We can say that this is one of the rare circumstances where roundoff errors help us!)

174   6 Eigenvalues and eigenvectors

Example 6.3 Consider the matrix $A(\alpha)$ of Example 6.1, with $\alpha = 16$. The eigenvector $\mathbf{x}_1$ of unit length associated with λ_1 is $(1/2, 1/2, 1/2, 1/2)^T$. Let us choose (on purpose!) the initial vector $(2, -2, 3, -3)^T$, which is orthogonal to $\mathbf{x}_1$. We report in Figure 6.3 the quantity $\cos(\theta^{(k)}) = (\mathbf{y}^{(k)})^T \mathbf{x}_1 / (\|\mathbf{y}^{(k)}\| \, \|\mathbf{x}_1\|)$. We can see that after about 30 iterations of the power method the cosine tends to -1 and the angle tends to π, while the sequence $\lambda^{(k)}$ approaches $\lambda_1 = 34$. The power method has therefore generated, thanks to the roundoff errors, a sequence of vectors $\mathbf{y}^{(k)}$ whose component along the direction of $\mathbf{x}_1$ is increasingly relevant. ∎

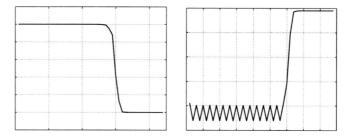

Fig. 6.3. The value of $(\mathbf{y}^{(k)})^T \mathbf{x}_1 / (\|\mathbf{y}^{(k)}\| \, \|\mathbf{x}_1\|)$ (*left*) and that of $\lambda^{(k)}$ (*right*), for $k = 1, \ldots, 44$

It is possible to prove that the power method converges even if λ_1 is a multiple root of $p_A(\lambda)$. On the contrary it does not converge when there exist two distinct eigenvalues both with maximum modulus. In that case the sequence $\lambda^{(k)}$ does not converge to any limit, rather it oscillates between two values.

See Exercises 6.1-6.3.

6.2 Generalization of the power method

A first possible generalization of the power method consists in applying it to the inverse of the matrix A (provided A is non singular!). Since the eigenvalues of A^{-1} are the reciprocals of those of A, the power method in that case allows us to approximate the eigenvalue of A of minimum modulus. In this way we obtain the so-called *inverse power method*:

given an initial vector $\mathbf{x}^{(0)}$, we set $\mathbf{y}^{(0)} = \mathbf{x}^{(0)} / \|\mathbf{x}^{(0)}\|$ and compute

$$\text{for } k = 1, 2, \ldots$$
$$\mathbf{x}^{(k)} = A^{-1} \mathbf{y}^{(k-1)}, \quad \mathbf{y}^{(k)} = \frac{\mathbf{x}^{(k)}}{\|\mathbf{x}^{(k)}\|}, \quad \mu^{(k)} = (\mathbf{y}^{(k)})^H A^{-1} \mathbf{y}^{(k)} \tag{6.8}$$

6.2 Generalization of the power method

If A admits linearly independent eigenvectors, and if also the eigenvalue λ_n of minimum modulus is distinct from the others, then

$$\lim_{k \to \infty} \mu^{(k)} = 1/\lambda_n,$$

i.e. $(\mu^{(k)})^{-1}$ tends to λ_n for $k \to \infty$.

At each step k we have to solve a linear system of the form $A\mathbf{x}^{(k)} = \mathbf{y}^{(k-1)}$. It is therefore convenient to generate the LU factorization of A (or its Cholesky factorization if A is symmetric and positive definite) once for all, and then solve two triangular systems at each iteration.

It is worth noticing that the lu command (in MATLAB and in Octave) can generate the LU decomposition even for complex matrices.

Example 6.4 When applied to the matrix A(30) of Example 6.1, after 7 iterations the inverse power method yields the value 3.5037. Thus the eigenvalue of A(30) of minimum modulus will be approximately equal to $1/3.5037 \simeq 0.2854$.
∎

A further generalization of the power method stems from the following consideration. Let λ_μ denote the (unknown) eigenvalue of A nearest to a given number (real or complex) μ. In order to approximate λ_μ, we can at first approximate the minimum length eigenvalue, say $\lambda_{min}(A_\mu)$, of the shifted matrix $A_\mu = A - \mu I$, and then set $\lambda_\mu = \lambda_{min}(A_\mu) + \mu$. We can therefore apply the inverse power method to A_μ to obtain an approximation of $\lambda_{min}(A_\mu)$. This technique is known as the *power method with shift*, and the number μ is called the *shift*.

In Program 6.2 we implement the inverse power method with shift. The inverse power method is recovered by simply setting $\mu = 0$. The first four input parameters are the same as in Program 6.1, while mu is the shift. Output parameters are the eigenvalue λ_μ of A, its associated eigenvector x and the actual number of iterations that have been carried out.

Program 6.2. invshift: inverse power method with shift

```
function [lambda,x,iter]=invshift(A,mu,tol,nmax,x0)
%INVSHIFT Numerically evaluate one eigenvalue of a
%   matrix.
%   LAMBDA=INVSHIFT(A) compute the  eigenvalue of A of
%   minimum modulus with the inverse power method.
%   LAMBDA=INVSHIFT(A,MU) computes the eigenvalue of A
%   closest to the given number (real or complex) MU.
%   LAMBDA=INVSHIFT(A,MU,TOL,NMAX,X0) uses an absolute
%   error tolerance TOL (the default is 1.e-6) and a
%   maximum number of iterations NMAX (the default is
%   100), starting from the initial vector X0.
%   [LAMBDA,V,ITER]=INVSHIFT(A,MU,TOL,NMAX,X0) also
%   returns the eigenvector V such that A*V=LAMBDA*V and
```

176   6 Eigenvalues and eigenvectors

```
%       the iteration number at which V was computed.
[n,m]=size(A);
if n ~= m, error('Only for square matrices'); end
if nargin == 1
    x0 = rand(n,1); nmax = 100; tol = 1.e-06; mu = 0;
elseif nargin == 2
    x0 = rand(n,1); nmax = 100; tol = 1.e-06;
end
[L,U]=lu(A-mu*eye(n));
if norm(x0) == 0
    x0 = rand(n,1);
end
x0=x0/norm(x0);
z0=L\x0;
pro=U\z0;
lambda=x0'*pro;
err=tol*abs(lambda)+1;         iter=0;
while err>tol*abs(lambda)&abs(lambda)~=0&iter<=nmax
    x = pro; x = x/norm(x);
    z=L\x;       pro=U\z;
    lambdanew = x'*pro;
    err = abs(lambdanew - lambda);
    lambda = lambdanew;
    iter = iter + 1;
end
lambda = 1/lambda + mu;
return
```

Example 6.5 For the matrix A(30) of Example 6.1 we seek the eigenvalue closest to the value 17. For that we use Program 6.2 with mu=17, tol =10^{-10} and x0=[1;1;1;1]. After 8 iterations the Program returns the value lambda=17.82079703055703. A less accurate knowledge of the *shift* would involve more iterations. For instance, if we set mu=13 the program returns the value 17.82079703064106 after 11 iterations. ∎

The value of the shift can be modified during the iterations, by setting $\mu = \lambda^{(k)}$. This yields a faster convergence; however the computational cost grows substantially since now at each iteration the matrix A_μ does change.

See Exercises 6.4-6.6.

6.3 How to compute the shift

In order to successfully apply the power method with shift we need to locate (more or less accurately) the eigenvalues of A in the complex plane. To this end let us introduce the following definition.

Let A be a square matrix of dimension n. The *Gershgorin circles* $C_i^{(r)}$ and $C_i^{(c)}$ associated with its i-th row and i-th column are respectively defined as

$$C_i^{(r)} = \{z \in \mathbb{C} : |z - a_{ii}| \leq \sum_{j=1, j \neq i}^{n} |a_{ij}|\},$$

$$C_i^{(c)} = \{z \in \mathbb{C} : |z - a_{ii}| \leq \sum_{j=1, j \neq i}^{n} |a_{ji}|\}.$$

$C_i^{(r)}$ is called the i-th *row circle* and $C_i^{(c)}$ the i-th *column circle*.

By the Program 6.3 we can visualize in two different windows (that are opened by the command figure) the row circles and the column circles of a matrix. The command hold on allows the overlapping of subsequent pictures (in our case, the different circles that have been computed in sequential mode). This command can be neutralized by the command hold off. The commands title, xlabel and ylabel have the scope of visualizing the title and the axis labels in the figure.

The command patch was used in order to color the circles, while the command axis sets scaling for the x- and y-axes on the current plot.

figure
hold on/off
title
xlabel
ylabel
patch
axis

Program 6.3. gershcircles: Gershgorin circles

```
function gershcircles(A)
%GERSHCIRCLES plots the Gershgorin circles
%   GERSHCIRCLES(A) draws the Gershgorin circles for
%   the square matrix A and its transpose.
n = size(A);
if n(1) ~= n(2)
    error('Only square matrices');
else
    n = n(1); circler = zeros(n,201); circlec = circler;
end
center = diag(A);
radiic = sum(abs(A-diag(center)));
radiir = sum(abs(A'-diag(center)));
one = ones(1,201); cosisin = exp(i*[0:pi/100:2*pi]);
figure(1); title('Row circles');
xlabel('Re'); ylabel('Im');
figure(2); title('Column circles');
xlabel('Re'); ylabel('Im');
for k = 1:n
    circlec(k,:) = center(k)*one + radiic(k)*cosisin;
    circler(k,:) = center(k)*one + radiir(k)*cosisin;
    figure(1);
    patch(real(circler(k,:)),imag(circler(k,:)),'red');
    hold on
    plot(real(circler(k,:)),imag(circler(k,:)),'k-',...
        real(center(k)),imag(center(k)),'kx');
    figure(2);
    patch(real(circlec(k,:)),imag(circlec(k,:)),'green');
    hold on
    plot(real(circlec(k,:)),imag(circlec(k,:)),'k-',...
        real(center(k)),imag(center(k)),'kx');
end
for k = 1:n
    figure(1);
    plot(real(circler(k,:)),imag(circler(k,:)),'k-',...
```

```
        real(center(k)),imag(center(k)),'kx');
    figure(2);
    plot(real(circlec(k,:)),imag(circlec(k,:)),'k-',...
        real(center(k)),imag(center(k)),'kx');
end
figure(1); axis image; hold off;
figure(2); axis image; hold off
return
```

Example 6.6 In Figure 6.4 we have plotted the Gershgorin circles associated with the matrix

$$A = \begin{bmatrix} 30 & 1 & 2 & 3 \\ 4 & 15 & -4 & -2 \\ -1 & 0 & 3 & 5 \\ -3 & 5 & 0 & -1 \end{bmatrix}.$$

The centers of the circles have been identified by a cross. ∎

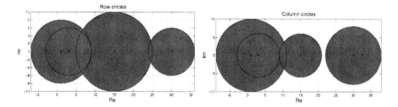

Fig. 6.4. Row circles (*left*) and column circles (*right*) for the matrix of Example 6.6

As previously anticipated, Gershgorin circles may be used to locate the eigenvalues of a matrix, as stated in the following proposition.

Proposition 6.1 *All eigenvalues of a given matrix $A \in \mathbb{C}^{n \times n}$ belong to the region of the complex plane which is the intersection of the two regions formed respectively by the union of the row circles and the union of the column circles.*
Moreover, should m row circles (or column circles), with $1 \leq m \leq n$, be disconnected from the union of the remaining $n - m$ circles, then their union contains exactly m eigenvalues.

There is no guarantee that a circle should contain eigenvalues, unless it is isolated from the others. The previous result can be applied in order to obtain a preliminary guess of the shift, as we show in the following example.

6.4 Computation of all the eigenvalues 179

Example 6.7 From the analysis of the row circles of the matrix A(30) of Example 6.1 we deduce that the real parts of the eigenvalues of A lie between -32 and 48. Thus we can use Program 6.2 to compute the maximum modulus eigenvalue by setting the value of the shift μ equal to 48. The convergence is achieved in 16 iterations, whereas 24 iterations would be required using the power method with the same initial guess `x0=[1;1;1;1]` and the same tolerance `tol=1.e-10`. ∎

Let us summarize

1. The power method is an iterative procedure to compute the eigenvalue of maximum modulus of a given matrix;
2. the inverse power method allows the computation of the eigenvalue of minimum modulus; it requires the factorization of the given matrix;
3. the power method with shift allows the computation of the eigenvalue closest to a given number; its effective application requires some a-priori knowledge of the location of the eigenvalues of the matrix, which can be achieved inspecting the Gershgorin circles.

See Exercises 6.7-6.8.

6.4 Computation of all the eigenvalues

Two square matrices A and B having the same dimension are called *similar* if there exists a non singular matrix P such that

$$P^{-1}AP = B.$$

Similar matrices share the same eigenvalues. Indeed, if λ is an eigenvalue of A and $\mathbf{x} \neq \mathbf{0}$ is an associated eigenvector, we have

$$BP^{-1}\mathbf{x} = P^{-1}A\mathbf{x} = \lambda P^{-1}\mathbf{x},$$

that is, λ is also an eigenvalue of B and its associated eigenvector is now $\mathbf{y} = P^{-1}\mathbf{x}$.

The methods which allow a simultaneous approximation of all the eigenvalues of a matrix are generally based on the idea of transforming A (after an infinite number of steps) into a similar matrix with diagonal or triangular form, whose eigenvalues are therefore given by the entries lying on its main diagonal.

Among these methods we mention the *QR method* which is implemented in MATLAB in the function `eig`. More precisely, the command `D=eig(A)` returns a vector D containing all the eigenvalues of A. However,

`eig`

by setting [X,D]=eig(A), we obtain two matrices: the diagonal matrix D formed by the eigenvalues of A, and a matrix X whose column vectors are the eigenvectors of A. Thus, A*X=X*D.

The method of QR iterations is called in this way since it makes a repeated use of the QR factorization introduced in Section 5.5 to compute the eigenvalues of the matrix A. Here we present the QR method only for real matrices and in its most elementary form (whose convergence is not always guaranteed). For a more complete description of this method we refer to [QSS06, Chapter 5], whereas for its extension to the complex case we refer to [GL96, Section 5.2.10] and [Dem97, Section 4.2.1].

The idea consists in building a sequence of matrices $A^{(k)}$, each of them similar to A. After setting $A^{(0)} = A$, at each $k = 1, 2, \ldots$, using the QR factorization we compute the matrices $Q^{(k+1)}$ and $R^{(k+1)}$ such that

$$Q^{(k+1)} R^{(k+1)} = A^{(k)},$$

whence we set $A^{(k+1)} = R^{(k+1)} Q^{(k+1)}$.

The matrices $A^{(k)}$, $k = 0, 1, 2, \ldots$ are all similar, thus they share with A their eigenvalues (see Exercise 6.9). Moreover, if $A \in \mathbb{R}^{n \times n}$ and its eigenvalues satisfy $|\lambda_1| > |\lambda_2| > \ldots > |\lambda_n|$, then

$$\lim_{k \to +\infty} A^{(k)} = T = \begin{bmatrix} \lambda_1 & t_{12} & \cdots & t_{1n} \\ 0 & \ddots & \ddots & \vdots \\ \vdots & & \lambda_{n-1} & t_{n-1,n} \\ 0 & \cdots & 0 & \lambda_n \end{bmatrix}. \tag{6.9}$$

The rate of decay to zero of the lower triangular coefficients, $a_{i,j}^{(k)}$ for $i > j$, when k tends to infinity, depends on $\max_i |\lambda_{i+1}/\lambda_i|$. In practice, the iterations are stopped when $\max_{i>j} |a_{i,j}^{(k)}| \leq \epsilon$, $\epsilon > 0$ being a given tolerance.

Under the further assumption that A is symmetric, the sequence $\{A^{(k)}\}$ converges to a diagonal matrix.

Program 6.4 implements the QR iteration method. The input parameters are the matrix A, the tolerance tol and the maximum number of iterations allowed, nmax.

Program 6.4. qrbasic: method of QR iterations

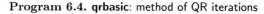

```
function D=qrbasic(A,tol,nmax)
%QRBASIC computes the eigenvalues of a matrix A.
%   D=QRBASIC(A,TOL,NMAX) computes by QR iterations all
%   the eigenvalues of A within a tolerance TOL and a
%   maximum number of iteration NMAX. The convergence of
%   this method is not always guaranteed.
[n,m]=size(A);
if n ~= m, error('The matrix must be squared'); end
```

6.4 Computation of all the eigenvalues

```
T = A; niter = 0; test = norm(tril(A,-1),inf);
while niter <= nmax & test >= tol
    [Q,R]=qr(T);     T = R*Q;
    niter = niter + 1;
    test = norm(tril(T,-1),inf);
end
if niter > nmax
  warning(['The method does not converge'
           'in the maximum number of iterations']);
else
  fprintf(['The method converges in ' ...
           '%i iterations\n'],niter);
end
D = diag(T);
return
```

Example 6.8 Let us consider the matrix A(30) of Example 6.1 and call Program 6.4 to compute its eigenvalues. We obtain
D=qrbasic(A(30),1.e-14,100)

The method converges in 56 iterations
D =
    39.3960
    17.8208
    -9.5022
     0.2854

These eigenvalues are in good agreement with those reported in Example 6.1, that were obtained with the command eig. The convergence rate decreases when there are eigenvalues whose moduli are almost the same. This is the case of the matrix corresponding to $\alpha = -30$: two eigenvalues have about the same modulus and the method requires as many as 1149 iterations to converge within the same tolerance
D=qrbasic(A(-30),1.e-14,2000)

The method converges in 1149 iterations
D =
    -30.6430
     29.7359
    -11.6806
      0.5878

■

A special case is the one of large sparse matrices. In this case, if A is stored in a sparse mode the command eigs(A,k) allows the computation of the k first eigenvalues of modulus larger than A.   eigs

Finally, let us mention how to compute the singular values of a rectangular matrix. Two MATLAB functions are available: svd and svds.   svd
The former computes all the singular values of a matrix, the latter only   svds
the first largest k. The integer k must be fixed as input (by default, k=6).

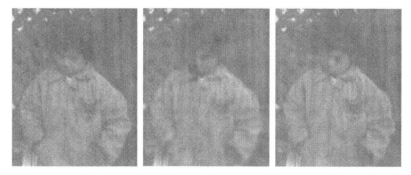

Fig. 6.5. The original image (*left*) and those obtained using the first 20 (*center*) and 40 (*right*) singular values, respectively

We refer to [ABB+99] for a thorough description of the algorithm that is actually used.

Example 6.9 (Image compression) With the MATLAB command A= imread('pout.tif') we upload a black and white image which is present in the MATLAB toolbox *Image Processing*. The variable A is a matrix of 291 by 240 eight bit integer numbers (uint8) that represent the intensity of gray.

imread
imshow

The command imshow(A) produces the image on the left hand of Figure 6.5. To compute the SVD of A we must first convert A in a double precision matrix (the floating-point numbers usually used by MATLAB), through the command A=double(A). Now, we set [U,S,V]=svd(A). In the middle of Figure 6.5 we report the image that is obtained by using only the first 20 singular values of S, through the commands

X=U(:,1:20)*S(1:20,1:20)*(V(:,1:20))'; imshow(uint8(X));

The image on the right-hand side of Figure 6.5 is obtained using the first 40 singular values. It requires the storage of 21280 coefficients (two matrices of 291 × 40 and 240 × 40 plus the first 40 singular values) instead of 69840 that would be required to store the whole original image. ∎

Octave 6.1 svds and eigs for computing the singular values and the eigenvalues of sparse matrices are not yet available in Octave. ∎

Let us summarize

1. The method of QR iterations allows the approximation of all the eigenvalues of a given matrix A;
2. in its basic version, this method is guaranteed to converge if A has real coefficients and distinct eigenvalues;
3. its asymptotic rate of convergence depends on the largest modulus of the ratio of two successive eigenvalues.

See Exercises 6.9-6.10.

6.5 What we haven't told you

We have not analyzed the issue of the condition number of the eigenvalue problem, which measures the sensitivity of the eigenvalues to the variation of the entries of the matrix. The interested reader is advised to refer to, for instance, [Wil65], [GL96] and [QSS06, Chapter 5].

Let us just remark that the eigenvalue computation is not necessarily an ill conditioned problem when the condition number of the matrix is large. An instance of this is provided by the Hilbert matrix (see Example 5.9): although its condition number is extremely large, the eigenvalue computation of the Hilbert matrix is well conditioned thanks to the fact that the matrix is symmetric and positive definite.

Besides the QR method, for computing simultaneously all the eigenvalues we can use the Jacobi method which transforms a symmetric matrix into a diagonal matrix, by eliminating, step-by-step, through similarity transformations, every off-diagonal element. This method does not terminate in a finite number of steps since, while a new off-diagonal element is set to zero, those previously treated can reassume non-zero values.

Other methods are the Lanczos method and the method which uses the so-called Sturm sequences. For a survey of all these methods see [Saa92].

The MATLAB library ARPACK (available through the command arpackc) can be used to compute the eigenvalues of large matrices. The MATLAB function eigs is a command that uses this library. arpackc

Let us mention that an appropriate use of the *deflation* technique (which consists in a successive elimination of the eigenvalues already computed) allows the acceleration of the convergence of the previous methods and hence the reduction of their computational cost.

6.6 Exercises

Exercise 6.1 Upon setting the tolerance equal to $\varepsilon = 10^{-10}$, use the power method to approximate the maximum modulus eigenvalue for the following matrices, starting from the initial vector $\mathbf{x}^{(0)} = (1, 2, 3)^T$:

$$A_1 = \begin{bmatrix} 1 & 2 & 0 \\ 1 & 0 & 0 \\ 0 & 1 & 0 \end{bmatrix}, \quad A_2 = \begin{bmatrix} 0.1 & 3.8 & 0 \\ 1 & 0 & 0 \\ 0 & 1 & 0 \end{bmatrix}, \quad A_3 = \begin{bmatrix} 0 & -1 & 0 \\ 1 & 0 & 0 \\ 0 & 1 & 0 \end{bmatrix}.$$

Then comment on the convergence behavior of the method in the three different cases.

Exercise 6.2 (Population dynamics) The features of a population of fishes are described by the following Leslie matrix introduced in Problem 6.2:

Age interval (months)	$\mathbf{x}^{(0)}$	m_i	s_i
0-3	6	0	0.2
3-6	12	0.5	0.4
6-9	8	0.8	0.8
9-12	4	0.3	---

Find the vector **x** of the normalized distribution of this population for different age intervals, according to what we have seen in Problem 6.2.

Exercise 6.3 Prove that the power method does not converge for matrices featuring an eigenvalue of maximum modulus $\lambda_1 = \gamma e^{i\vartheta}$ and another eigenvalue $\lambda_2 = \gamma e^{-i\vartheta}$, where $i = \sqrt{-1}$ and $\gamma, \vartheta \in \mathbb{R}$.

Exercise 6.4 Show that the eigenvalues of A^{-1} are the reciprocals of those of A.

Exercise 6.5 Verify that the power method is unable to compute the maximum modulus eigenvalue of the following matrix, and explain why:

$$A = \begin{bmatrix} \frac{1}{3} & \frac{2}{3} & 2 & 3 \\ 1 & 0 & -1 & 2 \\ 0 & 0 & -\frac{5}{3} & -\frac{2}{3} \\ 0 & 0 & 1 & 0 \end{bmatrix}.$$

Exercise 6.6 By using the power method with shift, compute the largest positive eigenvalue and the largest negative eigenvalue of

$$A = \begin{bmatrix} 3 & 1 & 0 & 0 & 0 & 0 & 0 \\ 1 & 2 & 1 & 0 & 0 & 0 & 0 \\ 0 & 1 & 1 & 1 & 0 & 0 & 0 \\ 0 & 0 & 1 & 0 & 1 & 0 & 0 \\ 0 & 0 & 0 & 1 & 1 & 1 & 0 \\ 0 & 0 & 0 & 0 & 1 & 2 & 1 \\ 0 & 0 & 0 & 0 & 0 & 1 & 3 \end{bmatrix}.$$

A is the so-called *Wilkinson matrix* and can be generated by the command `wilkinson(7)`.

Exercise 6.7 By using the Gershgorin circles, provide an estimate of the maximum number of the complex eigenvalues of the following matrices:

$$A = \begin{bmatrix} 2 & -\frac{1}{2} & 0 & -\frac{1}{2} \\ 0 & 4 & 0 & 2 \\ -\frac{1}{2} & 0 & 6 & \frac{1}{2} \\ 0 & 0 & 1 & 9 \end{bmatrix}, B = \begin{bmatrix} -5 & 0 & \frac{1}{2} & \frac{1}{2} \\ \frac{1}{2} & 2 & \frac{1}{2} & 0 \\ 0 & 1 & 0 & \frac{1}{2} \\ 0 & \frac{1}{4} & \frac{1}{2} & 3 \end{bmatrix}.$$

Exercise 6.8 Use the result of Proposition 6.1 to find a suitable shift for the computation of the maximum modulus eigenvalue of

$$A = \begin{bmatrix} 5 & 0 & 1 & -1 \\ 0 & 2 & 0 & -\frac{1}{2} \\ 0 & 1 & -1 & 1 \\ -1 & -1 & 0 & 0 \end{bmatrix}.$$

Then compare the number of iterations as well the computational cost of the power method both with and without shift by setting the tolerance equal to 10^{-14}.

Exercise 6.9 Show that the matrices $A^{(k)}$ generated by the QR iteration method are all similar to the matrix A.

Exercise 6.10 Use the command eig to compute all the eigenvalues of the two matrices given in Exercise 6.7. Then check how accurate are the conclusions drawn on the basis of Proposition 6.1.

7
Ordinary differential equations

A differential equation is an equation involving one or more derivatives of an unknown function. If all derivatives are taken with respect to a single independent variable we call it an *ordinary differential equation*, whereas we have a *partial differential equation* when partial derivatives are present.

The differential equation (ordinary or partial) has *order p* if p is the maximum order of differentiation that is present. The next chapter will be devoted to the study of partial differential equations, whereas in the present chapter we will deal with ordinary differential equations of first order.

Ordinary differential equations describe the evolution of many phenomena in various fields, as we can see from the following four examples.

Problem 7.1 (Thermodynamics) Consider a body having internal temperature T which is set in an environment with constant temperature T_e. Assume that its mass m is concentrated in a single point. Then the heat transfer between the body and the external environment can be described by the Stefan-Boltzmann law

$$v(t) = \epsilon \gamma S (T^4(t) - T_e^4),$$

where t is the time variable, ϵ the Boltzmann constant (equal to $5.6 \cdot 10^{-8}$ J/m^{2}K^{4}s where J stands for Joule, K for Kelvin and, obviously, m for meter, s for second), γ is the emissivity constant of the body, S the area of its surface and v is the rate of the heat transfer. The rate of variation of the energy $E(t) = mCT(t)$ (where C denotes the specific heat of the material constituting the body) equals, in absolute value, the rate v. Consequently, setting $T(0) = T_0$, the computation of $T(t)$ requires the solution of the ordinary differential equation

$$\frac{dT}{dt} = -\frac{v(t)}{mC}. \tag{7.1}$$

See Exercise 7.15. ∎

Problem 7.2 (Population dynamics) Consider a population of bacteria in a confined environment in which no more than B elements can coexist. Assume that, at the initial time, the number of individuals is equal to $y_0 \ll B$ and the growth rate of the bacteria is a positive constant C. In this case the rate of change of the population is proportional to the number of existing bacteria, under the restriction that the total number cannot exceed B. This is expressed by the differential equation

$$\frac{dy}{dt} = Cy\left(1 - \frac{y}{B}\right), \tag{7.2}$$

whose solution $y = y(t)$ denotes the number of bacteria at time t.

Assuming that two populations y_1 and y_2 be in competition, instead of (7.2) we would have

$$\begin{aligned}\frac{dy_1}{dt} &= C_1 y_1 \left(1 - b_1 y_1 - d_2 y_2\right), \\ \frac{dy_2}{dt} &= -C_2 y_2 \left(1 - b_2 y_2 - d_1 y_1\right),\end{aligned} \tag{7.3}$$

where C_1 and C_2 represent the growth rates of the two populations. The coefficients d_1 and d_2 govern the type of interaction between the two populations, while b_1 and b_2 are related to the available quantity of nutrients. The above equations (7.3) are called the Lotka-Volterra equations and form the basis of various applications. For their numerical solution, see Example 7.7. ∎

Problem 7.3 (Baseball trajectory) We want to simulate the trajectory of a ball from the pitcher to the catcher. By adopting the reference frame of Figure 7.1, the equations describing the ball motion are (see [Ada90], [Gio97])

$$\frac{d\mathbf{x}}{dt} = \mathbf{v}, \qquad \frac{d\mathbf{v}}{dt} = \mathbf{F},$$

where $\mathbf{x}(t) = (x(t), y(t), z(t))^T$ designates the position of the ball at time t, $\mathbf{v} = (v_x, v_y, v_z)^T$ its velocity, while $\mathbf{F}$ is the vector whose components are

$$\begin{aligned}F_x &= -F(v)vv_x + B\omega(v_z \sin\phi - v_y \cos\phi), \\ F_y &= -F(v)vv_y + B\omega v_x \cos\phi, \\ F_z &= -g - F(v)vv_z - B\omega v_x \sin\phi.\end{aligned} \tag{7.4}$$

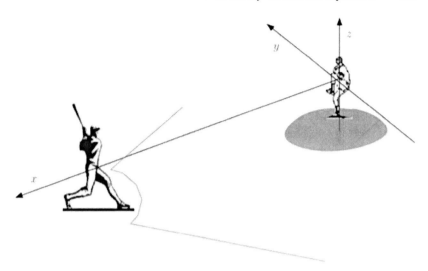

Fig. 7.1. The reference frame adopted for Problem 7.3

v is the modulus of **v**, $B = 4.1 \; 10^{-4}$, ϕ is the pitching angle, ω is the modulus of the angular velocity impressed to the ball from the pitcher. $F(v)$ is a friction coefficient, normally defined as

$$F(v) = 0.0039 + \frac{0.0058}{1 + e^{(v-35)/5}}.$$

The solution of this system of ordinary differential equations is postponed to Exercise 7.20. ■

Problem 7.4 (Electrical circuits) Consider the electrical circuit of Figure 7.2. We want to compute the function $v(t)$ representing the potential drop at the ends of the capacitor C starting from the initial time $t = 0$ at which the switch I has been turned off. Assume that the inductance L can be expressed as an explicit function of the current intensity i, that is $L = L(i)$. The Ohm law yields

$$e - \frac{d(i_1 L(i_1))}{dt} = i_1 R_1 + v,$$

where R_1 is a resistance. By assuming the current fluxes to be directed as indicated in Figure 7.2, upon differentiating with respect to t both sides of the Kirchoff law $i_1 = i_2 + i_3$ and noticing that $i_3 = C dv/dt$ and $i_2 = v/R_2$, we find the further equation

$$\frac{di_1}{dt} = C\frac{d^2 v}{dt^2} + \frac{1}{R_2}\frac{dv}{dt}.$$

We have therefore found a system of two differential equations whose solution allows the description of the time variation of the two unknowns i_1 and v. The second equation has order two. For its solution see Example 7.8. ∎

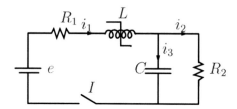

Fig. 7.2. The electrical circuit of Problem 7.4

7.1 The Cauchy problem

We confine ourselves to first order differential equations, as an equation of order $p > 1$ can always be reduced to a system of p equations of order 1. The case of first order systems will be addressed in Section 7.8.

An ordinary differential equation in general admits an infinite number of solutions. In order to fix one of them we must impose a further condition which prescribes the value taken by this solution at a given point of the integration interval. For instance, the equation (7.2) admits the family of solutions $y(t) = B\psi(t)/(1 + \psi(t))$ with $\psi(t) = e^{Ct+K}$, K being an arbitrary constant. If we impose the condition $y(0) = 1$, we pick up the unique solution corresponding to the value $K = \ln[1/(B - 1)]$.

We will therefore consider the solution of the so-called *Cauchy problem* which takes the following form:

find $y : I \to \mathbb{R}$ such that

$$\begin{cases} y'(t) = f(t, y(t)) & \forall t \in I, \\ y(t_0) = y_0, \end{cases} \quad (7.5)$$

where I is an interval of $\mathbb{R}$, $f : I \times \mathbb{R} \to \mathbb{R}$ is a given function and y' denotes the derivative of y with respect to t. Finally, t_0 is a point of I and y_0 a given value which is called the *initial data*.

In the following proposition we report a classical result of Analysis.

Proposition 7.1 *Assume that the function $f(t, y)$ is*

1. *continuous with respect to both arguments;*
2. *Lipschitz-continuous with respect to its second argument, that is, there exists a positive constant L such that*

$$|f(t, y_1) - f(t, y_2)| \leq L|y_1 - y_2|, \ \forall t \in I, \ \forall y_1, y_2 \in \mathbb{R}.$$

Then the solution $y = y(t)$ of the Cauchy problem (7.5) exists, is unique and belongs to $C^1(I)$.

Unfortunately, explicit solutions are available only for very special types of ordinary differential equations. In some other cases, the solution is available only in implicit form. This is, for instance, the case with the equation $y' = (y - t)/(y + t)$ whose solution satisfies the implicit relation

$$\frac{1}{2}\ln(t^2 + y^2) + \text{arctg}\frac{y}{t} = C,$$

where C is an arbitrary constant. In some other circumstances the solution is not even representable in implicit form, as in the case of the equation $y' = e^{-t^2}$ whose general solution can only be expressed through a series expansion. For all these reasons, we seek numerical methods capable of approximating the solution of *every* family of ordinary differential equations for which solutions do exist.

The common strategy of all these methods consists of subdividing the integration interval $I = [t_0, T]$, with $T < +\infty$, into N_h intervals of length $h = (T - t_0)/N_h$; h is called the *discretization step*. Then, at each node t_n ($0 \leq n \leq N_h - 1$) we seek the unknown value u_n which approximates $y_n = y(t_n)$. The set of values $\{u_0 = y_0, u_1, \ldots, u_{N_h}\}$ is our *numerical solution*.

7.2 Euler methods

A classical method, the *forward Euler* method, generates the numerical solution as follows

$$\boxed{u_{n+1} = u_n + h f_n, \qquad n = 0, \ldots, N_h - 1} \qquad (7.6)$$

where we have used the shorthand notation $f_n = f(t_n, u_n)$. This method is obtained by considering the differential equation (7.5) at every node t_n, $n = 1, \ldots, N_h$ and replacing the exact derivative $y'(t_n)$ by means of the incremental ratio (4.4).

In a similar way, using this time the incremental ratio (4.8) to approximate $y'(t_{n+1})$, we obtain the *backward Euler* method

$$\boxed{u_{n+1} = u_n + hf_{n+1}, \qquad n = 0, \ldots, N_h - 1} \qquad (7.7)$$

Both methods provide an instance of a *one-step method* since for computing the numerical solution u_{n+1} at the node t_{n+1} we only need the information related to the previous node t_n. More precisely, in the forward Euler method u_{n+1} depends exclusively on the value u_n previously computed, whereas in the backward Euler method it depends also on itself through the value f_{n+1}. For this reason the first method is called the *explicit* Euler method and the second the *implicit* Euler method.

For instance, the discretization of (7.2) by the forward Euler method requires at every step the simple computation of

$$u_{n+1} = u_n + hCu_n \left(1 - u_n/B\right),$$

whereas using the backward Euler method we must solve the nonlinear equation

$$u_{n+1} = u_n + hCu_{n+1} \left(1 - u_{n+1}/B\right).$$

Thus, implicit methods are more costly than explicit methods, since at every time-level t_{n+1} we must solve a nonlinear problem to compute u_{n+1}. However, we will see that implicit methods enjoy better stability properties than explicit ones.

The forward Euler method is implemented in the Program 7.1; the integration interval is tspan = [t0,tfinal], odefun is a string which contains the function $f(t,y(t))$ which depends on the variables t and y, or an inline function whose first two arguments stand for t and y.

Program 7.1. feuler: forward Euler method

```
function [t,y]=feuler(odefun,tspan,y,Nh,varargin)
%FEULER Solve differential equations using the forward
%   Euler method.
%   [T,Y]=FEULER(ODEFUN,TSPAN,Y0,NH) with TSPAN=[T0,TF]
%   integrates the system of differential equations
%   y'=f(t,y) from time T0 to TF with initial condition
%   Y0 using the forward Euler method on an equispaced
%   grid of NH intervals.Function ODEFUN(T,Y) must return
%   a column vector corresponding to f(t,y). Each row in
%   the solution array Y corresponds to a time returned
%   in the   column vector T.
%   [T,Y] = FEULER(ODEFUN,TSPAN,Y0,NH,P1,P2,...) passes
%   the additional parameters P1,P2,... to the function
%   ODEFUN as ODEFUN(T,Y,P1,P2...).
h=(tspan(2)-tspan(1))/Nh;
tt=linspace(tspan(1),tspan(2),Nh+1);
for t = tt(1:end-1)
   y=[y;y(end,:)+h*feval(odefun,t,y(end,:),varargin{:})];
end
t=tt;
return
```

7.2 Euler methods

The backward Euler method is implemented in the Program 7.2. Note that we have used the function fsolve for the solution of the non-linear problem at each step. As initial data for fsolve we use the last computed value of the numerical solution.

Program 7.2. beuler: backward Euler method

```
function [t,u]=beuler(odefun,tspan,y0,Nh,varargin)
%BEULER Solve differential equations using the backward
%   Euler method.
%   [T,Y]=BEULER(ODEFUN,TSPAN,Y0,NH) with TSPAN=[T0,TF]
%   integrates the system of differential equations
%   y'=f(t,y) from time T0 to TF with initial condition
%   Y0 using the backward Euler method on an equispaced
%   grid of NH intervals. Function ODEFUN(T,Y) must return
%   a column vector corresponding to f(t,y). Each row in
%   the solution array Y corresponds to a time returned
%   in the   column vector T.
%   [T,Y] = BEULER(ODEFUN,TSPAN,Y0,NH,P1,P2,...) passes
%   the additional parameters P1,P2,... to the function
%   ODEFUN as ODEFUN(T,Y,P1,P2...).
tt=linspace(tspan(1),tspan(2),Nh+1);
y=y0(:); % always create a vector column
u=y.';
global glob_h glob_t glob_y glob_odefun;
glob_h=(tspan(2)-tspan(1))/Nh;
glob_y=y;
glob_odefun=odefun;
glob_t=tt(2);

if ( ~exist('OCTAVE_VERSION') )
options=optimset;
options.Display='off';
options.TolFun=1.e-06;
options.MaxFunEvals=10000;
end

for glob_t=tt(2:end)
if ( exist('OCTAVE_VERSION') )
   [w info] = fsolve('beulerfun',glob_y);
else
   w = fsolve(@(w) beulerfun(w),glob_y,options);
end
   u = [u; w.'];
   glob_y = w;
end
t=tt;
clear glob_h glob_t glob_y glob_odefun;
end

function [z]=beulerfun(w)
   global glob_h glob_t glob_y glob_odefun;
   z=w-glob_y-glob_h*feval(glob_odefun,glob_t,w);
end
```

7.2.1 Convergence analysis

A numerical method is *convergent* if

$$\forall n = 0, \ldots, N_h, \quad |y_n - u_n| \leq C(h) \tag{7.8}$$

where $C(h)$ is infinitesimal with respect to h when h tends to zero. If $C(h) = \mathcal{O}(h^p)$ for some $p > 0$, then we say that the method converges with *order p*. In order to verify that the forward Euler method converges, we write the error as follows:

$$e_n = y_n - u_n = (y_n - u_n^*) + (u_n^* - u_n), \tag{7.9}$$

where

$$u_n^* = y_{n-1} + hf(t_{n-1}, y_{n-1})$$

denotes the numerical solution at time t_n which we would obtain starting from the exact solution at time t_{n-1}; see Figure 7.3. The term $y_n - u_n^*$ in (7.9) represents the error produced by a single step of the forward Euler method, whereas the term $u_n^* - u_n$ represents the propagation from t_{n-1} to t_n of the error accumulated at the previous time-level t_{n-1}. The method converges provided both terms tend to zero as $h \to 0$. Assuming that the second order derivative of y exists and is continuous, thanks to (4.6) we find

$$y_n - u_n^* = \frac{h^2}{2} y''(\xi_n), \text{ for a suitable } \xi_n \in (t_{n-1}, t_n). \tag{7.10}$$

The quantity

$$\tau_n(h) = (y_n - u_n^*)/h$$

is named *local truncation error* of the forward Euler method. More in general, the local truncation error of a given method represents the error that would be generated by forcing the exact solution to satisfy that specific numerical scheme, whereas the *global truncation error* is defined as

$$\tau(h) = \max_{n=0,\ldots,N_h} |\tau_n(h)|.$$

In view of (7.10), the truncation error for the forward Euler method takes the following form

$$\tau(h) = Mh/2, \tag{7.11}$$

where $M = \max_{t \in [t_0, T]} |y''(t)|$.

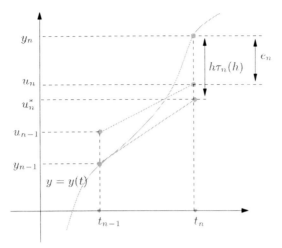

Fig. 7.3. Geometrical representation of a step of the forward Euler method

From (7.10) we deduce that $\lim_{h \to 0} \tau(h) = 0$, and a method for which this happens is said to be *consistent*. Further, we say that it is consistent with order p if $\tau(h) = \mathcal{O}(h^p)$ for a suitable integer $p \geq 1$.

Consider now the other term in (7.9). We have

$$u_n^* - u_n = e_{n-1} + h\left[f(t_{n-1}, y_{n-1}) - f(t_{n-1}, u_{n-1})\right]. \quad (7.12)$$

Since f is Lipschitz continuous with respect to its second argument, we obtain

$$|u_n^* - u_n| \leq (1 + hL)|e_{n-1}|.$$

If $e_0 = 0$, the previous relations yield

$$\begin{aligned}
|e_n| &\leq |y_n - u_n^*| + |u_n^* - u_n| \\
&\leq h|\tau_n(h)| + (1 + hL)|e_{n-1}| \\
&\leq \left[1 + (1 + hL) + \ldots + (1 + hL)^{n-1}\right] h\tau(h) \\
&= \frac{(1 + hL)^n - 1}{L}\tau(h) \leq \frac{e^{L(t_n - t_0)} - 1}{L}\tau(h).
\end{aligned}$$

We have used the identity

$$\sum_{k=0}^{n-1}(1 + hL)^k = [(1 + hL)^n - 1]/hL,$$

the inequality $1 + hL \leq e^{hL}$ and we have observed that $nh = t_n - t_0$. Therefore we find

$$|e_n| \le \frac{e^{L(t_n-t_0)}-1}{L}\frac{M}{2}h, \qquad \forall n = 0,\ldots, N_h, \qquad (7.13)$$

and thus we can conclude that *the forward Euler method converges with order 1*. We can note that the order of this method coincides with the order of its local truncation error. This property is shared by many numerical methods for the numerical solution of ordinary differential equations.

The convergence estimate (7.13) is obtained by simply requiring f to be Lipschitz continuous. A better estimate, precisely

$$|e_n| \le Mh(t_n - t_0)/2, \qquad (7.14)$$

holds if $\partial f/\partial y$ exists and satisfies the further requirement $\partial f(t,y)/\partial y \le 0$ for all $t \in [t_0, T]$ and all $-\infty < y < \infty$. Indeed, in that case, using Taylor expansion, from (7.12) we obtain

$$u_n^* - u_n = (1 + h\partial f/\partial y(t_{n-1}, \eta_n))e_{n-1},$$

where η_n belongs to the interval whose extrema are y_{n-1} and u_{n-1}, thus $|u_n^* - u_n| \le |e_{n-1}|$, provided the inequality

$$h < 2/\max_{t \in [t_0, T]} |\partial f/\partial y(t, y(t))| \qquad (7.15)$$

holds. Then $|e_n| \le |y_n - u_n^*| + |e_{n-1}| \le nh\tau(h) + |e_0|$, whence (7.14) owing to (7.11) and to the fact that $e_0 = 0$. The limitation (7.15) on the step h is in fact a stability restriction, as we will see in the sequel.

Remark 7.1 (Consistency) The property of consistency is necessary in order to get convergence. Actually, should it be violated, at each step the numerical method would generate an error which is not infinitesimal with respect to h. The accumulation with the previous errors would inhibit the global error to converge to zero when $h \to 0$. •

For the backward Euler method the local truncation error reads

$$\tau_n(h) = \frac{1}{h}[y_n - y_{n-1} - hf(t_n, y_n)].$$

Still using the Taylor expansion one obtains

$$\tau_n(h) = -\frac{h}{2}y''(\xi_n)$$

for a suitable $\xi_n \in (t_{n-1}, t_n)$, provided $y \in C^2$. Thus also the backward Euler method converges with order 1 with respect to h.

Example 7.1 Consider the Cauchy problem

7.3 The Crank-Nicolson method

$$\begin{cases} y'(t) = \cos(2y(t)) & t \in (0,1], \\ y(0) = 0, \end{cases} \quad (7.16)$$

whose solution is $y(t) = \frac{1}{2}\arcsin((e^{4t}-1)/(e^{4t}+1))$. We solve it by the forward Euler method (Program 7.1) and the backward Euler method (Program 7.2). By the following commands we use different values of h, $1/2$, $1/4, 1/8, \ldots, 1/512$:

```
tspan=[0,1]; y0=0; f=inline('cos(2*y)','t','y');
u=inline('0.5*asin((exp(4*t)-1)./(exp(4*t)+1))','t');
Nh=2;
for k=1:10
    [t,ufe]=feuler(f,tspan,y0,Nh);
    fe(k)=abs(ufe(end)-feval(u,t(end)));
    [t,ube]=beuler(f,tspan,y0,Nh);
    be(k)=abs(ube(end)-feval(u,t(end)));
    Nh = 2*Nh;
end
```

The errors committed at the point $t = 1$ are stored in the variable **fe** (forward Euler) and **be** (backward Euler), respectively. Then we apply formula (1.12) to estimate the order of convergence. Using the following commands

```
p=log(abs(fe(1:end-1)./fe(2:end)))/log(2); p(1:2:end)
```

```
1.2898   1.0349   1.0080   1.0019   1.0005
```

```
p=log(abs(be(1:end-1)./be(2:end)))/log(2); p(1:2:end)
```

```
0.90703   0.97198   0.99246   0.99808   0.99952
```

we can verify that both methods are convergent with order 1. ∎

Remark 7.2 The error estimate (7.13) was derived by assuming that the numerical solution $\{u_n\}$ is obtained in exact arithmetic. Should we account for the (inevitable) roundoff-errors, the error might blow up like $\mathcal{O}(1/h)$ as h approaches 0 (see, e.g., [Atk89]). This circumstance suggests that it might be unreasonable to go below a certain threshold h^* (which is actually extremely tiny) in practical computations. •

See the Exercises 7.1-7.3.

7.3 The Crank-Nicolson method

Adding together the generic steps of the forward and backward Euler methods we find the so-called *Crank-Nicolson method*

$$u_{n+1} = u_n + \frac{h}{2}[f_n + f_{n+1}], \quad n = 0, \ldots, N_h - 1 \quad (7.17)$$

It can also be derived by applying the fundamental theorem of integration (which we recalled in Section 1.4.3) to the Cauchy problem (7.5), obtaining

$$y_{n+1} = y_n + \int_{t_n}^{t_{n+1}} f(t, y(t))\, dt, \qquad (7.18)$$

and then approximating the integral on $[t_n, t_{n+1}]$ by the trapezoidal rule (4.19).

The local truncation error of the Crank-Nicolson method satisfies

$$\tau_n(h) = \frac{1}{h}[y(t_n) - y(t_{n-1})] - \frac{1}{2}[f(t_n, y(t_n)) + f(t_{n-1}, y(t_{n-1}))]$$

$$= \frac{1}{h}\int_{t_{n-1}}^{t_n} f(t, y(t))\, dt - \frac{1}{2}[f(t_n, y(t_n)) + f(t_{n-1}, y(t_{n-1}))].$$

The last equality follows from (7.18) and expresses the error associated with the trapezoidal rule for numerical integration (4.19). If we assume that $y \in C^3$ and use (4.20), we deduce that

$$\tau_n(h) = -\frac{h^2}{12} y'''(\xi_n) \text{ for a suitable } \xi_n \in (t_{n-1}, t_n). \qquad (7.19)$$

Thus the Crank-Nicolson method is consistent with order 2, i.e. its local truncation error tends to 0 as h^2. Using a similar approach to that followed for the forward Euler method, we can show that the Crank-Nicolson method is convergent with order 2 with respect to h.

The Crank-Nicolson method is implemented in the Program 7.3. Input and output parameters are the same as in the Euler methods.

Program 7.3. cranknic: Crank-Nicolson method

```
function [t,u]=cranknic(odefun,tspan,y0,Nh,varargin)
%CRANKNIC  Solve differential equations using the
%         Crank-Nicolson method.
%         [T,Y]=CRANKNIC(ODEFUN,TSPAN,Y0,NH) with TSPAN=[T0,TF]
%         integrates the system of differential equations
%         y'=f(t,y) from time T0 to TF with initial condition
%         Y0 using the Crank-Nicolson method on an equispaced
%         grid of NH intervals. Function ODEFUN(T,Y) must return
%         a column vector corresponding to f(t,y). Each row in
%         the solution array Y corresponds to a time returned
%         in the column vector T.
%         [T,Y] = CRANKNIC(ODEFUN,TSPAN,Y0,NH,P1,P2,...) passes
%         the additional parameters P1,P2,... to the function
%         ODEFUN as ODEFUN(T,Y,P1,P2...).
tt=linspace(tspan(1),tspan(2),Nh+1);
y=y0(:); % always create a vector column
u=y.';
global glob_h glob_t glob_y glob_odefun;
```

```
glob_h=(tspan(2)-tspan(1))/Nh;
glob_y=y;
glob_odefun=odefun;
if( ~exist('OCTAVE_VERSION') )
 options=optimset;
 options.Display='off';
 options.TolFun=1.e-06;
 options.MaxFunEvals=10000;
end

for glob_t=tt(2:end)
if ( exist('OCTAVE_VERSION') )
  [w info msg] = fsolve('cranknicfun',glob_y);
else
  w = fsolve(@(w) cranknicfun(w),glob_y,options);
end
  u = [u; w.'];
  glob_y = w;
end
t=tt;
clear glob_h glob_t glob_y glob_odefun;
end

function z=cranknicfun(w)
  global glob_h glob_t glob_y glob_odefun;
  z=w - glob_y - ...
     0.5*glob_h*(feval(glob_odefun,glob_t,w) + ...
     feval(glob_odefun,glob_t,glob_y));
end
```

Example 7.2 Let us solve the Cauchy problem (7.16) by using the Crank-Nicolson method with the same values of h as used in Example 7.1. As we can see, the results confirm that the estimated error tends to zero with order $p = 2$:

```
y0=0;  tspan=[0 1]; N=2; f=inline('cos(2*y)','t','y');
y='0.5*asin((exp(4*t)-1)./(exp(4*t)+1))';
for k=1:10
   [tt,u]=cranknic(f,tspan,y0,N);
   t=tt(end); e(k)=abs(u(end)-eval(y)); N=2*N;
end
p=log(abs(e(1:end-1)./e(2:end)))/log(2); p(1:2:end)

    1.7940    1.9944    1.9997    2.0000    2.0000
```

■

7.4 Zero-stability

There is a concept of stability, called zero-stability, which guarantees that, in a fixed bounded interval, small perturbations of data yield bounded perturbations of the numerical solution when $h \to 0$.

More precisely, a numerical method for the approximation of problem (7.5), where $I = [t_0, T]$, is *zero-stable* if $\exists h_0 > 0$, $\exists C > 0$ such that $\forall h \in (0, h_0], \forall \varepsilon > 0$ sufficiently small, if $|\rho_n| \leq \varepsilon, 0 \leq n \leq N_h$, then

$$|z_n - u_n| \leq C\varepsilon, \qquad 0 \leq n \leq N_h, \tag{7.20}$$

where C is a constant which might depend on the length of the integration interval I, z_n is the solution that would be obtained by applying the numerical method at hand to a *perturbed* problem, ρ_n denotes the size of the perturbation introduced at the n-th step and ε indicates the maximum size of the perturbation. Obviously, ε must be small enough to guarantee that the perturbed problem still has a unique solution on the interval of integration.

For instance, in the case of the forward Euler method u_n satisfies

$$\begin{cases} u_{n+1} = u_n + hf(t_n, u_n), \\ u_0 = y_0, \end{cases} \tag{7.21}$$

whereas z_n satisfies

$$\begin{cases} z_{n+1} = z_n + h\left[f(t_n, z_n) + \rho_{n+1}\right], \\ z_0 = y_0 + \rho_0 \end{cases} \tag{7.22}$$

for $0 \leq n \leq N_h - 1$, under the assumption that $|\rho_n| \leq \varepsilon$, $0 \leq n \leq N_h$.

For a consistent one-step method it can be proved that zero-stability is a consequence of the fact that f is Lipschitz-continuous with respect to its second argument (see, e.g. [QSS06]). In that case, the constant C that appears in (7.20) depends on $\exp((T - t_0)L)$, where L is the Lipschitz constant.

However, this is not necessarily true for other families of methods. Assume for instance that the numerical method can be written in the general form

$$u_{n+1} = \sum_{j=0}^{p} a_j u_{n-j} + h \sum_{j=0}^{p} b_j f_{n-j} + h b_{-1} f_{n+1}, \quad n = p, p+1, \ldots \tag{7.23}$$

for suitable coefficients $\{a_k\}$ and $\{b_k\}$ and for an integer $p \geq 0$. This is a linear *multistep method* and $p + 1$ denotes the number of steps. The initial values $u_0, u_1, \ldots, u_p$ must be provided. Apart from u_0, which is equal to y_0, the other values $u_1, \ldots, u_p$ can be generated by suitable accurate methods such as e.g., the Runge-Kutta methods that we will address in Section 7.6.

We will see some examples of multistep methods in Section 7.6. The polynomial

7.4 Zero-stability

$$\pi(r) = r^{p+1} - \sum_{j=0}^{p} a_j r^{p-j}$$

is called the *first characteristic polynomial* associated with the numerical method (7.23), and we denote its roots by r_j, $j = 0, \ldots, p$. The method (7.23) is zero-stable iff the following *root condition* is satisfied:

$$\begin{cases} |r_j| \leq 1 \text{ for all } j = 0, \ldots, p, \\ \text{furthermore } \pi'(r_j) \neq 0 \text{ for those } j \text{ such that } |r_j| = 1. \end{cases} \quad (7.24)$$

For example, for the forward Euler method we have $p = 0$, $a_0 = 1$, $b_{-1} = 0$, $b_0 = 1$. For the backward Euler method we have $p = 0$, $a_0 = 1$, $b_{-1} = 1$, $b_0 = 0$ and for the Crank-Nicolson method we have $p = 0$, $a_0 = 1$, $b_{-1} = 1/2$, $b_0 = 1/2$. In all cases there is only one root of $\pi(r)$ which is equal to 1 and therefore all these methods are zero-stable.

The following property, known as Lax-Ritchmyer *equivalence theorem*, is most crucial in the theory of numerical methods (see, e.g., [IK66]), and highlights the fundamental role played by the property of zero-stability:

$$\boxed{\text{Any consistent method is convergent iff it is zero-stable.}} \quad (7.25)$$

Coherently with what done before, the local truncation error for the multistep method (7.23) is defined as follows

$$\tau_n(h) = \frac{1}{h} \left\{ y_{n+1} - \sum_{j=0}^{p} a_j y_{n-j} \right. \\ \left. - h \sum_{j=0}^{p} b_j f(t_{n-j}, y_{n-j}) - h b_{-1} f(t_{n+1}, y_{n+1}) \right\}. \quad (7.26)$$

The method is said to be consistent if $\tau(h) = \max |\tau_n(h)|$ tends to zero when h tends to zero. We can prove that this condition is equivalent to require that

$$\boxed{\sum_{j=0}^{p} a_j = 1, \quad -\sum_{j=0}^{p} j a_j + \sum_{j=-1}^{p} b_j = 1} \quad (7.27)$$

which in turns amounts to say that $r = 1$ is a root of the polynomial $\pi(r)$ (see, e.g., [QSS06, Chapter 11]).

See the Exercises 7.4-7.5.

7.5 Stability on unbounded intervals

In the previous section we considered the solution of the Cauchy problem on bounded intervals. In that context, the number N_h of subintervals becomes infinite only if h goes to zero.

On the other hand, there are several situations in which the Cauchy problem needs to be integrated on very large (virtually infinite) time intervals. In this case, even if h is fixed, N_h tends to infinity, and then results like (7.13) become meaningless as the right hand side of the inequality contains an unbounded quantity. We are therefore interested in methods that are able to approximate the solution for arbitrarily long time-intervals, even with a step-size h relatively "large".

Unfortunately, the economical forward Euler method does not enjoy this property. To see this, let us consider the following *model problem*

$$\begin{cases} y'(t) = \lambda y(t), & t \in (0, \infty), \\ y(0) = 1, \end{cases} \quad (7.28)$$

where λ is a negative real number. The exact solution is $y(t) = e^{\lambda t}$, which tends to 0 as t tends to infinity. Applying the forward Euler method to (7.28) we find that

$$u_0 = 1, \qquad u_{n+1} = u_n(1 + \lambda h) = (1 + \lambda h)^{n+1}, \qquad n \geq 0. \quad (7.29)$$

Thus $\lim_{n \to \infty} u_n = 0$ iff

$$\boxed{-1 < 1 + h\lambda < 1, \quad \text{i.e.} \quad h < 2/|\lambda|} \quad (7.30)$$

This condition expresses the requirement that, for *fixed* h, the numerical solution should reproduce the behavior of the exact solution when t_n tends to infinity. If $h > 2/|\lambda|$, then $\lim_{n \to \infty} |u_n| = +\infty$; thus (7.30) is a stability condition. The property that $\lim_{n \to \infty} u_n = 0$ is called *absolute stability*.

Example 7.3 Let us apply the forward Euler method to solve problem (7.28) with $\lambda = -1$. In that case we must have $h < 2$ for absolute stability. In Figure 7.4 we report the solutions obtained on the interval $[0, 30]$ for 3 different values of h: $h = 30/14$ (which violates the stability condition), $h = 30/16$ (which satisfies, although by a little amount only, the stability condition) and $h = 1/2$. We can see that in the first two cases the numerical solution oscillates. However only in the first case (which violates the stability condition) the absolute value of the numerical solution does not vanish at infinity (and actually it diverges). ∎

Similar conclusions hold when λ is either a complex number (see Section 7.5.1) or a negative function of t in (7.28). However in this case, $|\lambda|$

7.5 Stability on unbounded intervals 203

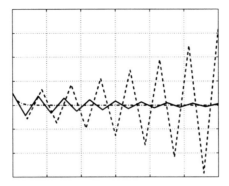

Fig. 7.4. Solutions of problem (7.28), with $\lambda = -1$, obtained by the forward Euler method, corresponding to $h = 30/14 (> 2)$ (*dashed line*), $h = 30/16 (< 2)$ (*solid line*) and $h = 1/2$ (*dashed-dotted line*)

must be replaced by $\max_{t\in[0,\infty)} |\lambda(t)|$ in the stability condition. This condition could however be relaxed to one which is less strict by using a *variable step-size* h_n which accounts for the local behavior of $|\lambda(t)|$ in every interval (t_n, t_{n+1}).

In particular, the following *adaptive* forward Euler method could be used:

choose $u_0 = y_0$ and $h_0 = 2\alpha/|\lambda(t_0)|$; then

for $n = 0, 1, \ldots$, do

$$\begin{aligned} t_{n+1} &= t_n + h_n, \\ u_{n+1} &= u_n + h_n \lambda(t_n) u_n, \\ h_{n+1} &= 2\alpha/|\lambda(t_{n+1})|, \end{aligned} \quad (7.31)$$

where α is a constant which must be less than 1 in order to have an absolutely stable method.

For instance, consider the problem

$$y'(t) = -(e^{-t} + 1)y(t), \quad t \in (0, 10),$$

with $y(0) = 1$. Since $|\lambda(t)|$ is decreasing, the most restrictive condition for absolute stability of the forward Euler method is $h < h_0 = 2/|\lambda(0)| = 1$. In Figure 7.5, left, we compare the solution of the forward Euler method with that of the adaptive method (7.31) for three values of α. Note that, although every $\alpha < 1$ is admissible for stability purposes, to get an accurate solution requires choosing α sufficiently small. In Figure 7.5, right, we also plot the behaviour of h_n on the interval $(0, 10]$ corresponding to the three values of α. This picture clearly shows that the sequence $\{h_n\}$ increases monotonically with n.

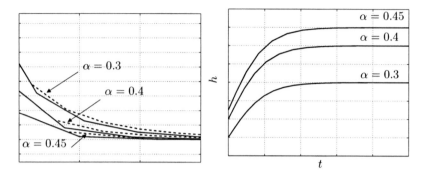

Fig. 7.5. Left: the numerical solution on the time interval $(0.5, 2)$ obtained by the forward Euler method with $h = \alpha h_0$ (*dashed line*) and by the adaptive variable stepping forward Euler method (7.31) (*solid line*) for three different values of α. Right: the behavior of the variable step-size h for the adaptive method (7.31)

In contrast to the forward Euler method, neither the backward Euler method nor the Crank-Nicolson method require limitations on h for absolute stability. In fact, with the backward Euler method we obtain $u_{n+1} = u_n + \lambda h u_{n+1}$ and therefore

$$u_{n+1} = \left(\frac{1}{1-\lambda h}\right)^{n+1}, \qquad n \geq 0,$$

which tends to zero as $n \to \infty$ for *all values* of $h > 0$. Similarly, with the Crank-Nicolson method we obtain

$$u_{n+1} = \left[\left(1 + \frac{h\lambda}{2}\right) \Big/ \left(1 - \frac{h\lambda}{2}\right)\right]^{n+1}, \qquad n \geq 0,$$

which still tends to zero as $n \to \infty$ for all possible values of $h > 0$. We can conclude that the forward Euler method is *conditionally absolutely stable*, while both the backward Euler and Crank-Nicolson methods are *unconditionally absolutely stable*.

7.5.1 The region of absolute stability

Let us suppose now that in (7.28) λ be a complex number with negative real part. In such a case, the solution $u(t) = e^{\lambda t}$ still tends to 0 when t tends to infinity. We call *region of absolute stability* $\mathcal{A}$ of a numerical method the set of complex numbers $z = h\lambda$ for which the method turns out to be absolutely stable (that is, $\lim_{n \to \infty} u_n = 0$). The region of absolute stability of forward Euler method is given by those numbers $h\lambda \in \mathbb{C}$ such that $|1 + h\lambda| < 1$, thus it coincides with the circle of radius one and with centre $(-1, 0)$. For the backward Euler method the property of absolute stability is instead satisfied by all values of $h\lambda$ which are exterior to the circle of radius one centered in $(1, 0)$ (see Figure 7.6). Finally, the region of absolute stability of Crank-Nicolson method coincides with the left hand complex plane of numbers with negative real part.

Methods that are unconditionally absolutely stable for all complex number λ in (7.28) with negative real part are called *A-stable*. Backward Euler and Crank-Nicolson method are therefore *A-stable*, and so are many other implicit methods. This property makes implicit methods attractive in spite of being computationally more expensive than explicit methods.

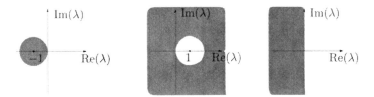

Fig. 7.6. The absolute stability regions (*in cyan*) of the forward Euler method (*left*), backward Euler method (*centre*) and Crank-Nicolson method (*right*)

Example 7.4 Let us compute the restriction on h when using the forward Euler method to solve the Cauchy problem $y'(t) = \lambda y$ with $\lambda = -1 + i$. This λ stands on the boundary of the absolute stability region $\mathcal{A}$ of the forward Euler method. Thus, any h such that $h \in (0, 1)$ will suffice to guarantee that $h\lambda \in \mathcal{A}$. If it were $\lambda = -2 + 2i$ we should choose $h \in (0, 1/2)$ in order to bring $h\lambda$ within the stability region $\mathcal{A}$. ∎

7.5.2 Absolute stability controls perturbations

Consider now the following *generalized model problem*

$$\begin{cases} y'(t) = \lambda(t)y(t) + r(t), & t \in (0, +\infty), \\ y(0) = 1, \end{cases} \quad (7.32)$$

where λ and r are two continuous functions and $-\lambda_{max} \leq \lambda(t) \leq -\lambda_{min}$ with $0 < \lambda_{min} \leq \lambda_{max} < +\infty$. In this case the exact solution does not necessarily tend to zero as t tends to infinity; for instance if both r and λ are constants we have

$$y(t) = \left(1 + \frac{r}{\lambda}\right) e^{\lambda t} - \frac{r}{\lambda}$$

whose limit when t tends to infinity is $-r/\lambda$. Thus, in general, it does not make sense to require a numerical method to be absolutely stable when applied to problem (7.32). However, we are going to show that a numerical method which is absolutely stable on the model problem (7.28), if applied to the generalized problem (7.32), guarantees that the perturbations are kept under control as t tends to infinity (possibly under a suitable constraint on the time-step h).

For the sake of simplicity we will confine our analysis to the forward Euler method; when applied to (7.32) it reads

$$\begin{cases} u_{n+1} = u_n + h(\lambda_n u_n + r_n), & n \geq 0, \\ u_0 = 1 \end{cases}$$

and its solution is (see Exercise 7.9)

$$u_n = u_0 \prod_{k=0}^{n-1}(1 + h\lambda_k) + h\sum_{k=0}^{n-1} r_k \prod_{j=k+1}^{n-1}(1 + h\lambda_j), \qquad (7.33)$$

where $\lambda_k = \lambda(t_k)$ and $r_k = r(t_k)$, with the convention that the last product is equal to one if $k+1 > n-1$. Let us consider the following "perturbed" method

$$\begin{cases} z_{n+1} = z_n + h(\lambda_n z_n + r_n + \rho_{n+1}), & n \geq 0, \\ z_0 = u_0 + \rho_0, \end{cases} \qquad (7.34)$$

where $\rho_0, \rho_1, \ldots$ are given perturbations which are introduced at every time step. This is a simple model in which ρ_0 and ρ_{n+1}, respectively, account for the fact that neither u_0 nor r_n can be determined exactly. (Should we account for *all* roundoff errors which are actually introduced at any step, our perturbed model would be far more involved and difficult to analyze.) The solution of (7.34) reads like (7.33) provided u_k is replaced by z_k and r_k by $r_k + \rho_{k+1}$, for all $k = 0, \ldots, n-1$. Then

$$z_n - u_n = \rho_0 \prod_{k=0}^{n-1}(1 + h\lambda_k) + h\sum_{k=0}^{n-1} \rho_{k+1} \prod_{j=k+1}^{n-1}(1 + h\lambda_j). \qquad (7.35)$$

The quantity $|z_n - u_n|$ is called the perturbation error at step n. It is worth noticing that this quantity does not depend on the function $r(t)$.

i. For the sake of exposition, let us consider first the special case where λ_k and ρ_k are two constants equal to λ and ρ, respectively. Assume that $h < h_0(\lambda) = 2/|\lambda|$, which is the condition on h that ensures the absolute stability of the forward Euler method applied to the model problem (7.28). Then, using the following identity for the geometric sum

7.5 Stability on unbounded intervals

$$\sum_{k=0}^{n-1} a^k = \frac{1-a^n}{1-a}, \qquad \text{if } |a| \neq 1, \tag{7.36}$$

we obtain

$$z_n - u_n = \rho \left\{ (1+h\lambda)^n \left(1 + \frac{1}{\lambda}\right) - \frac{1}{\lambda} \right\}. \tag{7.37}$$

It follows that the perturbation error satisfies (see Exercise 7.10)

$$|z_n - u_n| \leq \varphi(\lambda)|\rho|, \tag{7.38}$$

with $\varphi(\lambda) = 1$ if $\lambda \leq -1$, while $\varphi(\lambda) = |1 + 2/\lambda|$ if $-1 \leq \lambda < 0$. The conclusion that can be drawn is that the perturbation error is bounded by $|\rho|$ times a constant which is independent of n and h. Moreover,

$$\lim_{n \to \infty} |z_n - u_n| = \frac{\rho}{|\lambda|}.$$

Figure 7.7 corresponds to the case where $\rho = 0.1$, $\lambda = -2$ (*left*) and $\lambda = -0.5$ (*right*). In both cases we have taken $h = h_0(\lambda) - 0.01$. Obviously, the perturbation error blows up when n increases if the stability limit $h < h_0(\lambda)$ is violated.

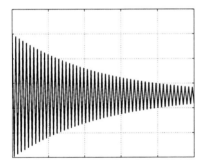
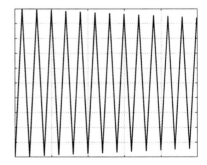

Fig. 7.7. The perturbation error when $\rho = 0.1$: $\lambda = -2$ (*left*) and $\lambda = -0.5$ (*right*). In both cases $h = h_0(\lambda) - 0.01$

ii. In the general case where λ and r are non-constant, let us require h to satisfy the restriction $h < h_0(\lambda)$, where this time $h_0(\lambda) = 2/\lambda_{max}$. Then,

$$|1 + h\lambda_k| \leq a(h) = \max\{|1 - h\lambda_{min}|, |1 - h\lambda_{max}|\}.$$

Since $a(h) < 1$, we can still use the identity (7.36) in (7.35) and obtain

7 Ordinary differential equations

$$|z_n - u_n| \leq \rho_{max} \left([a(h)]^n + h \frac{1 - [a(h)]^n}{1 - a(h)} \right), \tag{7.39}$$

where $\rho_{max} = \max |\rho_k|$. Notice that $a(h) = |1 - h\lambda_{min}|$ if $h \leq h^*$ while $a(h) = |1 - h\lambda_{max}|$ if $h^* \leq h < h_0(\lambda)$, having set $h^* = 2/(\lambda_{min} + \lambda_{max})$. When $h \leq h^*$, $a(h) > 0$ and it follows that

$$|z_n - u_n| \leq \frac{\rho_{max}}{\lambda_{min}} [1 - [a(h)]^n (1 - \lambda_{min})], \tag{7.40}$$

thus

$$\limsup_{n \to \infty} |z_n - u_n| \leq \frac{\rho_{max}}{\lambda_{min}}, \tag{7.41}$$

from which we still conclude that the perturbation error is bounded by ρ_{max} times a constant which is independent of n and h (although the oscillations are no longer damped as in the previous case).

In fact, similar conclusion holds also when $h^* \leq h \leq h_0(\lambda)$, although this does not follow from our upper bound (7.40) which is too pessimistic in this case.

In Figure 7.8 we report the perturbation errors computed on the problem (7.32), where $\lambda_k = \lambda(t_k) = -2 - \sin(t_k)$, $\rho_k = \rho(t_k) = 0.1 \sin(t_k)$ with $h < h^*$ (left) and with $h^* \leq h < h_0(\lambda)$ (right).

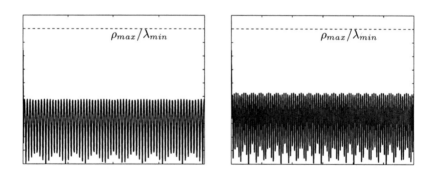

Fig. 7.8. The perturbation error when $\rho(t) = 0.1\sin(t)$ and $\lambda(t) = -2 - \sin(t)$ for $t \in (0, nh)$ with $n = 500$: the step-size is $h = h^* - 0.1 = 0.4$ (left) and $h = h^* + 0.1 = 0.6$ (right)

iii. We consider now the general Cauchy problem (7.5). We claim that this problem can be related to the generalized model problem (7.32), in those cases where

$$-\lambda_{max} < \partial f/\partial y(t, y) < -\lambda_{min}, \forall t \geq 0, \ \forall y \in (-\infty, \infty),$$

for suitable values λ_{min}, $\lambda_{max} \in (0,+\infty)$. To this end, for every t in the generic interval (t_n, t_{n+1}), we subtract (7.6) from (7.22) to obtain the following equation for the perturbation error:

$$z_n - u_n = (z_{n-1} - u_{n-1}) + h\{f(t_{n-1}, z_{n-1}) - f(t_{n-1}, u_{n-1})\} + h\rho_n.$$

By applying the mean-value theorem we obtain

$$f(t_{n-1}, z_{n-1}) - f(t_{n-1}, u_{n-1}) = \lambda_{n-1}(z_{n-1} - u_{n-1}),$$

where $\lambda_{n-1} = f_y(t_{n-1}, \xi_{n-1})$, $f_y = \partial f/\partial y$ and ξ_{n-1} is a suitable point in the interval whose endpoints are u_{n-1} and z_{n-1}. Thus

$$z_n - u_n = (1 + h\lambda_{n-1})(z_{n-1} - u_{n-1}) + h\rho_n.$$

By a recursive application of this formula we obtain the identity (7.35), from which we derive the same conclusions drawn in $ii.$, provided the stability restriction $0 < h < 2/\lambda_{max}$ holds.

Example 7.5 Let us consider the Cauchy problem

$$y'(t) = \arctan(3y) - 3y + t,\ t > 0,\ y(0) = 1. \tag{7.42}$$

Since $f_y = 3/(1+9y^2) - 3$ is negative, we can choose $\lambda_{max} = \max |f_y| = 3$ and set $h < 2/3$. Thus, we can expect that the perturbations on the forward Euler method are kept under control provided that $h < 2/3$. This is confirmed by the results which are reported in Figure 7.9. Note that in this example, taking $h = 2/3 + 0.01$ (thus violating the previous stability limit) the perturbation error blows up as t increases. ∎

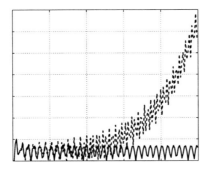

Fig. 7.9. The perturbation errors when $\rho(t) = \sin(t)$ with $h = 2/\lambda_{max} - 0.01$ (*thick line*) and $h = 2/\lambda_{max} + 0.01$ (*thin line*) for the Cauchy problem (7.42)

Example 7.6 We seek a limit on h that guarantees stability for the forward Euler method applied to approximate the Cauchy problem

$$y' = 1 - y^2, \quad t > 0, \tag{7.43}$$

with $y(0) = \dfrac{e-1}{e+1}$. The exact solution is $y(t) = (e^{2t+1} - 1)/(e^{2t+1} + 1)$ and $f_y = -2y$. Since $f_y \in (-2, -0.9)$ for all $t > 0$, we can take h less than $h_0 = 1$. In Figure 7.10, left, we report the solutions obtained on the interval $(0, 35)$ with $h = 0.95$ (*thick line*) and $h = 1.05$ (*thin line*). In both cases the solution oscillates, but remains bounded. Moreover in the first case, which satisfies the stability constraint, the oscillations are damped and the numerical solution tends to the exact one as t increases. In Figure 7.10, right, we report the perturbation errors corresponding to $\rho(t) = \sin(t)$ with $h = 0.95$ (*thick line*) and $h = h^* + 0.1$ (*thin line*). In both cases the perturbation errors remain bounded; moreover, in the former case the upper bound (7.41) is satisfied. ∎

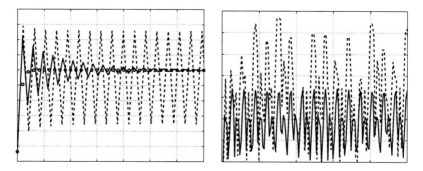

Fig. 7.10. On the left, numerical solutions of problem (7.43) obtained by the forward Euler method with $h = 20/19$ (*thin line*) and $h = 20/21$ (*thick line*). The values of the exact solution are indicated by circles. On the right, perturbation errors corresponding to $\rho(t) = \sin(t)$ with $h = 0.95$ (*thick line*) and $h = h^*$ (*thin line*)

In those cases where no information on y is available, finding the value $\lambda_{max} = \max |f_y|$ is not a simple matter. A more heuristic approach could be pursued in these situations, by adopting a variable stepping procedure. Precisely, one could take $t_{n+1} = t_n + h_n$, where

$$h_n < 2\frac{\alpha}{|f_y(t_n, u_n)|},$$

for suitable values of α strictly less than 1. Note that the denominator depends on the value u_n which is known. In Figure 7.11 we report the perturbation errors corresponding to the Example 7.6 for two different values of α.

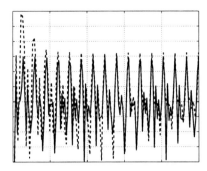

Fig. 7.11. The perturbation errors corresponding to $\rho(t) = \sin(t)$ with $\alpha = 0.8$ (*thick line*) and $\alpha = 0.9$ (*thin line*) for the Example 7.6, using the adaptive strategy

The previous analysis can be carried out also for other kind of one-step methods, in particular for the backward Euler and Crank-Nicolson methods. For these methods which are A-stable, the same conclusions about the perturbation error can be drawn without requiring any limitation on the time-step. In fact, in the previous analysis one should replace each term $1 + h\lambda_n$ by $(1 - h\lambda_n)^{-1}$ in the backward Euler case and by $(1 + h\lambda_n/2)/(1 - h\lambda_n/2)$ in the Crank-Nicolson case.

Let us summarize

1. An absolutely stable method is one which generates a solution u_n of the model problem (7.28) which tends to zero as t_n tends to infinity;
2. a method is said *A-stable* if it is absolutely stable for any possible choice of the time-step h (otherwise a method is called conditionally stable, and h should be lower than a constant depending on λ);
3. when an absolutely stable method is applied to a generalized model problem (like (7.32)), the perturbation error (that is the absolute value of the difference between the perturbed and unperturbed solution) is uniformly bounded (with respect to h). In short we can say that absolutely stable methods keep the perturbation controlled;
4. the analysis of absolute stability for the linear model problem can be exploited to find stability conditions on the time-step when considering the nonlinear Cauchy problem (7.5) with a function f satisfying $\partial f/\partial y < 0$. In that case the stability restriction requires the step-size to be chosen as a function of $\partial f/\partial y$. Precisely, the new integration interval $[t_n, t_{n+1}]$ is chosen in such a way that $h_n = t_{n+1} - t_n$ satisfies $h_n < 2\alpha/|\partial f(t_n, u_n)/\partial y|$ for a suitable $\alpha \in (0, 1)$.

See the Exercises 7.6-7.13.

7.6 High order methods

All methods presented so far are elementary examples of one-step methods. More sophisticated schemes, which allow the achievement of a higher order of accuracy, are the *Runge-Kutta methods* and the *multistep methods*. (whose general form was already introduced in (7.23)). Runge-Kutta (briefly, RK) methods are still one-step methods; however, they involve several evaluations of the function $f(t,y)$ on every interval $[t_n, t_{n+1}]$. In its most general form, a RK method can be written as

$$u_{n+1} = u_n + h\sum_{i=1}^{s} b_i K_i, \qquad n \geq 0 \qquad (7.44)$$

where

$$K_i = f(t_n + c_i h, u_n + h\sum_{j=1}^{s} a_{ij} K_j), \quad i = 1, 2, \ldots, s$$

and s denotes the number of *stages* of the method. The coefficients $\{a_{ij}\}$, $\{c_i\}$ and $\{b_i\}$ fully characterize a RK method and are usually collected in the so-called *Butcher array*

$$\begin{array}{c|c} \mathbf{c} & A \\ \hline & \mathbf{b}^T \end{array},$$

where $A = (a_{ij}) \in \mathbb{R}^{s \times s}$, $\mathbf{b} = (b_1, \ldots, b_s)^T \in \mathbb{R}^s$ and $\mathbf{c} = (c_1, \ldots, c_s)^T \in \mathbb{R}^s$. If the coefficients a_{ij} in A are equal to zero for $j \geq i$, with $i = 1, 2, \ldots, s$, then each K_i can be explicitly computed in terms of the $i-1$ coefficients $K_1, \ldots, K_{i-1}$ that have already been determined. In such a case the RK method is *explicit*. Otherwise, it is *implicit* and solving a nonlinear system of size s is necessary for computing the coefficients K_i.

One of the most celebrated Runge-Kutta methods reads

$$\boxed{u_{n+1} = u_n + \frac{h}{6}(K_1 + 2K_2 + 2K_3 + K_4)} \qquad (7.45)$$

where

$$K_1 = f_n,$$
$$K_2 = f(t_n + \tfrac{h}{2}, u_n + \tfrac{h}{2}K_1),$$
$$K_3 = f(t_n + \tfrac{h}{2}, u_n + \tfrac{h}{2}K_2),$$
$$K_4 = f(t_{n+1}, u_n + hK_3),$$

$$\begin{array}{c|cccc} 0 & & & & \\ \tfrac{1}{2} & \tfrac{1}{2} & & & \\ \tfrac{1}{2} & 0 & \tfrac{1}{2} & & \\ 1 & 0 & 0 & 1 & \\ \hline & \tfrac{1}{6} & \tfrac{1}{3} & \tfrac{1}{3} & \tfrac{1}{6} \end{array}.$$

This method can be derived from (7.18) by using the Simpson quadrature rule (4.23) to evaluate the integral between t_n and t_{n+1}. It is explicit, of fourth order with respect to h; at each time step, it involves four new evaluations of the function f. Other Runge-Kutta methods, either explicit or implicit, with arbitrary order can be constructed. For instance, an implicit RK method of order 4 with 2 stages is defined by the following Butcher array

$$\begin{array}{c|cc} \frac{3-\sqrt{3}}{6} & \frac{1}{4} & \frac{3-2\sqrt{3}}{12} \\ \frac{3+\sqrt{3}}{6} & \frac{3+2\sqrt{3}}{12} & \frac{1}{4} \\ \hline & \frac{1}{2} & \frac{1}{2} \end{array}.$$

The absolute stability region $\mathcal{A}$ of the RK methods, including explicit RK methods, can grow in surface with the order: an example is provided by the left graph in Figure 7.13, where $\mathcal{A}$ has been reported for some explicit RK methods of increasing order: RK1 is the forward Euler method, RK2 is the improved Euler method, (7.52), RK3 represents the following Butcher array

$$\begin{array}{c|ccc} 0 & & & \\ \frac{1}{2} & \frac{1}{2} & & \\ 1 & -1 & 2 & \\ \hline & \frac{1}{6} & \frac{2}{3} & \frac{1}{6} \end{array} \tag{7.46}$$

and RK4 represents method (7.45) introduced previously.

The RK methods stand at the base of a family of MATLAB programs whose names contain the root ode followed by numbers and letters. In particular, ode45 is based on a pair of explicit Runge-Kutta methods (the so-called Dormand-Prince pair) of order 4 and 5, respectively. ode23 is the implementation of another pair of explicit Runge-Kutta methods (the Bogacki and Shampine pair). In these methods the integration step varies in order to guarantee that the error remains below a given tolerance (the default scalar relative error tolerance RelTol is equal to 10^{-3}). The program ode23tb is an implementation of an implicit Runge-Kutta formula whose first stage is the trapezoidal rule, while the second stage is a backward differentiation formula of order two (see (7.49)).

ode
ode45
ode23

ode23tb

Multistep methods (see (7.23)) achieve a high order of accuracy by involving the values $u_n, u_{n-1}, \ldots, u_{n-p}$ for the determination of u_{n+1}. They can be derived by applying first the formula (7.18) and then approximating the integral by a quadrature formula which involves the interpolant of f at a suitable set of nodes. A notable example of multistep method is the three-step ($p = 2$), third order (explicit) Adams-Bashforth formula (AB3)

$$u_{n+1} = u_n + \frac{h}{12}(23f_n - 16f_{n-1} + 5f_{n-2}) \tag{7.47}$$

214   7 Ordinary differential equations

which is obtained by replacing f in (7.18) by its interpolating polynomial of degree two at the nodes t_{n-2}, t_{n-1}, t_n. Another important example is the three-step, fourth order (implicit) Adams-Moulton formula (AM4)

$$u_{n+1} = u_n + \frac{h}{24}(9f_{n+1} + 19f_n - 5f_{n-1} + f_{n-2}) \tag{7.48}$$

which is obtained by replacing f in (7.18) by its interpolating polynomial of degree three at the nodes $t_{n-2}, t_{n-1}, t_n, t_{n+1}$.

Another family of multistep methods can be obtained by writing the differential equation at time t_{n+1} and replacing $y'(t_{n+1})$ by a one-sided incremental ratio of high order. An instance is provided by the two-step, second order (implicit) *backward difference formula* (BDF2)

$$u_{n+1} = \frac{4}{3}u_n - \frac{1}{3}u_{n-1} + \frac{2h}{3}f_{n+1} \tag{7.49}$$

or by the following three-step, third order (implicit) *backward difference formula* (BDF3)

$$u_{n+1} = \frac{18}{11}u_n - \frac{9}{11}u_{n-1} + \frac{2}{11}u_{n-2} + \frac{6h}{11}f_{n+1} \tag{7.50}$$

All these methods can be recasted in the general form (7.23). It is easy to verify that for all of them the relations (7.27) are satisfied, thus they are consistent. Moreover, they are zero-stable. Indeed, in both cases (7.47) and (7.48), the first characteristic polynomial is $\pi(r) = r^3 - r^2$ and its roots are $r_0 = 1$, $r_1 = r_2 = 0$, while the first characteristic polynomial of (7.50) is $\pi(r) = r^3 - 18/11r^2 + 9/11r - 2/11$ and its roots are $r_0 = 1$, $r_1 = 0.3182 + 0.2839i$, $r_2 = 0.3182 - 0.2839i$, where i is the imaginary unit. In all cases, the root condition (7.24) is satisfied.

When applied to the model problem (7.28), AB3 is absolutely stable if $h < 0.545/|\lambda|$, while AM4 is absolutely stable if $h < 3/|\lambda|$. The method BDF3 is unconditionally absolutely stable (i.e., A-stable) for all real negative λ. However, this is no longer true if $\lambda \in \mathbb{C}$ (with negative real part). In other words, BDF3 fails to be A-stable (see, Figure 7.13). More generally, according to the second Dahlquist barrier there is no multistep A-stable method of order strictly greater than two.

In Figures 7.12 the regions of absolute stability of several Adams-Bashfort and Adams-Moulton methods are drawn. Note that their size reduces as far as the order increases. In the right-hand side graphs of Figure 7.13 we report the (unlimited) absolute stability regions of some BDF methods: these cover a surface in the complex plane which reduces when the order increases, as opposed to those of the Runge-Kutta methods (reported on the left) which increase in surface when the order increases.

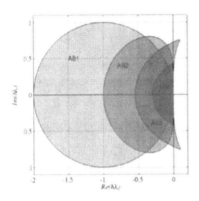
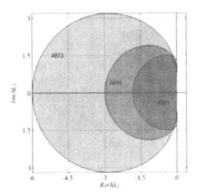

Fig. 7.12. The absolute stability regions of several Adams-Basforth (*left*) and Adams-Moulton (*right*) methods

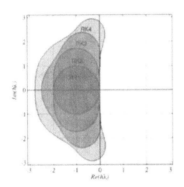
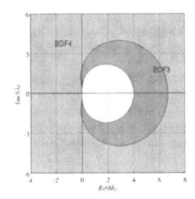

Fig. 7.13. The absolute stability regions of several explicit RK (*left*) and BDF methods (*right*). In this case the regions are unlimited and span in the direction shown by the arrows

Remark 7.3 (Computing absolute stability regions) It is possible to compute the boundary $\partial \mathcal{A}$ of the absolute stability region $\mathcal{A}$ of a multistep method with a simple trick. The boundary is in fact composed by the complex numbers $h\lambda$ such that

$$h\lambda = \left(r^{p+1} - \sum_{j=0}^{p} a_j r^{p-j} \right) \Big/ \left(\sum_{j=-1}^{p} b_j r^{p-j} \right), \qquad (7.51)$$

with r as a complex number of module one. Therefore, to obtain with MATLAB an approximate representation of $\partial \mathcal{A}$ it is sufficient to evaluate the second member of (7.51) with different values of r on the unit circle (for instance, by setting `r = exp(i*pi*(0:2000)/1000)`, where `i` is the imaginary unit). The graphs in Figures 7.12 and 7.13 have been obtained in this way. •

According to the first Dahlquist barrier the maximum order q of a $p+1$-step method satisfying the root condition is $q = p+1$ for explicit methods and, for implicit methods $q = p+2$ if $p+1$ is odd, $q = p+3$ if $p+1$ is even.

Remark 7.4 (Cyclic composite methods) It is possible to overcome the Dahlquist barriers by appropriately combining several multistep methods. For instance, the two following methods

$$u_{n+1} = -\frac{8}{11}u_n + \frac{19}{11}u_{n-1} + \frac{h}{33}(30f_{n+1} + 57f_n + 24f_{n-1} - f_{n-2}),$$

$$u_{n+1} = \frac{449}{240}u_n + \frac{19}{30}u_{n-1} - \frac{361}{240}u_{n-2}$$
$$+ \frac{h}{720}(251f_{n+1} + 456f_n - 1347f_{n-1} - 350f_{n-2}),$$

have order five, but are unstable. However, by using them in a combined way (the former if n is even, the latter if n is odd) they produce an A-stable 3-step method of order five. ●

Multistep methods are implemented in several MATLAB programs, for instance in ode15s.

ode15s

Octave 7.1 ode23 and ode45 are also available in Octave-forge. The optional arguments however differ from MATLAB. Note that ode45 in Octave-forge offers two possible strategies: the default one based on the Dormand and Prince method produces generally more accurate results than the other option that is based on the Fehlberg method. ■

7.7 The predictor-corrector methods

In Section 7.2 it was pointed out that implicit methods yield at each step a nonlinear problem for the unknown value u_{n+1}. For its solution we can use one of the methods introduced in Chapter 2, or else apply the function fsolve as we have done with the Programs 7.2 and 7.3.

Alternatively, we can carry out fixed point iterations at every time-step. For example, for the Crank-Nicolson method (7.17), for $k = 0, 1, \ldots$, we compute until convergence

$$u_{n+1}^{(k+1)} = u_n + \frac{h}{2}\left[f_n + f(t_{n+1}, u_{n+1}^{(k)})\right].$$

It can be proved that if the initial guess $u_{n+1}^{(0)}$ is chosen conveniently, a single iteration suffices in order to obtain a numerical solution $u_{n+1}^{(1)}$ whose accuracy is of the same order as the solution u_{n+1} of the original implicit method. More precisely, if the original implicit method has order p, then the initial guess $u_{n+1}^{(0)}$ must be generated by an explicit method of order (at least) $p - 1$.

7.7 The predictor-corrector methods

For instance, if we use the first order (explicit) forward Euler method to initialize the Crank-Nicolson method, we get the *Heun method* (also called *improved Euler method*), which is a second order explicit Runge-Kutta method:

$$
\begin{aligned}
u_{n+1}^* &= u_n + h f_n, \\
u_{n+1} &= u_n + \frac{h}{2} \left[f_n + f(t_{n+1}, u_{n+1}^*) \right]
\end{aligned}
\tag{7.52}
$$

The explicit step is called a *predictor*, whereas the implicit one is called a *corrector*. Another example combines the (AB3) method (7.47) as predictor with the (AM4) method (7.48) as corrector. These kinds of methods are therefore called *predictor-corrector* methods. They enjoy the order of accuracy of the corrector method. However, being explicit, they undergo a stability restriction which is typically the same as that of the predictor method (see, for instance, the regions of absolute stability of Figure 7.14). Thus they are not adequate to integrate a Cauchy problem on unbounded intervals.

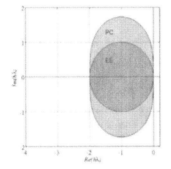 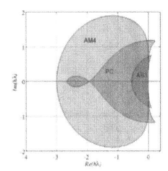

Fig. 7.14. The absolute stability regions of the predictor-corrector methods obtained by combining the explicit Euler (EE) and Crank-Nicolson methods (*left*) and AB3 and AM4 (*right*). Notice the reduced surface of the region when compared to the corresponding implicit methods (in the first case the region of the Crank-Nicolson method hasn't been reported as it coincides with all the complex half-plane $Re(h\lambda) < 0$)

In Program 7.4 we implement a general predictor-corrector method. The strings `predictor` and `corrector` identify the type of method that is chosen. For instance, if we use the functions `eeonestep` and `cnonestep`, which are defined in Program 7.5, we can call `predcor` as follows:

```
>> [t,u]=predcor(t0,y0,T,N,f,'eeonestep','cnonestep');
```

218   7 Ordinary differential equations

and obtain the Heun method.

Program 7.4. predcor: predictor-corrector method

```
function [t,u]=predcor(odefun,tspan,y,Nh,...
                predictor,corrector,varargin)
%PREDCOR  Solve differential equations using a
%  predictor-corrector method
%  [T,Y]=PREDCOR(ODEFUN,TSPAN,Y0,NH,PRED,CORR) with
%  TSPAN=[T0 TF] integrates the system of differential
%  equations y' = f(t,y) from time T0 to TF with initial
%  condition Y0 using a general predictor corrector
%  method on an equispaced grid of NH intervals.
%  Function ODEFUN(T,Y) must return a column-vector
%  corresponding to f(t,y). Each row in the solution
%  array Y corresponds to a time returned in the column
%  vector T. Functions PRED and CORR identify the type
%  of method that is chosen.
%  [T,Y]=PREDCOR(ODEFUN,TSPAN,Y0,NH,PRED,CORR,P1,..)
%  passes the additional parameters P1,... to the
%  functions ODEFUN,PRED and CORR as ODEFUN(T,Y,P1,...),
%  PRED(T,Y,P1,P2...), CORR(T,Y,P1,P2...).
h=(tspan(2)-tspan(1))/Nh;  tt=[tspan(1):h:tspan(2)];
u=y; [n,m]=size(u); if n < m, u=u'; end
for t=tt(1:end-1)
    y = u(:,end); fn = feval(odefun,t,y,varargin{:});
    upre = feval(predictor,t,y,h,fn);
    ucor = feval(corrector,t+h,y,upre,h,odefun,...
           fn,varargin{:});
    u = [u, ucor];
end
t = tt;
end
```

Program 7.5. onestep: one step of forward Euler (eeonestep), one step of backward Euler (eionestep), one step of Crank-Nicolson (cnonestep)

```
function [u]=feonestep(t,y,h,f)
u = y + h*f;
return

function [u]=beonestep(t,u,y,h,f,fn,varargin)
u = u + h*feval(f,t,y,varargin{:});
return

function [u]=cnonestep(t,u,y,h,f,fn,varargin)
u = u + 0.5*h*(feval(f,t,y,varargin{:})+fn);
return
```

ode113   The MATLAB program ode113 implements a combined Adams-Moulton-Bashforth scheme with variable step-size.

See the Exercises 7.14-7.17.

7.8 Systems of differential equations

Let us consider the following system of first-order ordinary differential equations whose unknowns are $y_1(t), \ldots, y_m(t)$:

$$\begin{cases} y_1' = f_1(t, y_1, \ldots, y_m), \\ \vdots \\ y_m' = f_m(t, y_1, \ldots, y_m), \end{cases}$$

where $t \in (t_0, T]$, with the initial conditions

$$y_1(t_0) = y_{0,1}, \ldots, y_m(t_0) = y_{0,m}.$$

For its solution we could apply to each individual equation one of the methods previously introduced for a scalar problem. For instance, the n-th step of the forward Euler method would read

$$\begin{cases} u_{n+1,1} = u_{n,1} + hf_1(t_n, u_{n,1}, \ldots, u_{n,m}), \\ \vdots \\ u_{n+1,m} = u_{n,m} + hf_m(t_n, u_{n,1}, \ldots, u_{n,m}). \end{cases}$$

By writing the system in vector form $\mathbf{y}'(t) = \mathbf{F}(t, \mathbf{y}(t))$, with obvious choice of notation, the extension of the methods previously developed for the case of a single equation to the vector case is straightforward. For instance, the method

$$\mathbf{u}_{n+1} = \mathbf{u}_n + h(\vartheta \mathbf{F}(t_{n+1}, \mathbf{u}_{n+1}) + (1 - \vartheta)\mathbf{F}(t_n, \mathbf{u}_n)), \qquad n \geq 0,$$

with $\mathbf{u}_0 = \mathbf{y}_0$, $0 \leq \vartheta \leq 1$, is the vector form of the forward Euler method if $\vartheta = 0$, the backward Euler method if $\vartheta = 1$ and the Crank-Nicolson method if $\vartheta = 1/2$.

Example 7.7 (Population dynamics) Let us apply the forward Euler method to solve the Lotka-Volterra equations (7.3) with $C_1 = C_2 = 1$, $b_1 = b_2 = 0$ and $d_1 = d_2 = 1$. In order to use Program 7.1 for a *system* of ordinary differential equations, let us create a function f which contains the component of the vector function $\mathbf{F}$, which we save in the file f.m. For our specific system we have:

```
function y = f(t,y)
C1=1; C2=1; d1=1; d2=1; b1=0; b2=0;
yy(1)=C1*y(1)*(1-b1*y(1)-d2*y(2));   % first equation
y(2)=-C2*y(2)*(1-b2*y(2)-d1*y(1));   % second equation
y(1)=yy(1);
return
```

Now we execute Program 7.1 with the following instructions

```
[t,u]=feuler('fsys',[0,0.1],[0 0],100);
```

They correspond to solving the Lotka-Volterra system on the time interval $[0, 10]$ with a time-step $h = 0.005$.

The graph in Figure 7.15, left, represents the time evolution of the two components of the solution. Note that they are periodic with period 2π. The second graph in Figure 7.15, right, shows the trajectory issuing from the initial value in the so-called *phase plane*, that is, the Cartesian plane whose coordinate axes are y_1 and y_2. This trajectory is confined within a bounded region of the (y_1, y_2) plane. If we start from the point $(1.2, 1.2)$, the trajectory would stay in an even smaller region surrounding the point $(1, 1)$. This can be explained as follows. Our differential system admits 2 *points of equilibrium* at which $y_1' = 0$ and $y_2' = 0$, and one of them is precisely $(1, 1)$ (the other being $(0, 0)$). Actually, they are obtained by solving the nonlinear system

$$\begin{cases} y_1' = y_1 - y_1 y_2 = 0, \\ y_2' = -y_2 + y_2 y_1 = 0. \end{cases}$$

If the initial data coincide with one of these points, the solution remains constant in time. Moreover, while $(0, 0)$ is an unstable equilibrium point, $(1, 1)$ is stable, that is, all trajectories issuing from a point near $(1, 1)$ stay bounded in the phase plane. ∎

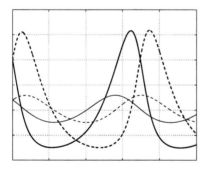
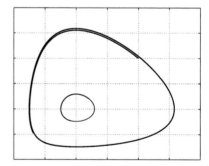

Fig. 7.15. Numerical solutions of system (7.3). On the left, we represent y_1 and y_2 on the time interval $(0, 10)$, the solid line refers to y_1, the dashed line to y_2. Two different initial data are considered: $(2, 2)$ (*thick lines*) and $(1.2, 1.2)$ (*thin lines*). On the right, we report the corresponding trajectories in the phase plane

When we use an explicit method, the step-size h should undergo a stability restriction similar to the one encountered in Section 7.5. When the real part of the eigenvalues λ_k of the Jacobian $A(t) = [\partial \mathbf{F}/\partial \mathbf{y}](t, \mathbf{y})$ of $\mathbf{F}$ are all negative, we can set $\lambda = -\max_t \rho(A(t))$, where $\rho(A(t))$ is the spectral radius of $A(t)$. This λ is a candidate to replace the one entering in the stability conditions (such as, e.g., (7.30)) that were derived for the scalar Cauchy problem.

Remark 7.5 The MATLAB programs (ode23, ode45, ...) that we have mentioned before can be used also for the solution of systems of ordinary differential equations. The syntax is odeXX('f',[t0 tf],y0), where y0 is the vector of the initial conditions, f is a function to be specified by the user and odeXX is one of the methods available in MATLAB. ●

Now consider the case of an ordinary differential equation of order m

$$y^{(m)}(t) = f(t, y, y', \ldots, y^{(m-1)}) \qquad (7.53)$$

for $t \in (t_0, T]$, whose solution (when existing) is a family of functions defined up to m arbitrary constants. The latter can be fixed by prescribing m initial conditions

$$y(t_0) = y_0, \ y'(t_0) = y_1, \ \ldots, \ y^{(m-1)}(t_0) = y_{m-1}.$$

Setting

$$w_1(t) = y(t), \ w_2(t) = y'(t), \ \ldots, \ w_m(t) = y^{(m-1)}(t),$$

the equation (7.53) can be transformed into a first-order system of m differential equations

$$\begin{cases} w_1' = w_2, \\ w_2' = w_3, \\ \vdots \\ w_{m-1}' = w_m, \\ w_m' = f(t, w_1, \ldots, w_m), \end{cases}$$

with initial conditions

$$w_1(t_0) = y_0, \ w_2(t_0) = y_1, \ \ldots, \ w_m(t_0) = y_{m-1}.$$

Thus we can always approximate the solution of a differential equation of order $m > 1$ by resorting to the equivalent system of m first-order equations, and then applying to this system a convenient discretization method.

Example 7.8 (Electrical circuits) Consider the circuit of Problem 7.4 and suppose that $L(i_1) = L$ is constant and that $R_1 = R_2 = R$. In this case v can be obtained by solving the following system of two differential equations:

$$\begin{cases} v'(t) = w(t), \\ w'(t) = -\dfrac{1}{LC}\left(\dfrac{L}{R} + RC\right)w(t) - \dfrac{2}{LC}v(t) + \dfrac{e}{LC}, \end{cases} \qquad (7.54)$$

with initial conditions $v(0) = 0$, $w(0) = 0$. The system has been obtained from the second-order differential equation

$$LC\frac{d^2v}{dt^2} + \left(\frac{L}{R_2} + R_1C\right)\frac{dv}{dt} + \left(\frac{R_1}{R_2} + 1\right)v = e. \qquad (7.55)$$

We set $L = 0.1$ Henry, $C = 10^{-3}$ Farad, $R = 10$ Ohm and $e = 5$ Volt, where Henry, Farad, Ohm and Volt are respectively the unit measure of inductance, capacitance, resistance and voltage. Now we apply the forward Euler method with $h = 0.01$ seconds in the time interval $[0, 0.1]$, by the Program 7.1:

```
[t,u]=feuler('fsys',[0,0.1],[0 0],100);
```

where fsys is contained in the file fsys.m:

```
function y=fsys(t,y)
L=0.1; C=1.e-03; R=10; e=5; LC = L*C;
yy=y(2); y(2)=-(L/R+R*C)/(LC)*y(2)-2/(LC)*y(1)+e/(LC);
y(1)=yy;
return
```

In Figure 7.16 we report the approximated values of v and w. As expected, $v(t)$ tends to $e/2 = 2.5$ Volt for large t. In this case the real part of the eigenvalues of $A(t) = [\partial \mathbf{F}/\partial \mathbf{y}](t, \mathbf{y})$ is negative and λ can be set equal to -141.4214. Then a condition for absolute stability is to take $h < 2/|\lambda| = 0.0282$. ∎

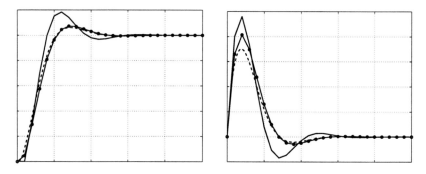

Fig. 7.16. Numerical solutions of system (7.54). The potential drop $v(t)$ is reported on the left, its derivative w on the right: the dashed line represents the solution obtained for $h = 0.001$ with the forward Euler method, the continuous line is for the one generated via the same method with $h = 0.004$, and the dotted line is for the one produced via the Newmark method (7.59) (with $\theta = 1/2$ and $\varsigma = 1/4$) with $h = 0.004$

Sometimes numerical approximations can be directly derived on the high order equation without passing through the equivalent first order system. Consider for instance the case of the 2nd order Cauchy problem

$$\begin{cases} y''(t) = f(t, y(t), y'(t)) & t \in (t_0, T], \\ y(t_0) = \alpha_0, \quad y'(t_0) = \beta_0. \end{cases} \qquad (7.56)$$

7.8 Systems of differential equations

A simple numerical scheme can be constructed as follows: find u_n for $1 \leq n \leq N_h$ such that

$$\frac{u_{n+1} - 2u_n + u_{n-1}}{h^2} = f(t_n, u_n, v_n) \tag{7.57}$$

with $u_0 = \alpha_0$ and $v_0 = \beta_0$. The quantity v_k represents a second order approximation of $y'(t_k)$ (since $(y_{n+1} - 2y_n + y_{n-1})/h^2$ is a second order approximation of $y''(t_n)$). One possibility is to take

$$v_n = \frac{u_{n+1} - u_{n-1}}{2h}, \quad \text{with } v_0 = \beta_0. \tag{7.58}$$

The *leap-frog method* (7.57)-(7.58) is accurate of order 2 with respect to h.

A more general method is the *Newmark method*, in which we build two sequences

$$u_{n+1} = u_n + hv_n + h^2 \left[\zeta f(t_{n+1}, u_{n+1}, v_{n+1}) + (1/2 - \zeta) f(t_n, u_n, v_n) \right], \tag{7.59}$$

$$v_{n+1} = v_n + h \left[(1-\theta) f(t_n, u_n, v_n) + \theta f(t_{n+1}, u_{n+1}, v_{n+1}) \right],$$

with $u_0 = \alpha_0$ and $v_0 = \beta_0$, where ζ and θ are two non-negative real numbers. This method is implicit unless $\zeta = \theta = 0$, second order if $\theta = 1/2$, whereas it is first order accurate if $\theta \neq 1/2$. The condition $\theta \geq 1/2$ is necessary to ensure stability. For $\theta = 1/2$ and $\zeta = 1/4$ we find a rather popular method that is unconditionally stable. However, this method is not suitable for simulations on long time intervals as it introduces oscillatory spurious solutions. For these simulations it is preferable to use $\theta > 1/2$ and $\zeta > (\theta + 1/2)^2/4$ even though the method degenerates to a first order one.

In Program 7.6 we implement the Newmark method. The vector param allows to specify the values of the coefficients (param(1)=ζ, param(2)=θ).

Program 7.6. newmark: Newmark method

```
function [tt,u]=newmark(odefun,tspan,y,Nh,param,varargin)
%NEWMARK Solve second order differential equations using
%    the Newmark method
%    [T,Y]=NEWMARK(ODEFUN,TSPAN,Y0,NH,PARAM) with TSPAN =
%    [T0 TF] integrates the system of differential
%    equations y''=f(t,y,y') from time T0 to TF with
%    initial conditions Y0=(y(t0),y'(t0)) using the
%    Newmark method on an equispaced grid of NH intervals.
%    Function ODEFUN(T,Y) must return a   scalar value
%    corresponding to f(t,y,y').
tt=linspace(tspan(1),tspan(2),Nh+1);
u(1,:)=y;

global glob_h glob_t glob_y glob_odefun;
global glob_zeta glob_theta glob_varargin glob_fn;
```

224   7 Ordinary differential equations

```
glob_h=(tspan(2)-tspan(1))/Nh;
glob_y=y;
glob_odefun=odefun;
glob_t=tt(2);
glob_zeta = param(1);
glob_theta = param(2);
glob_varargin=varargin;

if ( ~exist( 'OCTAVE_VERSION' ) )
 options=optimset;
 options.TolFun=1.e-12;
 options.MaxFunEvals=10000;
end

glob_fn =feval(odefun,tt(1),u(1,:),varargin{:});
for glob_t=tt(2:end)
if ( exist( 'OCTAVE_VERSION' ) )
  w = fsolve('newmarkfun', glob_y )
else
  w = fsolve(@(w) newmarkfun(w),glob_y,options);
end
  glob_fn =feval(odefun,glob_t,w,varargin{:});
  u = [u; w];
  y = w;
end
t=tt;
clear glob_h glob_t glob_y glob_odefun;
clear glob_zeta glob_theta glob_varargin glob_fn;
end

function z=myfun(w)
 global glob_h glob_t glob_y glob_odefun;
 global glob_zeta glob_theta glob_varargin glob_fn;
 fn1 = feval(glob_odefun,glob_t,glob_w,glob_varargin{:});
 z=w - glob_y -...
    glob_h*[glob_y(1,2), ...
             (1-glob_theta)*glob_fn+glob_theta*fn1]-...
    glob_h^2*[glob_zeta*fn1+(0.5-glob_zeta)*glob_fn ,0];
end
```

Example 7.9 (Electrical circuits) We consider again the circuit of Problem 7.4 and we solve the second order equation (7.55) with the Newmark scheme. In Figure 7.16 we compare the numerical approximations of the function v computed using the Euler scheme (*dashed line* and *continuous line*) and the Newmark scheme with $\theta = 1/2$ and $\zeta = 1/4$ (*dotted line*), with the time-step $h = 0.04$. The better accuracy of the latter solution is due to the fact that the method (7.57)-(7.58) is second order accurate with respect to h. ∎

See the Exercises 7.18-7.20.

7.9 Some examples

We end this chapter by considering and solving three non-trivial examples of systems of ordinary differential equations.

7.9.1 The spherical pendulum

The motion of a point $\mathbf{x}(t) = (x_1(t), x_2(t), x_3(t))^T$ with mass m subject to the gravity force $\mathbf{F} = (0, 0, -gm)^T$ (with $g = 9.8$ m/s^2) and constrained to move on the spherical surface of equation $\Phi(\mathbf{x}) = x_1^2 + x_2^2 + x_3^2 - 1 = 0$ is described by the following system of ordinary differential equations

$$\ddot{\mathbf{x}} = \frac{1}{m}\left(\mathbf{F} - \frac{m\dot{\mathbf{x}}^T \mathrm{H}\dot{\mathbf{x}} + \nabla\Phi^T \mathbf{F}}{|\nabla\Phi|^2}\nabla\Phi\right) \quad \text{for } t > 0. \tag{7.60}$$

We denote by $\dot{\mathbf{x}}$ the first derivative with respect to t, with $\ddot{\mathbf{x}}$ the second derivative, with $\nabla\Phi$ the spatial gradient of Φ, equal to $2\mathbf{x}^T$, with H the Hessian matrix of Φ whose components are $\mathrm{H}_{ij} = \partial^2\Phi/\partial x_i \partial x_j$ for $i, j = 1, 2, 3$. In our case H is a diagonal matrix with coefficients equal to 2. System (7.60) must be provided with the initial conditions $\mathbf{x}(0) = \mathbf{x}_0$ and $\dot{\mathbf{x}}(0) = \mathbf{v}_0$.

To numerically solve (7.60) let us transform it into a system of differential equations of order 1 in the new variable $\mathbf{y}$, a vector with 6 components. Having set $y_i = x_i$ and $y_{i+3} = \dot{x}_i$ with $i = 1, 2, 3$, and

$$\lambda = \left(m(y_4, y_5, y_6)^T \mathrm{H}(y_4, y_5, y_6) + \nabla\Phi^T \mathbf{F}\right)/|\nabla\Phi|^2,$$

we obtain, for $i = 1, 2, 3$,

$$\begin{aligned}\dot{y}_i &= y_{3+i}, \\ \dot{y}_{3+i} &= \frac{1}{m}\left(F_i - \lambda\frac{\partial\Phi}{\partial y_i}\right).\end{aligned} \tag{7.61}$$

We apply the Euler and Crank-Nicolson methods. Initially it is necessary to define a MATLAB *function* (fvinc in Program 7.7) which yields the expressions of the right-hand terms (7.61). Furthermore, let us suppose that the initial conditions are given by vector y0=[0,1,0,.8,0,1.2] and that the integration interval is tspan=[0,25]. We recall the explicit Euler method in the following way

[t,y]=feuler('fvinc',tspan,y0,nt);

(and analogously for the backward Euler beuler and Crank-Nicolson cranknic methods), where nt is the number of intervals (of constant width) used to discretize the interval [tspan(1),tspan(2)]. In the graphs in Figure 7.17 we report the trajectories obtained with 10000

226   7 Ordinary differential equations

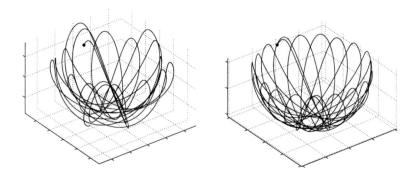

Fig. 7.17. The trajectories obtained with the explicit Euler method with $h = 0.0025$ (*on the left*) and $h = 0.00025$ (*on the right*). The blackened point shows the initial datum

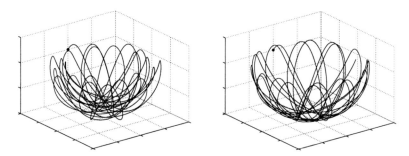

Fig. 7.18. The trajectories obtained using the implicit Euler method with $h = 0.00125$ (*on the left*) and using the Crank-Nicolson method with $h = 0.025$ (*on the right*)

and 100000 discretization nodes. In the second case, the solution looks reasonably accurate. As a matter of fact, although we do not know the exact solution to the problem, we can have an idea of the accuracy by noticing that the solution satisfies $r(\mathbf{y}) \equiv y_1^2 + y_2^2 + y_3^2 - 1 = 0$ and by consequently measuring the maximal value of the residual $r(\mathbf{y}_n)$ when n varies, $\mathbf{y}_n$ being the approximation of the exact solution generated at time t_n. By using 10000 discretization nodes we find $r = 1.0578$, while with 100000 nodes we have $r = 0.1111$, in accordance with the theory requiring the explicit Euler method to converge with order 1.

By using the implicit Euler method with 20000 steps we obtain the solution reported in Figure 7.18, while the Crank-Nicolson method (of order 2) with only 2000 steps provides the solution reported in the same figure on the right, which is undoubtedly more accurate. Indeed, we find $r = 0.5816$ for the implicit Euler method and $r = 0.0966$ for the Crank-Nicolson method.

7.9 Some examples

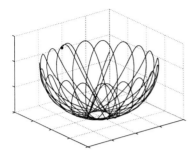 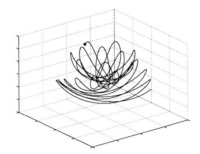

Fig. 7.19. The trajectories obtained using methods ode23 (*left*) and ode45 (*right*) with the same accuracy criteria. In the second case the error control fails and the solution obtained is less accurate

As a comparison, let us solve the same problem using the explicit adaptive methods of type Runge-Kutta ode23 and ode45, featured in MATLAB. These (unless differently specified) modify the integration step in order to guarantee that the relative error on the solution is less than 10^{-3} and the absolute error is less than 10^{-6}. We run them using the following commands

```
[t1,y1]=ode23('fvinc',tspan,y0');
[t2,y2]=ode45('fvinc',tspan,y0');
```

obtaining the solutions in Figure 7.19.

The two methods used 783, respectively 537, non-uniformly distributed discretization nodes. The residual r is equal to 0.0238 for ode23 and 3.2563 for ode45. Surprisingly, the result obtained with the highest-order method is thus less accurate and this warns us as to using the ode programs available in MATLAB. An explanation of this behavior is in the fact that the error estimator implemented in ode45 is less constraining than that in ode23. By slightly decreasing the relative tolerance (it is sufficient to set options=odeset('RelTol',1.e-04)) and renaming the program to [t,y]=ode45(@fvinc,tspan,y0,options); we can in fact find comparable results.

Program 7.7. fvinc: forcing term for the spherical pendulum problem

```
function [f]=fvinc(t,y)
[n,m]=size(y); phix='2*y(1)';
phiy='2*y(2)'; phiz='2*y(3)'; H=2*eye(3);
mass=1;   % Mass
F1='0*y(1)'; F2='0*y(2)'; F3='-mass*9.8'; % Weight
f=zeros(n,m); xpunto=zeros(3,1); xpunto(1:3)=y(4:6);
F=[eval(F1);eval(F2);eval(F3)];
G=[eval(phix);eval(phiy);eval(phiz)];
lambda=(m*xpunto'*H*xpunto+F'*G)/(G'*G);
f(1:3)=y(4:6);
for k=1:3; f(k+3)=(F(k)-lambda*G(k))/mass; end
return
```

228   7 Ordinary differential equations

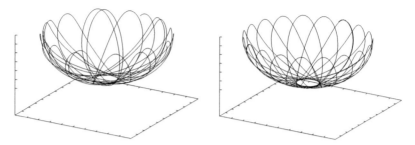

Fig. 7.20. The trajectories obtained using methods ode23 (*left*) and ode45 (*right*) with the same accuracy criteria.

Octave 7.2 ode23 requires 924 steps while ode45 requires 575 steps for the same accuracy.

Note that ode45 gives results similar to ode23 as opposed to ode45 in MATLAB, see Figure 7.20. ∎

7.9.2 The three-body problem

We want to compute the evolution of a system composed by three bodies, knowing their initial positions and velocities and their masses under the influence of their reciprocal gravitational attraction. The problem can be formulated by using Newton's laws of motion. However, as opposed to the case of two bodies, there are no known closed form solutions. We suppose that one of the three bodies has considerably larger mass than the two remaining, and in particular we study the case of the Sun-Earth-Mars system, a problem studied by celeber mathematicians such as Lagrange in the eighteenth century, Poincaré towards the end of the nineteenth century and Levi-Civita in the twentieth century.

We denote by M_s the mass of the Sun, by M_e that of the Earth and by M_m that of Mars. The Sun's mass being about 330000 times that of the Earth and the mass of Mars being about one tenth of the Earth's, we can imagine that the center of gravity of the three bodies approximately coincides with the center of the Sun (which will therefore remain still in this model) and that the three objects remain in the plane described by their initial positions. In such case the total force exerted on the Earth will be for instance

$$\mathbf{F}_e = \mathbf{F}_{es} + \mathbf{F}_{em} = M_e \frac{d^2 \mathbf{x}_e}{dt^2}, \qquad (7.62)$$

where $\mathbf{x}_e = (x_e, y_e)^T$ denote the Earth's position, while $\mathbf{F}_{es}$ and $\mathbf{F}_{em}$ denote the force exerted by the Sun and Mars, respectively, on the Earth.

7.9 Some examples

By applying the universal gravitational law, (7.62) becomes ($\mathbf{x}_m$ denotes the position of Mars)

$$M_e \frac{d^2\mathbf{x}_e}{dt^2} = -GM_eM_s \frac{\mathbf{x}_e}{|\mathbf{x}_e|^3} + GM_eM_m \frac{\mathbf{x}_m - \mathbf{x}_e}{|\mathbf{x}_m - \mathbf{x}_e|^3}.$$

By adimensionalizing the equations and scaling the lengths with respect to the length of the Earth orbit's semi-major axis, the following equation is obtained

$$M_e \frac{d^2\mathbf{x}_e}{dt^2} = 4\pi^2 \left(\frac{M_m}{M_s} \frac{\mathbf{x}_m - \mathbf{x}_e}{|\mathbf{x}_m - \mathbf{x}_e|^3} - \frac{\mathbf{x}_e}{|\mathbf{x}_e|^3} \right). \tag{7.63}$$

The analogous equation for planet Mars can be obtained with a similar computation

$$M_m \frac{d^2\mathbf{x}_m}{dt^2} = 4\pi^2 \left(\frac{M_e}{M_s} \frac{\mathbf{x}_e - \mathbf{x}_m}{|\mathbf{x}_e - \mathbf{x}_m|^3} - \frac{\mathbf{x}_m}{|\mathbf{x}_m|^3} \right). \tag{7.64}$$

The second-order system (7.63)-(7.64) immediately reduces to a system of eight equations of order one. Program 7.8 allows to evaluate a *function* containing the right-hand side terms of system (7.63)-(7.64).

Program 7.8. threebody: forcing term for the simplified three body system

```
function f=threebody(t,y)
f=zeros(8,1);
Ms=330000;
Me=1;
Mm=0.1;
D1 = ((y(5)-y(1))^2+(y(7)-y(3))^2)^(3/2);
D2 = (y(1)^2+y(3)^2)^(3/2);
f(1)=y(2);
f(2)=4*pi^2*(Me/Ms*(y(5)-y(1))/D1-y(1)/D2);
f(3)=y(4);
f(4)=4*pi^2*(Me/Ms*(y(7)-y(3))/D1-y(3)/D2);
D2 = (y(5)^2+y(7)^2)^(3/2);
f(5)=y(6);
f(6)=4*pi^2*(Mm/Ms*(y(1)-y(5))/D1-y(5)/D2);
f(7)=y(8);
f(8)=4*pi^2*(Mm/Ms*(y(3)-y(7))/D1-y(7)/D2);
return
```

Let us compare the Crank-Nicolson method (implicit) and the adaptive Runge-Kutta method implemented in `ode23` (explicit). Having set the Earth to be 1 unit away from the Sun, Mars will be located at about 1.52 units: the initial position will therefore be $(1,0)$ for the Earth and $(1.52,0)$ for Mars. Let us further suppose that the two planets initially have null horizontal velocity and vertical velocity equal to -5.1 units (Earth) and -4.6 units (Mars): this way they should move along reasonably stable orbits around the Sun. For the Crank-Nicolson method we choose 2000 discretization steps.

230   7 Ordinary differential equations

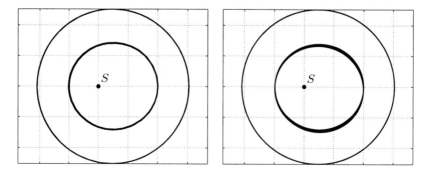

Fig. 7.21. The Earth's (*inmost*) and Mars's orbit with respect to the Sun as computed with the adaptive method ode23 (*on the left*) (with 564 steps) and with the Crank-Nicolson method (*on the right*) (with 2000 steps)

```
[t23,u23]=ode23('threebody',[0 10],...
         [1.52 0 0 -4.6 1 0 0 -5.1]);
[tcn,ucn]=cranknic('threebody',[0 10],...
         [1.52 0 0 -4.6 1 0 0 -5.1],2000);
```

The graphs in Figure 7.21 show that the two methods are both able to reproduce the elliptical orbits of the two planets around the Sun. Method ode23 only required 543 (non-uniform) steps to generate a more accurate solution than that generated by an implicit method with the same order of accuracy, but which does not use step adaptivity.

Octave 7.3 ode23 requires 847 steps to generate a solution with a tolerance of 1e-6.   ∎

7.9.3 Some stiff problems

Let us consider the following differential problem, proposed by [Gea71], as a variant of the model problem (7.28):

$$\begin{cases} y'(t) = \lambda(y(t) - g(t)) + g'(t), & t > 0, \\ y(0) = y_0, \end{cases} \quad (7.65)$$

where g is a regular function and $\lambda \ll 0$, whose solution is

$$y(t) = (y_0 - g(0))e^{\lambda t} + g(t), \quad t \geq 0. \quad (7.66)$$

It has two components, $(y_0 - g(0))e^{\lambda t}$ and $g(t)$, the first being negligible with respect to the second one for t large enough. In particular, we set $g(t) = t$, $\lambda = -100$ and solve problem (7.65) over the interval $(0, 100)$ using the explicit Euler method: since in this case $f(t, y) = \lambda(y(t) - g(t)) + g'(t)$ we have $\partial f / \partial y = \lambda$, and the stability

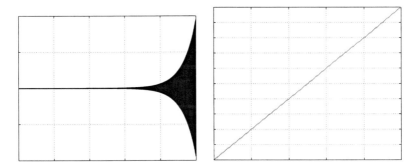

Fig. 7.22. Solutions obtained using method (7.47) for problem (7.65) violating the stability condition ($h = 0.0055$, *left*) and respecting it ($h = 0.0054$, *right*)

analysis performed in Section 7.4 suggests we choose $h < 2/100$. This restriction is dictated by the presence of the component behaving like e^{-100t} and appears completely unjustified when we think of its weight with respect to the whole solution (to get an idea, for $t = 1$ we have $e^{-100} \approx 10^{-44}$). The situation gets worse using a higher order explicit method, such as for instance the Adams-Bashforth (7.47) method of order 3: the absolute stability region reduces (see Figure 7.12) and, consequently, the restriction on h becomes even stricter, $h < 0.00545$. Violating – even slightly – such restriction produces completely unacceptable solutions (as shown in Figure 7.22 on the left).

We thus face an apparently simple problem, but one that becomes difficult to solve with an explicit method (and more generally with a method which is not A-stable) due to the presence in the solution of two components having a dramatically different behavior for t tending to infinity: such a problem is said to be *stiff*.

More precisely, we say that a system of differential equations of the form

$$\mathbf{y}'(t) = A\mathbf{y}(t) + \boldsymbol{\varphi}(t), \qquad A \in \mathbb{R}^{n \times n}, \quad \boldsymbol{\varphi}(t) \in \mathbb{R}^n, \qquad (7.67)$$

where A has n distinct eigenvalues λ_j, $j = 1, \ldots, n$, with $\mathrm{Re}(\lambda_j) < 0$, $j = 1, \ldots, n$, is stiff if

$$r_s = \frac{\max_j |\mathrm{Re}(\lambda_j)|}{\min_j |\mathrm{Re}(\lambda_j)|} \gg 1.$$

The exact solution to (7.67) is

$$\mathbf{y}(t) = \sum_{j=1}^{n} C_j e^{\lambda_j t} \mathbf{v}_j + \boldsymbol{\psi}(t), \qquad (7.68)$$

where $C_1, \ldots, C_n$ are n constants and $\{\mathbf{v}_j\}$ is a base formed by the eigenvectors of A, while $\boldsymbol{\psi}(t)$ is a given solution of the differential equation.

232   7 Ordinary differential equations

If $r_s \gg 1$ we observe once again the presence of components of the solution **y** which tend to zero with different speed. The component which tends to zero fastest for t tending to infinity (the one associated to the eigenvalue having maximal value) will be the one involving the strictest restriction on the integration step, unless of course we use a method which is absolutely stable under any condition.

Example 7.10 Let us consider the system $\mathbf{y}' = A\mathbf{y}$ with $t \in (0, 100)$ with initial condition $\mathbf{y}(0) = \mathbf{y}_0$, where $\mathbf{y} = (y_1, y_2)^T$, $\mathbf{y}_0 = (y_{1,0}, y_{2,0})^T$ and

$$A = \begin{bmatrix} 0 & 1 \\ -\lambda_1\lambda_2 & \lambda_1 + \lambda_2 \end{bmatrix},$$

where λ_1 and λ_2 are two different negative numbers such that $|\lambda_1| \gg |\lambda_2|$. Matrix A has eigenvalues λ_1 and λ_2 and eigenvectors $\mathbf{v}_1 = (1, \lambda_1)^T$, $\mathbf{v}_2 = (1, \lambda_2)^T$. Thanks to (7.68) the system's solution is

$$\mathbf{y}(t) = \begin{pmatrix} C_1 e^{\lambda_1 t} + C_2 e^{\lambda_2 t} \\ C_1 \lambda_1 e^{\lambda_1 t} + C_2 \lambda_2 e^{\lambda_2 t} \end{pmatrix}^T. \tag{7.69}$$

The constants C_1 and C_2 are obtained by fulfilling the initial condition:

$$C_1 = \frac{\lambda_2 y_{1,0} - y_{2,0}}{\lambda_2 - \lambda_1}, \quad C_2 = \frac{y_{2,0} - \lambda_1 y_{1,0}}{\lambda_2 - \lambda_1}.$$

Based on the remarks made earlier, the integration step of an explicit method used for the resolution of such a system will depend uniquely on the eigenvalue having maximal module, λ_1. Let us assess this experimentally using the explicit Euler method and choosing $\lambda_1 = -100$, $\lambda_2 = -1$, $y_{1,0} = y_{2,0} = 1$. In Figure 7.23 we report the solutions computed by violating (*left*) or respecting (*right*) the stability condition $h < 1/50$. ∎

The definition of stiff problem can be extended, by exerting some precautions, to the nonlinear case (see for instance [QSS06, Chapter 11]). One of the most studied nonlinear *stiff* problems is given by the *Van der Pol equation*

$$\frac{d^2 x}{dt^2} = \mu(1 - x^2)\frac{dx}{dt} - x, \tag{7.70}$$

proposed in 1920 and used in the study of circuits containing thermoionic valves, the so-called vacuum tubes, such as cathodic tubes in television sets or magnetrons in microwave ovens.

If we set $\mathbf{y} = (x, y)^T$, (7.70) is equivalent to the following nonlinear first order system

$$\mathbf{y}' = \begin{bmatrix} 0 & 1 \\ -1 & \mu(1 - x^2) \end{bmatrix} \mathbf{y}. \tag{7.71}$$

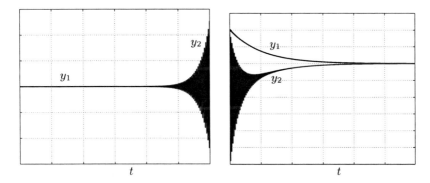

Fig. 7.23. Solutions to the problem in Example 7.10 for $h = 0.0207$ (*left*) and $h = 0.0194$ (*right*). In the first case the condition $h < 2/|\lambda_1| = 0.02$ is violated and the method is unstable. Consider the totally different scale in the two graphs

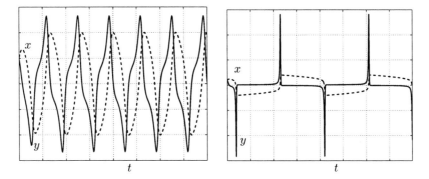

Fig. 7.24. Behavior of the components of the solutions **y** to system (7.71) for $\mu = 1$ (*left*) and $\mu = 10$ (*right*)

Such system becomes increasingly stiff with the increase of the μ parameter. In the solution we find in fact two components which denote totally different dynamics with the increase of μ. The one having the fastest dynamics imposes a limitation on the integration step which gets more and more prohibitive with the increase of μ's value.

If we solve (7.70) using ode23 and ode45, we realize that these are too costly when μ is large. With $\mu = 100$ and initial condition $\mathbf{y} = (1, 1)^T$, ode23 requires 7835 steps and ode45 23473 steps to integrate between $t = 0$ and $t = 100$. Reading the MATLAB *help* we discover that these methods are not recommended for stiff problems: for these, other procedures are suggested, such as for instance the implicit methods ode23s or ode15s. The difference in terms of number of steps is remarkable, as shown in Table 7.1. Notice however that the number of steps for ode23s

μ	ode23	ode45	ode23s	ode15s
0.1	471	509	614	586
1	775	1065	838	975
10	1220	2809	1005	1077
100	7835	23473	299	305
1000	112823	342265	183	220

Table 7.1. Behavior of the number of integration steps for various approximation methods with growing μ parameter

is smaller than that for ode23 only for large enough values of μ (thus for very stiff problems).

7.10 What we haven't told you

For a complete derivation of the whole family of the Runge-Kutta methods we refer to [But87], [Lam91] and [QSS06, Chapter 11].

For derivation and analysis of multistep methods we refer to [Arn73] and [Lam91].

7.11 Exercises

Exercise 7.1 Apply the backward Euler and forward Euler methods for the solution of the Cauchy problem

$$y' = \sin(t) + y, \ t \in (0, 1], \text{ with } y(0) = 0, \qquad (7.72)$$

and verify that both converge with order 1.

Exercise 7.2 Consider the Cauchy problem

$$y' = -te^{-y}, \ t \in (0, 1], \text{ with } y(0) = 0. \qquad (7.73)$$

Apply the forward Euler method with $h = 1/100$ and estimate the number of exact significant digits of the approximate solution at $t = 1$ (use the property that the value of the exact solution is included between -1 and 0).

Exercise 7.3 The backward Euler method applied to problem (7.73) requires at each step the solution of the nonlinear equation: $u_{n+1} = u_n - ht_{n+1}e^{-u_{n+1}} = \phi(u_{n+1})$. The solution u_{n+1} can be obtained by the following fixed-point iteration: for $k = 0, 1, \ldots$, compute $u_{n+1}^{(k+1)} = \phi(u_{n+1}^{(k)})$, with $u_{n+1}^{(0)} = u_n$. Find under which restriction on h these iterations converge.

Exercise 7.4 Repeat Exercise 7.1 for the Crank-Nicolson method.

Exercise 7.5 Verify that the Crank-Nicolson method can be derived from the following integral form of the Cauchy problem (7.5)

$$y(t) - y_0 = \int_{t_0}^{t} f(\tau, y(\tau)) d\tau$$

provided that the integral is approximated by the trapezoidal formula (4.19).

Exercise 7.6 Solve the model problem (7.28) with $\lambda = -1+i$ by the forward Euler method and find the values of h for which we have absolute stability.

Exercise 7.7 Show that the Heun method defined in (7.52) is consistent. Write a MATLAB program to implement it for the solution of the Cauchy problem (7.72) and verify experimentally that the method has order of convergence equal to 2 with respect to h.

Exercise 7.8 Prove that the Heun method (7.52) is absolutely stable if $-2 \leq h\lambda \leq 0$ where λ is real and negative.

Exercise 7.9 Prove formula (7.33).

Exercise 7.10 Prove the inequality (7.38).

Exercise 7.11 Prove the inequality (7.39).

Exercise 7.12 Verify the consistency of the method (7.46). Write a MATLAB program to implement it for the solution of the Cauchy problem (7.72) and verify experimentally that the method has order of convergence equal to 3 with respect to h. The methods (7.52) and (7.46) stand at the base of the MATLAB program ode23 for the solution of ordinary differential equations.

Exercise 7.13 Prove that the method (7.46) is absolutely stable if $-2.5 \leq h\lambda \leq 0$ where λ is real and negative.

Exercise 7.14 The *modified Euler method* is defined as follows:

$$u_{n+1}^* = u_n + hf(t_n, u_n), \quad u_{n+1} = u_n + hf(t_{n+1}, u_{n+1}^*). \quad (7.74)$$

Find under which condition on h this method is absolutely stable.

Exercise 7.15 (Thermodynamics) Solve equation (7.1) by the Crank-Nicolson method and the Heun method when the body in question is a cube with side equal to 1 m and mass equal to 1 Kg. Assume that $T_0 = 180K$, $T_e = 200K$, $\gamma = 0.5$ and $C = 100 J/(Kg/K)$. Compare the results obtained by using $h = 20$ and $h = 10$, for t ranging from 0 to 200 seconds.

Exercise 7.16 Use MATLAB to compute the region of absolute stability of the Heun method.

Exercise 7.17 Solve the Cauchy problem (7.16) by the Heun method and verify its order.

Exercise 7.18 The displacement $x(t)$ of a vibrating system represented by a body of a given weight and a spring, subjected to a resistive force proportional to the velocity, is described by the second-order differential equation $x'' + 5x' + 6x = 0$. Solve it by the Heun method assuming that $x(0) = 1$ and $x'(0) = 0$, for $t \in [0, 5]$.

Exercise 7.19 The motion of a frictionless Foucault pendulum is described by the system of two equations
$$x'' - 2\omega \sin(\Psi) y' + k^2 x = 0, \quad y'' + 2\omega \cos(\Psi) x' + k^2 y = 0,$$
where Ψ is the latitude of the place where the pendulum is located, $\omega = 7.29 \cdot 10^{-5}$ sec^{-1} is the angular velocity of the Earth, $k = \sqrt{g/l}$ with $g = 9.8$ m/sec^2 and l is the length of the pendulum. Apply the forward Euler method to compute $x = x(t)$ and $y = y(t)$ for t ranging between 0 and 300 seconds and $\Psi = \pi/4$.

Exercise 7.20 (Baseball trajectory) Using ode23, solve Problem 7.3 by assuming that the initial velocity of the ball be $\mathbf{v}(0) = v_0 (\cos(\theta), 0, \sin(\theta))^T$, with $v_0 = 38$ m/s, $\theta = 1$ degree and an angular velocity equal to $180 \cdot 1.047198$ radiants per second. If $\mathbf{x}(0) = \mathbf{0}$, after how many seconds (approximately) will the ball touch the ground (i.e., $z = 0$)?

8
Numerical methods for (initial-)boundary-value problems

Boundary-value problems are differential problems set in an interval (a, b) of the real line or in an open multidimensional region $\Omega \subset \mathbb{R}^d$ ($d = 2, 3$) for which the value of the unknown solution (or its derivatives) is prescribed at the end-points a and b of the interval, or on the boundary $\partial \Omega$ of the multidimensional region.

In the multidimensional case the differential equation will involve *partial derivatives* of the exact solution with respect to the space coordinates. Equations depending on time (denoted with t), like the heat equation and the wave equation, are called initial-boundary-value problems. In that case initial conditions at $t = 0$ need to be prescribed as well.

Some examples of boundary-value problems are reported below.

1. *Poisson equation*:
$$-u''(x) = f(x), \; x \in (a, b), \qquad (8.1)$$

or (in several dimensions)
$$-\Delta u(\mathbf{x}) = f(\mathbf{x}), \; \mathbf{x} = (x_1, \ldots, x_d)^T \in \Omega, \qquad (8.2)$$

where f is a given function and Δ is the so-called *Laplace operator*:
$$\Delta u = \sum_{i=1}^{d} \frac{\partial^2 u}{\partial x_i^2}.$$

The symbol $\partial \cdot / \partial x_i$ denotes partial derivative with respect to the x_i variable, that is, for every point $\mathbf{x}^0$
$$\frac{\partial u}{\partial x_i}(\mathbf{x}^0) = \lim_{h \to 0} \frac{u(\mathbf{x}^0 + h \mathbf{e}_i) - u(\mathbf{x}^0)}{h}, \qquad (8.3)$$

where $\mathbf{e}_i$ is i-th unitary vector of $\mathbb{R}^d$.

2. *Heat equation*:
$$\frac{\partial u(x,t)}{\partial t} - \mu \frac{\partial^2 u(x,t)}{\partial x^2} = f(x,t),\ x \in (a,b),\ t > 0, \quad (8.4)$$

or (in several dimensions)
$$\frac{\partial u(\mathbf{x},t)}{\partial t} - \mu \Delta u(\mathbf{x},t) = f(\mathbf{x},t),\ \mathbf{x} \in \Omega,\ t > 0, \quad (8.5)$$

where $\mu > 0$ is a given coefficient representing the thermal conductivity, and f is again a given function.

3. *Wave equation*:
$$\frac{\partial^2 u(x,t)}{\partial t^2} - c\frac{\partial^2 u(x,t)}{\partial x^2} = 0,\ x \in (a,b),\ t > 0,$$

or (in several dimensions)
$$\frac{\partial^2 u(\mathbf{x},t)}{\partial t^2} - c\Delta u(\mathbf{x},t) = 0,\ \mathbf{x} \in \Omega,\ t > 0,$$

where c is a given positive constant.

For more general partial differential equations, the reader is referred for instance to [QV94], [EEHJ96] or [Lan03].

Problem 8.1 (Hydrogeology) The study of filtration in groundwater can lead, in some cases, to an equation like (8.2). Consider a portion Ω occupied by a porous medium (like ground or clay). According to the Darcy law, the water velocity filtration $\mathbf{q} = (q_1, q_2, q_3)^T$ is equal to the variation of the water level ϕ in the medium, precisely

$$\mathbf{q} = -K\nabla\phi, \quad (8.6)$$

where K is the constant hydraulic conductivity of the porous medium and $\nabla\phi$ denotes the spatial gradient of ϕ. Assume that the fluid density is constant; then the mass conservation principle yields the equation div$\mathbf{q} = 0$, where div$\mathbf{q}$ is the *divergence* of the vector $\mathbf{q}$ and is defined as

$$\text{div}\mathbf{q} = \sum_{i=1}^{3} \frac{\partial q_i}{\partial x_i}.$$

Thanks to (8.6) we therefore find that ϕ satisfies the Poisson problem $\Delta\phi = 0$ (see Exercise 8.9). ∎

8 Numerical methods for (initial-)boundary-value problems   239

Problem 8.2 (Thermodynamics) Let $\Omega \subset \mathbb{R}^d$ be a volume occupied by a fluid. Denoting by $\mathbf{J}(\mathbf{x},t)$ and $T(\mathbf{x},t)$ the heat flux and the flow temperature, respectively, the Fourier law states that heat flux is proportional to the variation of the temperature T, that is

$$\mathbf{J}(\mathbf{x},t) = -k\nabla T(\mathbf{x},t),$$

where k is a positive constant expressing the thermal conductivity coefficient. Imposing the conservation of energy, that is, the rate of change of energy of a volume equals the rate at which heat flows into it, we obtain the heat equation

$$\rho c \frac{\partial T}{\partial t} = k \Delta T, \qquad (8.7)$$

where ρ is the mass density of the fluid and c is the specific heat capacity (per unit mass). If, in addition, heat is produced at the rate $f(\mathbf{x},t)$ by some other means (e.g., electrical heating), (8.7) becomes

$$\rho c \frac{\partial T}{\partial t} = k \Delta T + f. \qquad (8.8)$$

For the solution of this problem see Example 8.4.   ■

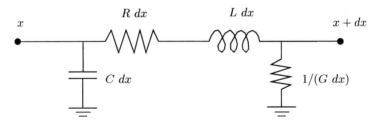

Fig. 8.1. An element of cable of length dx

Problem 8.3 (Communications) We consider a telegraph wire with resistance R and self-inductance L per unit length. Assuming that the current can drain away to ground through a capacitance C and a conductance G per unith length (see Figure 8.1), the equation for the voltage v is

$$\frac{\partial^2 v}{\partial t^2} - c^2 \frac{\partial^2 v}{\partial x^2} = -\alpha \frac{\partial v}{\partial t} - \beta v, \qquad (8.9)$$

where $c^2 = 1/(LC)$, $\alpha = R/L + G/C$ and $\beta = RG/(LC)$. Equation (8.9) is an example of a second order hyperbolic equation. The solution of this problem is given in Example 8.7.   ■

8.1 Approximation of boundary-value problems

The differential equations presented so far feature an infinite number of solutions. With the aim of obtaining a unique solution we must impose suitable conditions on the boundary $\partial \Omega$ of Ω and, for the time-dependent equations, suitable initial conditions at time $t = 0$.

In this section we consider the Poisson equations (8.1) or (8.2). In the one-dimensional case (8.1), to fix the solution one possibility is to prescribe the value of u at $x = a$ and $x = b$, obtaining

$$\begin{aligned} -u''(x) &= f(x) \text{ for } x \in (a,b), \\ u(a) &= \alpha, \quad u(b) = \beta \end{aligned} \qquad (8.10)$$

where α and β are two given real numbers. This is a *Dirichlet boundary-value problem*, and is precisely the problem that we will face in the next section.

Performing double integration it is easily seen that if $f \in C^0([a,b])$, the solution u exists and is unique; moreover it belongs to $C^2([a,b])$.

Although (8.10) is an ordinary differential problem, it cannot be cast in the form of a Cauchy problem for ordinary differential equations since the value of u is prescribed at two different points.

In the two-dimensional case, the Dirichlet boundary-value problem takes the following form: being given two functions $f = f(\mathbf{x})$ and $g = g(\mathbf{x})$, find a function $u = u(\mathbf{x})$ such that

$$\begin{aligned} -\Delta u(\mathbf{x}) &= f(\mathbf{x}) & \text{for } \mathbf{x} \in \Omega, \\ u(\mathbf{x}) &= g(\mathbf{x}) & \text{for } \mathbf{x} \in \partial \Omega \end{aligned} \qquad (8.11)$$

Alternatively to the boundary condition on (8.11), we can prescribe a value for the partial derivative of u with respect to the normal direction to the boundary $\partial \Omega$, in which case we will get a *Neumann boundary-value problem*.

It can be proven that if f and g are two continuous functions and the region Ω is regular enough, then the Dirichlet boundary-value problem (8.11) has a unique solution (while the solution of the Neumann boundary-value problem is unique up to an additive constant).

The numerical methods which are used for its solution are based on the same principles used for the approximation of the one-dimensional boundary-value problem. This is the reason why in Sections 8.1.1 and 8.1.2 we will make a digression on the numerical solution of problem (8.10).

With this aim we introduce on $[a,b]$ a partition into intervals $I_j = [x_j, x_{j+1}]$ for $j = 0, \ldots, N$ with $x_0 = a$ and $x_{N+1} = b$. We assume for simplicity that all intervals have the same length h.

8.1.1 Approximation by finite differences

The differential equation must be satisfied in particular at any point x_j (which we call *nodes* from now on) internal to (a, b), that is

$$-u''(x_j) = f(x_j), \qquad j = 1, \ldots, N.$$

We can approximate this set of N equations by replacing the second derivative with a suitable finite difference as we have done in Chapter 4 for the first derivatives. In particular, we observe that if $u : [a, b] \to \mathbb{R}$ is a sufficiently smooth function in a neighborhood of a generic point $\bar{x} \in (a, b)$, then the quantity

$$\delta^2 u(\bar{x}) = \frac{u(\bar{x} + h) - 2u(\bar{x}) + u(\bar{x} - h)}{h^2} \qquad (8.12)$$

provides an approximation to $u''(\bar{x})$ of order 2 with respect to h (see Exercise 8.3). This suggests the use of the following approximation to problem (8.10): find $\{u_j\}_{j=1}^N$ such that

$$-\frac{u_{j+1} - 2u_j + u_{j-1}}{h^2} = f(x_j), \qquad j = 1, \ldots, N \qquad (8.13)$$

with $u_0 = \alpha$ and $u_{N+1} = \beta$. Equations (8.13) provide a linear system

$$\mathbf{A}\mathbf{u}_h = h^2 \mathbf{f}, \qquad (8.14)$$

where $\mathbf{u}_h = (u_1, \ldots, u_N)^T$ is the vector of unknowns, $\mathbf{f} = (f(x_1) + \alpha/h^2, f(x_2), \ldots, f(x_{N-1}), f(x_N) + \beta/h^2)^T$, and A is the tridiagonal matrix

$$A = \text{tridiag}(-1, 2, -1) = \begin{bmatrix} 2 & -1 & 0 & \cdots & 0 \\ -1 & 2 & \ddots & & \vdots \\ 0 & \ddots & \ddots & -1 & 0 \\ \vdots & & -1 & 2 & -1 \\ 0 & \cdots & 0 & -1 & 2 \end{bmatrix}. \qquad (8.15)$$

This system admits a unique solution since A is symmetric and positive definite (see Exercise 8.1). Moreover, it can be solved by the Thomas algorithm introduced in Section 5.4. We note however that, for small values of h (and thus for large values of N), A is ill-conditioned. Indeed, $K(A) = \lambda_{max}(A)/\lambda_{min}(A) = Ch^{-2}$, for a suitable constant C independent of h (see Exercise 8.2). Consequently, the numerical solution of system (8.14), by either direct or iterative methods, requires special care. In particular, when using iterative methods a suitable preconditioner ought to be employed.

It is possible to prove (see, e.g., [QSS06, Chapter 12]) that if $f \in C^2([a,b])$ then

$$\max_{j=0,\ldots,N+1} |u(x_j) - u_j| \leq \frac{h^2}{96} \max_{x \in [a,b]} |f''(x)| \qquad (8.16)$$

that is, the finite difference method (8.13) converges with order two with respect to h.

In Program 8.1 we solve the boundary-value problem

$$\begin{cases} -u''(x) + \delta u'(x) + \gamma u(x) = f(x) \text{ for } x \in (a,b), \\ u(a) = \alpha \qquad\qquad\qquad u(b) = \beta, \end{cases} \qquad (8.17)$$

which is a generalization of problem (8.10). For this problem the finite difference method, which generalizes (8.13), reads:

$$\begin{cases} -\dfrac{u_{j+1} - 2u_j + u_{j-1}}{h^2} + \delta \dfrac{u_{j+1} - u_{j-1}}{2h} + \gamma u_j = f(x_j), \, j = 1, \ldots, N, \\ u_0 = \alpha, \qquad\qquad\qquad\qquad\qquad\qquad u_{N+1} = \beta. \end{cases}$$

The input parameters of Program 8.1 are the end-points a and b of the interval, the number N of internal nodes, the constant coefficients δ and γ and the function bvpfun specifying the function f. Finally, ua and ub represent the values that the solution should attain at x=a and x=b, respectively. Output parameters are the vector of nodes x and the computed solution uh. Notice that the solutions can be affected by spurious oscillations if $h \geq 2/|\delta|$ (see Exercise 8.6).

Program 8.1. bvp: approximation of a two-point boundary-value problem by the finite difference method

```
function [x,uh]=bvp(a,b,N,delta,gamma,bvpfun,ua,ub,...
                    varargin)
%BVP Solve two-point boundary value problems.
%   [X,UH]=BVP(A,B,N,DELTA,GAMMA,BVPFUN,UA,UB) solves
%   with the centered finite difference method the
%   boundary-value problem
%            -D(DU/DX)/DX+DELTA*DU/DX+GAMMA*U=BVPFUN
%   on the interval (A,B) with boundary conditions
%   U(A)=UA and U(B)=UB. BVPFUN can be an inline
%   function.
h = (b-a)/(N+1);
z = linspace(a,b,N+2);
e = ones(N,1);
h2 = 0.5*h*delta;
A = spdiags([-e-h2 2*e+gamma*h^2 -e+h2],-1:1,N,N);
x = z(2:end-1);
f = h^2*feval(bvpfun,x,varargin{:});
f=f';     f(1) = f(1) + ua;    f(end) = f(end) + ub;
uh = A\f;
uh=[ua; uh; ub];
x = z;
```

8.1.2 Approximation by finite elements

The *finite element method* represents an alternative to the finite difference method and is derived from a suitable reformulation of the differential problem.

Let us consider again (8.10) and multiply both sides of the differential equation by a generic function $v \in C^1([a, b])$. Integrating the corresponding equality on the interval (a, b) and using integration by parts we obtain

$$\int_a^b u'(x)v'(x)\,dx - [u'(x)v(x)]_a^b = \int_a^b f(x)v(x)\,dx.$$

By making the further assumption that v vanishes at the end-points $x = a$ and $x = b$, problem (8.10) becomes: find $u \in C^1([a, b])$. such that $u(a) = \alpha$, $u(b) = \beta$ and

$$\int_a^b u'(x)v'(x)\,dx = \int_a^b f(x)v(x)\,dx \qquad (8.18)$$

for each $v \in C^1([a, b])$ such that $v(a) = v(b) = 0$. This is called *weak formulation* of problem (8.10). (Indeed, both u and the test function v can be less regular than $C^1([a, b])$, see, e.g. [QSS06], [QV94].)

Its finite element approximation is defined as follows:

$$\boxed{\text{find } u_h \in V_h \text{ such that } u_h(a) = \alpha, u_h(b) = \beta \text{ and} \\ \sum_{j=0}^N \int_{x_j}^{x_{j+1}} u_h'(x)v_h'(x)\,dx = \int_a^b f(x)v_h(x)\,dx, \qquad \forall v_h \in V_h^0} \qquad (8.19)$$

where

$$V_h = \left\{ v_h \in C^0([a, b]) : v_h|_{I_j} \in \mathbb{P}_1, j = 0, \ldots, N \right\},$$

i.e. V_h is the space of continuous functions on (a, b) whose restrictions on every sub-interval I_j are linear polynomials. Moreover, V_h^0 is the subspace of V_h of those functions vanishing at the end-points a and b. V_h is called space of finite elements of degree 1.

The functions in V_h^0 are piecewise linear polynomials (see Figure 8.2, left). In particular, every function v_h of V_h^0 admits the representation

$$v_h(x) = \sum_{j=1}^N v_h(x_j)\varphi_j(x),$$

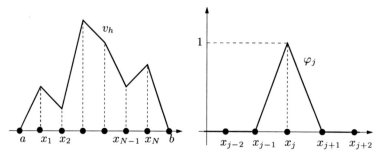

Fig. 8.2. To the left, a generic function $v_h \in V_h^0$. To the right, the basis function of V_h^0 associated with the k-th node

where for $j = 1, \ldots, N$,

$$\varphi_j(x) = \begin{cases} \dfrac{x - x_{j-1}}{x_j - x_{j-1}} & \text{if } x \in I_{j-1}, \\ \dfrac{x - x_{j+1}}{x_j - x_{j+1}} & \text{if } x \in I_j, \\ 0 & \text{otherwise.} \end{cases}$$

Thus, φ_j is null at every node x_i except at x_j where $\varphi_j(x_j) = 1$ (see Figure 8.2, right). The functions φ_j, $j = 1, \ldots, N$ are called *shape functions* and provide a basis for the vector space V_h^0.

Consequently, we can limit ourselves to fulfill (8.19) only for the shape functions φ_j, $j = 1, \ldots, N$. By exploiting the fact that φ_j vanishes outside the intervals I_{j-1} and I_j, from (8.19) we obtain

$$\int_{I_{j-1} \cup I_j} u_h'(x) \varphi_j'(x)\, dx = \int_{I_{j-1} \cup I_j} f(x) \varphi_j(x)\, dx, \quad j = 1, \ldots, N. \quad (8.20)$$

On the other hand, we can write $u_h(x) = \sum_{j=1}^{N} u_j \varphi_j(x) + \alpha \varphi_0(x) + \beta \varphi_{N+1}(x)$, where $u_j = u_h(x_j)$, $\varphi_0(x) = (a + h - x)/h$ for $a \leq x \leq a + h$, and $\varphi_{N+1}(x) = (x - b + h)/h$ for $b - h \leq x \leq b$, while both $\varphi_0(x)$ and $\varphi_{N+1}(x)$ are zero otherwise. By substituting this expression in (8.20), we find that for all $j = 1, \ldots, N$

$$u_{j-1} \int_{I_{j-1}} \varphi_{j-1}'(x) \varphi_j'(x)\, dx + u_j \int_{I_{j-1} \cup I_j} \varphi_j'(x) \varphi_j'(x)\, dx$$
$$+ u_{j+1} \int_{I_j} \varphi_{j+1}'(x) \varphi_j'(x)\, dx = \int_{I_{j-1} \cup I_j} f(x) \varphi_j(x)\, dx + B_{1,j} + B_{N,j},$$

where

8.1 Approximation of boundary-value problems

$$B_{1,j} = \begin{cases} -\alpha \int_{I_0} \varphi'_0(x)\varphi'_1(x) \, dx = -\dfrac{\alpha}{x_1 - a} & \text{if } j = 1, \\ 0 \text{ otherwise,} \end{cases}$$

while

$$B_{N,j} = \begin{cases} -\beta \int_{I_N} \varphi'_{N+1}(x)\varphi'_j(x) \, dx = -\dfrac{\beta}{b - x_N} & \text{if } j = N, \\ 0 \text{ otherwise.} \end{cases}$$

In the special case where all intervals have the same length h, then $\varphi'_{j-1} = -1/h$ in I_{j-1}, $\varphi'_j = 1/h$ in I_{j-1} and $\varphi'_j = -1/h$ in I_j, $\varphi'_{j+1} = 1/h$ in I_j. Consequently, we obtain for $j = 1, \ldots, N$

$$-u_{j-1} + 2u_j - u_{j+1} = h \int_{I_{j-1} \cup I_j} f(x)\varphi_j(x) \, dx + B_{1,j} + B_{N,j}.$$

This linear system has the same matrix as the finite difference system (8.14), but a different right-hand side (and a different solution too, in spite of coincidence of notation). Finite difference and finite element solutions share however the same accuracy with respect to h when the nodal maximum error is computed.

Obviously the finite element approach can be generalized to problems like (8.17) (also in the case when δ and γ depend on x). A further generalization consists of using piecewise polynomials of degree greater than 1, allowing the achievement of higher convergence orders. In these cases, the finite element matrix does not coincide anymore with that of finite differences, and the convergence order is greater than when using piecewise linear polynomials.

See Exercises 8.1-8.8.

8.1.3 Approximation by finite differences of two-dimensional problems

Let us consider a partial differential equation, for instance equation (8.2), in a two-dimensional region Ω.

The idea behind finite differences relies on approximating the partial derivatives that are present in the PDE again by incremental ratios computed on a suitable grid (called the computational grid) made of a finite number of nodes. Then the solution u of the PDE will be approximated only at these nodes.

The first step therefore consists of introducing a computational grid. Assume for simplicity that Ω is the rectangle $(a, b) \times (c, d)$. Let us introduce a partition of $[a, b]$ in subintervals (x_k, x_{k+1}) for $k = 0, \ldots, N_x$,

8 Numerical methods for (initial-)boundary-value problems

with $x_0 = a$ and $x_{N_x+1} = b$. Let us denote by $\Delta_x = \{x_0, \ldots, x_{N_x+1}\}$ the set of end-points of such intervals and by $h_x = \max_{k=0,\ldots,N_x}(x_{k+1} - x_k)$ their maximum length.

In a similar manner we introduce a discretization of the y-axis $\Delta_y = \{y_0, \ldots, y_{N_y+1}\}$ with $y_0 = c$ and $y_{N_y+1} = d$. The cartesian product $\Delta_h = \Delta_x \times \Delta_y$ provides the computational grid on Ω (see Figure 8.3), and $h = \max\{h_x, h_y\}$ is a characteristic measure of the grid-size. We are looking for values $u_{i,j}$ which approximate $u(x_i, y_j)$. We will assume for the sake of simplicity that the nodes be uniformly spaced, that is, $x_i = x_0 + i h_x$ for $i = 0, \ldots, N_x+1$ and $y_j = y_0 + j h_y$ for $j = 0, \ldots, N_y+1$.

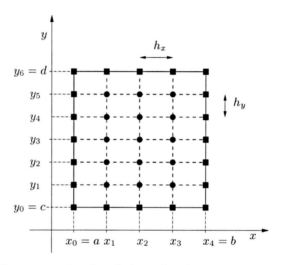

Fig. 8.3. The computational grid Δ_h with only 15 internal nodes on a rectangular domain

The second order partial derivatives of a function can be approximated by a suitable incremental ratio, as we did for ordinary derivatives. In the case of a function of two variables, we define the following incremental ratios:

$$\delta_x^2 u_{i,j} = \frac{u_{i-1,j} - 2u_{i,j} + u_{i+1,j}}{h_x^2},$$

$$\delta_y^2 u_{i,j} = \frac{u_{i,j-1} - 2u_{i,j} + u_{i,j+1}}{h_y^2}. \quad (8.21)$$

They are second order accurate with respect to h_x and h_y, respectively, for the approximation of $\partial^2 u/\partial x^2$ and $\partial^2 u/\partial y^2$ at the node (x_i, y_j). If we replace the second order partial derivatives of u with the formula

8.1 Approximation of boundary-value problems

(8.21), by requiring that the PDE is satisfied at all internal nodes of Δ_h, we obtain the following set of equations:

$$-(\delta_x^2 u_{i,j} + \delta_y^2 u_{i,j}) = f_{i,j}, \quad i = 1, \ldots, N_x, \ j = 1, \ldots, N_y. \quad (8.22)$$

We have set $f_{i,j} = f(x_i, y_j)$. We must add the equations that enforce the Dirichlet data at the boundary, which are

$$u_{i,j} = g_{i,j} \ \forall i, j \text{ such that } (x_i, y_j) \in \partial \Delta_h, \quad (8.23)$$

where $\partial \Delta_h$ indicates the set of nodes belonging to the boundary $\partial \Omega$ of Ω. These nodes are indicated by small squares in Figure 8.3. If we make the further assumption that the computational grid is uniform in both cartesian directions, that is, $h_x = h_y = h$, instead of (8.22) we obtain

$$-\frac{1}{h^2}(u_{i-1,j} + u_{i,j-1} - 4u_{i,j} + u_{i,j+1} + u_{i+1,j}) = f_{i,j},$$
$$i = 1, \ldots, N_x, \ j = 1, \ldots, N_y \quad (8.24)$$

The system given by equations (8.24) (or (8.22)) and (8.23) allows the computation of the nodal values $u_{i,j}$ at all nodes of Δ_h. For every fixed pair of indices i and j, equation (8.24) involves five unknown nodal values as we can see in Figure 8.4. For that reason this finite difference scheme is called *the five-point scheme* for the Laplace operator. We note that the unknowns associated with the boundary nodes can be eliminated using (8.23) (or (8.22)), and therefore (8.24) involves only $N = N_x N_y$ unknowns.

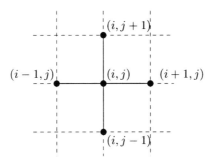

Fig. 8.4. The stencil of the five point scheme for the Laplace operator

The resulting system can be written in a more interesting form if we adopt the *lexicographic* order according to which the nodes (and, correspondingly, the unknown components) are numbered by proceeding from left to right, from the top to the bottom. We obtain a system of the

248   8 Numerical methods for (initial-)boundary-value problems

form (8.14), with a matrix $A \in \mathbb{R}^{N \times N}$ which takes the following block tridiagonal form:

$$A = \text{tridiag}(D, T, D). \tag{8.25}$$

There are N_y rows and N_y columns, and every entry (denoted by a capital letter) consists of a $N_x \times N_x$ matrix. In particular, $D \in \mathbb{R}^{N_x \times N_x}$ is a diagonal matrix whose diagonal entries are $-1/h_y^2$, while $T \in \mathbb{R}^{N_x \times N_x}$ is a symmetric tridiagonal matrix

$$T = \text{tridiag}(-\frac{1}{h_x^2}, \frac{2}{h_x^2} + \frac{2}{h_y^2}, -\frac{1}{h_x^2}).$$

A is symmetric since all diagonal blocks are symmetric. It is also positive definite, that is $\mathbf{v}^T A \mathbf{v} > 0 \; \forall \mathbf{v} \in \mathbb{R}^N, \mathbf{v} \neq \mathbf{0}$. Actually, by partitioning $\mathbf{v}$ in N_y vectors $\mathbf{v}_i$ of length N_x we obtain

$$\mathbf{v}^T A \mathbf{v} = \sum_{k=1}^{N_y} \mathbf{v}_k^T T \mathbf{v}_k - \frac{2}{h_y^2} \sum_{k=1}^{N_y-1} \mathbf{v}_k^T \mathbf{v}_{k+1}. \tag{8.26}$$

We can write $T = 2/h_y^2 I + 1/h_x^2 K$ where K is the (symmetric and positive definite) matrix given in (8.15). Consequently, (8.26) becomes

$$(\mathbf{v}_1^T K \mathbf{v}_1 + \mathbf{v}_2^T K \mathbf{v}_2 + \ldots + \mathbf{v}_{N_y}^T K \mathbf{v}_{N_y})/h_x^2$$

which is a strictly positive real number since K is positive definite and at least one vector $\mathbf{v}_i$ is non-null.

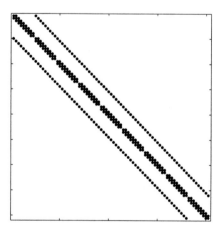

Fig. 8.5. Pattern of the matrix associated with the five-point scheme using the lexicographic ordering of the unknowns

8.1 Approximation of boundary-value problems

Having proven that A is non-singular we can conclude that the finite difference system admits a unique solution $\mathbf{u}_h$.

The matrix A is *sparse*; as such, it will be stored in the format sparse of MATLAB (see Section 5.4). In Figure 8.5 (obtained by using the command spy(A)) we report the structure of the matrix corresponding to a uniform grid of 11×11 nodes, after having eliminated the rows and columns associated to the nodes of $\partial \Delta_h$. It can be noted that the only nonzero elements lie on five diagonals.

Since A is symmetric and positive definite, the associated system can be solved efficiently by either direct or iterative methods, as illustrated in Chapter 5. Finally, it is worth pointing out that A shares with its one-dimensional analog the property of being ill-conditioned: indeed, its condition number grows like h^{-2} as h tends to zero, where $h = \max(h_x, h_y)$.

In the Program 8.2 we construct and solve the system (8.22)-(8.23) (using the command \, see Section 5.6). The input parameters a, b, c and d denote the corners of the rectangular domain $\Omega = (a,c) \times (b,d)$, while nx and ny denote the values of N_x and N_y (the case $N_x \neq N_y$ is admitted). Finally, the two strings fun and bound represent the right-hand side $f = f(x,y)$ (otherwise called the source term) and the boundary data $g = g(x,y)$. The output is a two-dimensional array u whose i,j-th entry is the nodal value $u_{i,j}$. The numerical solution can be visualized by the command mesh(x,y,u). The (optional) string uex represents the exact solution of the original problem for those cases (of theoretical interest) where this solution is known. In such cases the output parameter error contains the nodal relative error between the exact and numerical solution, which is computed as follows:

$$\texttt{error} = \max_{i,j}|u(x_i,y_j) - u_{i,j}|/\max_{i,j}|u(x_i,y_j)|.$$

Program 8.2. poissonfd: approximation of the Poisson problem with Dirichlet data by the five-point finite difference method

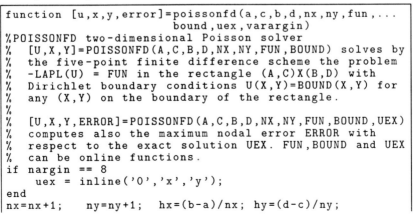

```
nx1=nx+1;    hx2=hx^2;  hy2=hy^2;
kii=2/hx2+2/hy2;         kix=-1/hx2;  kiy=-1/hy2;
dim=(nx+1)*(ny+1);       K=speye(dim,dim);
rhs=zeros(dim,1);
y = c;
for m = 2:ny
    x = a; y = y + hy;
    for n = 2:nx
        i = n+(m-1)*(nx+1);
        x = x + hx;
        rhs(i) = feval(fun,x,y,varargin{:});
        K(i,i) = kii;              K(i,i-1) = kix;
        K(i,i+1) = kix;            K(i,i+nx1) = kiy;
        K(i,i-nx1) = kiy;
    end
end
rhs1 = zeros(dim,1);
x = [a:hx:b];
rhs1(1:nx1) = feval(bound,x,c,varargin{:});
rhs1(dim-nx:dim) = feval(bound,x,d,varargin{:});
y = [c:hy:d];
rhs1(1:nx1:dim-nx) = feval(bound,a,y,varargin{:});
rhs1(nx1:nx1:dim) = feval(bound,b,y,varargin{:});
rhs = rhs - K*rhs1;
nbound = [[1:nx1],[dim-nx:dim],...
          [1:nx1:dim-nx],[nx1:nx1:dim]];
ninternal = setdiff([1:dim],nbound);
K = K(ninternal,ninternal);
rhs = rhs(ninternal);
utemp = K\rhs;
uh = rhs1;
uh(ninternal) = utemp;
k = 1; y = c;
for j = 1:ny+1
    x = a;
    for i = 1:nx1
        u(i,j) = uh(k);
        k = k + 1;
        ue(i,j) = feval(uex,x,y,varargin{:});
        x = x + hx;
    end
    y = y + hy;
end
x = [a:hx:b];
y = [c:hy:d];
if nargout == 4
  if nargin == 8
     warning('Exact solution not available');
     error = [ ];
  else
     error = max(max(abs(u-ue)))/max(max(abs(ue)));
  end
end
return
```

Example 8.1 The transverse displacement u of an elastic membrane from a reference plane $\Omega = (0,1)^2$ under a load whose intensity is $f(x,y) = 8\pi^2 \sin(2\pi x) \cos(2\pi y)$ satisfies a Poisson problem like (8.2) in the domain Ω.

8.1 Approximation of boundary-value problems 251

The Dirichlet value of the displacement is prescribed on $\partial\Omega$ as follows: $g = 0$ on the sides $x = 0$ and $x = 1$, and $g(x,0) = g(x,1) = \sin(2\pi x)$, $0 < x < 1$. This problem admits the exact solution $u(x,y) = \sin(2\pi x)\cos(2\pi y)$. In Figure 8.6 we show the numerical solution obtained by the five-point finite difference scheme on a uniform grid. Two different values of h have been used: $h = 1/10$ (*left*) and $h = 1/20$ (*right*). When h decreases the numerical solution improves, and actually the nodal relative error is 0.0292 for $h = 1/10$ and 0.0081 for $h = 1/20$. ∎

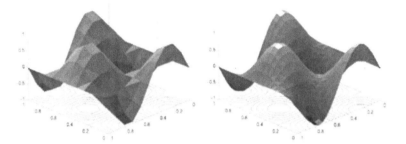

Fig. 8.6. Transverse displacement of an elastic membrane computed on two uniform grids. On the horizontal plane we report the isolines of the numerical solution. The triangular partition of Ω only serves the purpose of the visualization of the results

Also the finite element method can be easily extended to the two-dimensional case. To this end the problem (8.2) must be reformulated in an integral form and the partition of the interval (a,b) in one dimension must be replaced by a decomposition of Ω by polygons (typically, triangles) called *elements*. The shape function φ_k will still be a continuous function, whose restriction on each element is a polynomial of degree 1 on each element, which is equal to 1 at the k-th vertex (or node) of the triangulation and 0 at all other vertices. For its implementation one can use the MATLAB toolbox pde.

pde

8.1.4 Consistency and convergence

In the previous section we have shown that the solution of the finite difference problem exists and is unique. Now we investigate the approximation error. We will assume for simplicity that $h_x = h_y = h$. If

$$\max_{i,j}|u(x_i,y_j) - u_{i,j}| \to 0 \text{ as } h \to 0 \qquad (8.27)$$

the method is called convergent.

As we have already pointed out, consistency is a necessary condition for convergence. A method is *consistent* if the residual that is obtained when the exact solution is plugged into the numerical scheme tends to zero when h tends to zero. If we consider the five point finite difference scheme, at every internal node (x_i, y_j) of Δ_h we define

$$\tau_h(x_i, y_j) = -f(x_i, y_j)$$

$$-\frac{1}{h^2}\left[u(x_{i-1}, y_j) + u(x_i, y_{j-1}) - 4u(x_i, y_j) + u(x_i, y_{j+1}) + u(x_{i+1}, y_j)\right].$$

This is the *local truncation error* at the node (x_i, y_j). By (8.2) we obtain

$$\tau_h(x_i, y_j) = \left\{\frac{\partial^2 u}{\partial x^2}(x_i, y_j) - \frac{u(x_{i-1}, y_j) - 2u(x_i, y_j) + u(x_{i+1}, y_j)}{h^2}\right\}$$
$$+ \left\{\frac{\partial^2 u}{\partial y^2}(x_i, y_j) - \frac{u(x_i, y_{j-1}) - 2u(x_i, y_j) + u(x_i, y_{j+1})}{h^2}\right\}.$$

Thanks to the analysis that was carried out in Section 8.1.3 we can conclude that both terms vanish as h tends to 0. Thus

$$\lim_{h \to 0} \tau_h(x_i, y_j) = 0, \ \forall (x_i, y_j) \in \Delta_h \setminus \partial \Delta_h,$$

that is, the five-point method is consistent. It is also convergent, as stated in the following Proposition (for its proof, see, e.g., [IK66]):

Proposition 8.1 *Assume that the exact solution $u \in C^4(\bar{\Omega})$, i.e. all its partial derivatives up to the fourth order are continuous in the closed domain $\bar{\Omega}$. Then there exists a constant $C > 0$ such that*

$$\max_{i,j}|u(x_i, y_j) - u_{i,j}| \leq CMh^2 \tag{8.28}$$

where M is the maximum absolute value attained by the fourth order derivatives of u in $\bar{\Omega}$.

Example 8.2 Let us verify that the five-point scheme applied to solve the Poisson problem of Example 8.1 converges with order two with respect to h. We start from $h = 1/4$ and, then we halve subsequently the value of h, until $h = 1/64$, through the following instructions:

```
a=0;b=1;c=0;d=1;
f=inline('8*pi^2*sin(2*pi*x).*cos(2*pi*y)','x','y');
g=inline('sin(2*pi*x).*cos(2*pi*y)','x','y');
uex=g; nx=4; ny=4;
for n=1:5
   [u,x,y,error(n)]=poissonfd(a,c,b,d,nx,ny,f,g,uex);
   nx = 2*nx; ny = 2*ny;
end
```

The vector containing the error is
```
format short e; error
```

```
    1.3565e-01   4.3393e-02   1.2308e-02   3.2775e-03   8.4557e-04
```

As we can verify using the following commands
```
p=log(abs(error(1:end-1)./error(2:end)))/log(2)
```

```
    1.6443e+00   1.8179e+00   1.9089e+00   1.9546e+00
```

this error decreases as h^2 when $h \to 0$. ∎

Let us summarize

1. Boundary-value problems are differential equations set in a spatial domain $\Omega \subset \mathbb{R}^d$ (which is an interval if $d = 1$) that require information on the solution on the domain boundary;
2. finite difference approximations are based on the discretization of the given differential equation at selected points (called nodes) where derivatives are replaced by finite difference formulae;
3. the finite difference method provides a nodal vector whose components converge to the corresponding nodal values of the exact solution quadratically with respect to the grid-size;
4. the finite element method is based on a suitable integral reformulation of the original differential equation, then on the assumption that the approximate solution is a piecewise polynomial;
5. matrices arising from both finite difference and finite element approximations are sparse and ill-conditioned.

8.2 Finite difference approximation of the heat equation

We consider the one-dimensional heat equation (8.4) with homogeneous Dirichlet boundary conditions $u(a,t) = u(b,t) = 0$ for any $t > 0$ and initial condition $u(x,0) = u_0(x)$ for $x \in [a,b]$.

To solve this equation numerically we have to discretize both the x and t variables. We can start by dealing with the x-variable, following the same approach as in Section 8.1.1. We denote by $u_j(t)$ an approximation of $u(x_j,t)$, $j = 0, \ldots, N$, and approximate the Dirichlet problem (8.4) by the scheme: for all $t > 0$

$$\frac{du_j}{dt}(t) - \frac{\mu}{h^2}(u_{j-1}(t) - 2u_j(t) + u_{j+1}(t)) = f_j(t), \quad j = 1, \ldots, N-1,$$
$$u_0(t) = u_N(t) = 0,$$

where $f_j(t) = f(x_j, t)$ and, for $t = 0$,

$$u_j(0) = u_0(x_j), \qquad j = 0, \ldots, N.$$

This is actually a *semi-discretization* of the heat equation, yielding a system of ordinary differential equations of the following form

$$\begin{cases} \dfrac{d\mathbf{u}}{dt}(t) = -\dfrac{\mu}{h^2} A \mathbf{u}(t) + \mathbf{f}(t), \ \forall t > 0, \\ \mathbf{u}(0) = \mathbf{u}_0, \end{cases} \qquad (8.29)$$

where $\mathbf{u}(t) = (u_1(t), \ldots, u_{N-1}(t))^T$ is the vector of unknowns, $\mathbf{f}(t) = (f_1(t), \ldots, f_{N-1}(t))^T$, $\mathbf{u}_0 = (u_0(x_1), \ldots, u_0(x_{N-1}))^T$ and A is the tridiagonal matrix introduced in (8.15). Note that for the derivation of (8.29) we have assumed that $u_0(x_0) = u_0(x_N) = 0$, which is coherent with the homogeneous Dirichlet boundary conditions.

A popular scheme for the integration of (8.29) with respect to time is the so-called θ–*method*. Let $\Delta t > 0$ be a constant time-step, and denote by v^k the value of a variable v referred at the time level $t^k = k\Delta t$. Then the θ-method reads

$$\dfrac{\mathbf{u}^{k+1} - \mathbf{u}^k}{\Delta t} = -\dfrac{\mu}{h^2} A(\theta \mathbf{u}^{k+1} + (1-\theta)\mathbf{u}^k) + \theta \mathbf{f}^{k+1} + (1-\theta)\mathbf{f}^k,$$

$$k = 0, 1, \ldots$$

$$\mathbf{u}^0 = \mathbf{u}_0$$

(8.30)

or, equivalently,

$$\left(I + \dfrac{\mu}{h^2} \theta \Delta t A\right) \mathbf{u}^{k+1} = \left(I - \dfrac{\mu}{h^2} \Delta t (1-\theta) A\right) \mathbf{u}^k + \mathbf{g}^{k+1}, \qquad (8.31)$$

where $\mathbf{g}^{k+1} = \Delta t(\theta \mathbf{f}^{k+1} + (1-\theta)\mathbf{f}^k)$ and I is the identity matrix of order $N - 1$.

For suitable values of the parameter θ, from (8.31) we can recover some familiar methods that have been introduced in Chapter 7. For example, if $\theta = 0$ the method (8.31) coincides with the forward Euler scheme and we can obtain $\mathbf{u}^{k+1}$ explicitly; otherwise, a linear system (with constant matrix $I + \mu \theta \Delta t A / h^2$) needs to be solved at each time-step.

Regarding stability, when $f = 0$ the exact solution $u(x,t)$ tends to zero for every x as $t \to \infty$. Then we would expect the discrete solution to have the same behaviour, in which case we would call our scheme (8.31) *asymptotically stable*, this being coherent with what we did in Section 7.5 for ordinary differential equations.

If $\theta = 0$, from (8.31) it follows that

8.2 Finite difference approximation of the heat equation

$$\mathbf{u}^k = (I - \mu \Delta t A/h^2)^k \mathbf{u}^0, \qquad k = 1, 2, \ldots$$

whence $\mathbf{u}^k \to \mathbf{0}$ as $k \to \infty$ iff

$$\rho(I - \mu \Delta t A/h^2) < 1. \tag{8.32}$$

On the other hand, the eigenvalues λ_j of A are given by (see Exercise 8.2) $\lambda_j = 2 - 2\cos(j\pi/N)$, $j = 1, \ldots, N-1$. Then (8.32) is satisfied iff

$$\Delta t < \frac{1}{2\mu} h^2.$$

As expected, the forward Euler method is conditionally stable, and the time-step Δt should decay as the square of the grid spacing h.

In the case of the backward Euler method ($\theta = 1$), we would have from (8.31)

$$\mathbf{u}^k = \left[(I + \mu \Delta t A/h^2)^{-1}\right]^k \mathbf{u}^0, \qquad k = 1, 2, \ldots$$

Since all the eigenvalues of the matrix $(I + \mu \Delta t A/h^2)^{-1}$ are real, positive and strictly less than 1 for every value of Δt, this scheme is unconditionally stable. More generally, the θ-scheme is unconditionally stable for all the values $1/2 \le \theta \le 1$, and conditionally stable if $0 \le \theta < 1/2$ (see, for instance, [QSS06, Chapter 13]).

As far as the accuracy of the θ-method is concerned, its local truncation error is of the order of $\Delta t + h^2$ if $\theta \ne \frac{1}{2}$ while it is of the order of $\Delta t^2 + h^2$ if $\theta = \frac{1}{2}$. The latter is the *Crank-Nicolson method* (see Section 7.3) and is therefore unconditionally stable and second-order accurate with respect to both Δt and h.

The same conclusions hold for the heat equation in a two-dimensional domain. In this case in the scheme (8.30) one must substitute to the matrix A/h^2 the finite difference matrix defined in (8.25).

Program 8.3 solves numerically the heat equation on the time interval $(0, T)$ and on the square domain $\Omega = (a, b) \times (c, d)$ using the θ-method. The input parameters are the vector xspan=[a,b], yspan=[c,d] and tspan=[0,T], the number of discretization intervals in space (nstep(1)) and in time (nstep(2)), the string fun which contains the function $f(t, x_1(t), x_2(t))$, g which contains the Dirichlet function and u0 that defines the initial function $u_0(x_1, x_2)$. Finally, the real number theta is the coefficient θ.

Program 8.3. heattheta: θ-method for the heat equation in a square domain

```
function [x,u]=heattheta(xspan,tspan,nstep,theta,mu,...
              u0,g,f,varargin)
%HEATTHETA  solve the heat equation with the
%    theta-method.
%    [X,U]=HEATTHETA(XSPAN,TSPAN,NSTEP,THETA,MU,U0,G,F)
%    solve the heat equation D U/DT - MU D^2U/DX^2 = F in
```

```
%   (XSPAN(1),XSPAN(2)) X (TSPAN(1),TSPAN(2)) using the
%   theta-method with initial condition U(X,0)=U0(X) and
%   Dirichlet boundary conditions U(X,T)=G(X,T) for
%   X=XSPAN(1) and X=XSPAN(2). MU is a positive constant,
%   F, G and U0 are inline functions. NSTEP(1) is the
%   number of space integration intervals, NSTEP(2)+1 is
%   the number of time-integration intervals.
h  = (xspan(2)-xspan(1))/nstep(1);
dt = (tspan(2)-tspan(1))/nstep(2);
N  = nstep(1)+1;
e  = ones(N,1);
D  = spdiags([-e 2*e -e],[-1,0,1],N,N);
I  = speye(N);
A  = I+mu*dt*theta*D/h^2;
An = I-mu*dt*(1-theta)*D/h^2;
A(1,:) = 0; A(1,1) = 1;
A(N,:) = 0; A(N,N) = 1;
x  = linspace(xspan(1),xspan(2),N);
x  = x';
fn = feval(f,x,tspan(1),varargin{:});
un = feval(u0,x,varargin{:});
[L,U]=lu(A);
for t = tspan(1)+dt:dt:tspan(2)
    fn1 = feval(f,x,t,varargin{:});
    rhs = An*un+dt*(theta*fn1+(1-theta)*fn);
    temp = feval(g,[xspan(1),xspan(2)],t,varargin{:});
    rhs([1,N]) = temp;
    u = L\rhs;
    u = U\u;
    fn = fn1;
    un = u;
end
return
```

Example 8.3 We consider the heat equation (8.4) in $(a,b) = (0,1)$ with $\mu = 1$, $f(x,t) = -\sin(x)\sin(t) + \sin(x)\cos(t)$, initial condition $u(x,0) = \sin(x)$ and boundary conditions $u(0,t) = 0$ and $u(1,t) = \sin(1)\cos(t)$. In this case the exact solution is $u(x,t) = \sin(x)\cos(t)$. In Figure 8.7 we compare the behavior of the errors $\max_{i=0,\ldots,N} |u(x_i,1) - u_i^M|$ with respect to the time-step on a uniform grid in space with $h = 0.002$. $\{u_i^M\}$ are the values of the finite difference solution computed at time $t^M = 1$. As expected, for $\theta = 0.5$ the θ-method is second order accurate until when the time-step is so small that the spatial error dominates over the error due to the temporal discretization.
∎

Example 8.4 (Thermodynamics) We consider an aluminum bar (whose density is $\rho = 2700$ Kg/m^3), of three meters length, with thermal conductivity $k = 273$ W/mK (Watt per meters-Kelvin). We are interested to the evolution of the temperature in the bar starting from the initial condition $T(x,0) = 500$ K if $x \in (1,2)$, 250 K otherwise and subject to the following Dirichlet boundary conditions: $T(0,t) = T(3,t) = 250$ K. In Figure 8.8 we report the evolution of the temperature starting from the initial data computed with the Euler method ($\theta = 1$, *left*) and the Crank-Nicolson method ($\theta = 0.5$, *right*). The

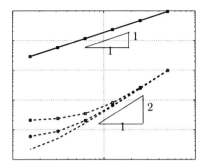

Fig. 8.7. The error versus Δt for the θ-method (for $\theta = 1$, *solid line*, and $\theta = 0.5$ *dashed line*), for three different values of h: 0.008 ($\square$), 0.004 ($\circ$) and 0.002 (*no symbols*)

results show that the Crank-Nicolson method suffers a clear instability due to the low smoothness of the initial datum (about this point, see also [QV94, Chapter 11]). On the contrary, the implicit Euler method provides a stable solution which decays correctly to 250 K as t grows since the source term f is null. ∎

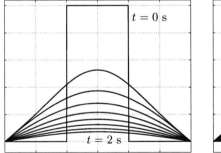
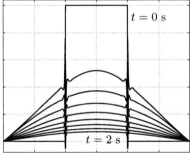

Fig. 8.8. Temperature profiles in an aluminum bar at different time-steps (from $t = 0$ to $t = 2$ seconds with steps of 0.25 seconds), obtained with the backward Euler method (*left*) and the Crank-Nicolson method (*right*)

8.3 The wave equation

We consider the second-order hyperbolic equation in one dimension

$$\boxed{\frac{\partial^2 u}{\partial t^2} - c\frac{\partial^2 u}{\partial x^2} = f} \tag{8.33}$$

When $f = 0$, the general solution of (8.33) is the d'Alembert traveling-wave solution

$$u(x,t) = \psi_1(\sqrt{c}t - x) + \psi_2(\sqrt{c}t + x), \qquad (8.34)$$

for arbitrary functions ψ_1 and ψ_2.

In the sequel we consider problem (8.33) for $x \in (a,b)$ and $t > 0$. Therefore, we complete the differential equation with the initial data

$$u(x,0) = u_0(x) \text{ and } \frac{\partial u}{\partial t}(x,0) = v_0(x), \quad x \in (a,b),$$

and the boundary data

$$u(a,t) = 0 \text{ and } u(b,t) = 0, \quad t > 0. \qquad (8.35)$$

In this case, u may represent the transverse displacement of an elastic vibrating string of length $b-a$, fixed at the endpoints, and c is a positive coefficient depending on the specific mass of the string and on its tension. The string is subjected to a vertical force of density f. The functions $u_0(x)$ and $v_0(x)$ denote respectively the initial displacement and the initial velocity of the string.

The change of variables

$$\omega_1 = \frac{\partial u}{\partial x}, \qquad \omega_2 = \frac{\partial u}{\partial t},$$

transforms (8.33) into the first-order system

$$\boxed{\frac{\partial \boldsymbol{\omega}}{\partial t} + \mathrm{A}\frac{\partial \boldsymbol{\omega}}{\partial x} = \mathbf{f}, \qquad x \in (a,b),\ t > 0} \qquad (8.36)$$

where

$$\boldsymbol{\omega} = \begin{bmatrix} \omega_1 \\ \omega_2 \end{bmatrix}, \mathrm{A} = \begin{bmatrix} 0 & -1 \\ -c & 0 \end{bmatrix}, \mathbf{f} = \begin{bmatrix} 0 \\ f \end{bmatrix},$$

and the initial conditions are $\omega_1(x,0) = u_0'(x)$ and $\omega_2(x,0) = v_0(x)$ for $x \in (a,b)$.

In general, we can consider systems of the form (8.36) where $\boldsymbol{\omega}, \mathbf{f} : \mathbb{R} \times [0,\infty) \to \mathbb{R}^p$ and $\mathrm{A} \in \mathbb{R}^{p \times p}$ is a matrix with constant coefficients. This system is said *hyperbolic* if A is diagonalizable and has real eigenvalues, that is, if there exists a nonsingular matrix $\mathrm{T} \in \mathbb{R}^{p \times p}$ such that

$$\mathrm{A} = \mathrm{T}\Lambda\mathrm{T}^{-1},$$

where $\Lambda = \mathrm{diag}(\lambda_1, ..., \lambda_p)$ is the diagonal matrix of the real eigenvalues of A, while $\mathrm{T} = (\boldsymbol{\omega}^1, \boldsymbol{\omega}^2, \ldots, \boldsymbol{\omega}^p)$ is the matrix whose column vectors are the right eigenvectors of A. Thus

8.3 The wave equation

$$A\boldsymbol{\omega}^k = \lambda_k \boldsymbol{\omega}^k, \qquad k = 1, \ldots, p.$$

Introducing the *characteristic variables* $\mathbf{w} = T^{-1}\boldsymbol{\omega}$, system (8.36) becomes

$$\frac{\partial \mathbf{w}}{\partial t} + \Lambda \frac{\partial \mathbf{w}}{\partial x} = \mathbf{g},$$

where $\mathbf{g} = T^{-1}\mathbf{f}$. This is a system of p independent scalar equations of the form

$$\frac{\partial w_k}{\partial t} + \lambda_k \frac{\partial w_k}{\partial x} = g_k, \qquad k = 1, \ldots, p.$$

When $g_k = 0$, its solution is given by $w_k(x,t) = w_k(x - \lambda_k t, 0)$, $k = 1, \ldots, p$ and thus the solution $\boldsymbol{\omega} = T\mathbf{w}$ of problem (8.36) with $\mathbf{f} = \mathbf{0}$ can be written as

$$\boldsymbol{\omega}(x,t) = \sum_{k=1}^{p} w_k(x - \lambda_k t, 0)\boldsymbol{\omega}^k.$$

The curve $(x_k(t), t)$ in the plane (x, t) that satisfies $x'_k(t) = \lambda_k$ is the k-th characteristic curve and w_k is constant along it. Then $\boldsymbol{\omega}(\overline{x}, \overline{t})$ depends only on the initial datum at the points $\overline{x} - \lambda_k \overline{t}$. For this reason, the set of p points that form the feet of the characteristics issuing from the point $(\overline{x}, \overline{t})$,

$$D(\overline{t}, \overline{x}) = \{x \in \mathbb{R} \; : \; x = \overline{x} - \lambda_k \overline{t} \; , \; k = 1, \ldots, p\}, \qquad (8.37)$$

is called the *domain of dependence* of the solution $\boldsymbol{\omega}(\overline{x}, \overline{t})$.

If (8.36) is set on a bounded interval (a, b) instead of on the whole real line, the inflow point for each characteristic variable w_k is determined by the sign of λ_k. Correspondingly, the number of positive eigenvalues determines the number of boundary conditions that can be assigned at $x = a$, whereas at $x = b$ it is admissible to assign a number of conditions which equals the number of negative eigenvalues.

Example 8.5 System (8.36) is hyperbolic since A is diagonalizable with matrix

$$T = \begin{bmatrix} -\frac{1}{\sqrt{c}} & \frac{1}{\sqrt{c}} \\ 1 & 1 \end{bmatrix}$$

and presents two distinct real eigenvalues $\pm\sqrt{c}$ (representing the propagation velocities of the wave). Moreover, one boundary condition needs to be prescribed at every end-point, as in (8.35). ■

Remark 8.1 Notice that replacing $\frac{\partial^2 u}{\partial t^2}$ by t^2, $\frac{\partial^2 u}{\partial x^2}$ by x^2 and f by one, the wave equation becomes $t^2 - cx^2 = 1$ which represents an hyperbola in the (x,t) plane. Proceeding analogously in the case of the heat equation (8.4), we end up with $t - \mu x^2 = 1$ which represents a parabola in the (x,t) plane. Finally, for the Poisson equation in two dimensions, replacing $\frac{\partial^2 u}{\partial x_1^2}$ by x_1^2, $\frac{\partial^2 u}{\partial x_2^2}$ by x_2^2 and f by one, we get $x_1^2 + x_2^2 = 1$ which represents an ellipse in the (x_1, x_2) plane. Due to the geometric interpretation above, the corresponding differential operators are classified as hyperbolic, parabolic and elliptic, respectively. •

8.3.1 Approximation by finite differences

To discretize in time the wave equation we use the Newmark method (7.59) proposed in Chapter 7. Still denoting by Δt the (uniform) time-step and using in space the classical finite difference method on a grid with nodes $x_j = x_0 + jh$, $j = 0, \ldots, N$, $x_0 = a$ and $x_N = b$, we obtain the following scheme: for any $n \geq 1$ find $\{u_j^n, v_j^n, j = 1, \ldots, N-1\}$ such that

$$u_j^{n+1} = u_j^n + \Delta t v_j^n$$
$$+ \Delta t^2 \left[\zeta(cw_j^{n+1} + f(t^{n+1}, x_j)) + (1/2 - \zeta)(cw_j^n + f(t^n, x_j)) \right], \quad (8.38)$$
$$v_j^{n+1} = v_j^n + \Delta t \left[(1-\theta)(cw_j^n + f(t^n, x_j)) + \theta(cw_j^{n+1} + f(t^{n+1}, x_j)) \right],$$

with $u_j^0 = u_0(x_j)$ and $v_j^0 = v_0(x_j)$ and $w_j^k = (u_{j+1}^k - 2u_j^k + u_{j-1}^k)/h^2$ for $k = n$ or $k = n+1$. System (8.38) must be completed imposing the boundary conditions (8.35).

This method is implemented in Program 8.4. The input parameters are the vectors xspan=[a,b] and tspan=[0,T], the number of discretization intervals in space (nstep(1)) and in time (nstep(2)), the string fun which contains the function $f(t, x(t))$ and the strings u0 and v0 to define the initial data. Finally, the vector param allows to specify the values of the coefficients (param(1)=θ, param(2)=ζ). The Newmark method is second order accurate with respect to Δt if $\theta = 1/2$, whereas it is first order if $\theta \neq 1/2$. Moreover, the condition $\theta \geq 1/2$ is necessary to ensure stability (see Section 7.8).

Program 8.4. newmarkwave: Newmark method for the wave equation

```
function [x,u]=newmarkwave(xspan,tspan,nstep,param,c,...
                u0,v0,g,f,varargin)
%NEWMARKWAVE solve the wave equation with the Newmark
% method.
% [X,U]=NEWMARKWAVE(XSPAN,TSPAN,NSTEP,PARAM,C,U0,VO,G,F)
% solve the wave equation D^2 U/DT^2 - C D^2U/DX^2 = F
% in (XSPAN(1),XSPAN(2)) X (TSPAN(1),TSPAN(2)) using the
% Newmark method with initial conditions U(X,0)=U0(X),
% DU/DX(X,0)=V0(X) and Dirichlet boundary conditions
% U(X,T)=G(X,T) for X=XSPAN(1) and X=XSPAN(2). C is a
% positive constant, F,G,U0 and V0 are inline functions.
```

```
% NSTEP(1) is the number of space integration intervals,
% NSTEP(2)+1 is the number of time-integration intervals.
% PARAM(1)=THETA and PARAM(2)=ZETA.
% [X,U]=NEWMARKWAVE(XSPAN,TSPAN,NSTEP,PARAM,C,U0,V0,G,F,
% P1,P2,...) passes the additional parameters P1,P2,...
% to the functions U0,V0,G,F.
h    = (xspan(2)-xspan(1))/nstep(1);
dt   = (tspan(2)-tspan(1))/nstep(2);
theta = param(1);    zeta = param(2);
N    = nstep(1)+1;
e    = ones(N,1);  D = spdiags([e -2*e e],[-1,0,1],N,N);
I    = speye(N);
lambda = dt/h;
A    = I-c*lambda^2*zeta*D;
An   = I+c*lambda^2*(0.5-zeta)*D;
A(1,:) = 0; A(1,1) = 1; A(N,:) = 0; A(N,N) = 1;
x    = linspace(xspan(1),xspan(2),N);
x    = x';
fn   = feval(f,x,tspan(1),varargin{:});
un   = feval(u0,x,varargin{:});
vn   = feval(v0,x,varargin{:});
[L,U]=lu(A);
alpha = dt^2*zeta; beta = dt^2*(0.5-zeta);
theta1 = 1-theta;
for t = tspan(1)+dt:dt:tspan(2)
    fn1 = feval(f,x,t,varargin{:});
    rhs = An*un+dt*I*vn+alpha*fn1+beta*fn;
    temp = feval(g,[xspan(1),xspan(2)],t,varargin{:});
    rhs([1,N]) = temp;
    u = L\rhs;    u = U\u;
    v = vn + dt*((1-theta)*(c*D*un/h^2+fn)+...
         theta*(c*D*u/h^2+fn1));
    fn = fn1;    un = u;    vn = v;
end
return
```

Example 8.6 Using Program 8.4 we study the evolution of the initial condition $u_0(x) = e^{-10x^2}$ for $x \in (-2, 2)$. We assume $v_0 = 0$ and homogeneous Dirichlet boundary conditions. In Figure 8.9 we compare the solutions obtained at time $t = 3$ using $h = 0.04$ and time-steps equal to 0.15 (*dashed line*), to 0.075 (*continuous line*) and to 0.0375 (*dashed-dotted line*). The parameters of the Newmark method are $\theta = 1/2$ and $\zeta = 0.25$, that ensure a second order unconditionally stable method. ∎

Example 8.7 (Communications) In this example we use the equation (8.9) to model how a telegraph wire transmits a pulse of voltage. The equation is a combination of diffusion and wave equations, and accounts for effects of finite velocity in a standard mass transport equation. In Figure 8.10 we compare the evolution of a sinusoidal pulse using the wave equation (8.33) (*dotted line*) and the telegraph equation (8.9) with $c = 1$, $\alpha = 2$ and $\beta = 1$ (*continuous line*). The presence of the diffusion effect is evident. ∎

An alternative approach to the Newmark method is to discretize the first order equivalent system (8.36). We consider for simplicity the case

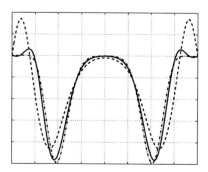

Fig. 8.9. Comparison between the solutions obtained using the Newmark method for a discretization with $h = 0.04$ and $\Delta t = 0.154$ (*dashed line*), $\Delta t = 0.075$ (*continuous line*) and $\Delta t = 0.0375$ (*dashed-dotted line*)

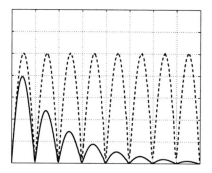

Fig. 8.10. Propagation of a pulse of voltage using the wave equation (*dotted line*) and the telegraph equation (*continuous line*)

$(a, b) = \mathbb{R}$ and $\mathbf{f} = \mathbf{0}$. Then, the half-plane $\{(x,t) : -\infty < x < \infty,\ t > 0\}$ is discretized by choosing a spatial grid size h, a temporal step Δt and the grid points (x_j, t^n) as follows

$$x_j = jh, \quad j \in \mathbb{Z}, \quad t^n = n\Delta t, \quad n \in \mathbb{N}.$$

By setting $\lambda = \Delta t/h$, some popular schemes for the discretization of (8.36) are:

1. the *upwind (or forward Euler/uncentred) method*

$$\begin{aligned}\omega_j^{n+1} &= \omega_j^n - \frac{\lambda}{2}\mathrm{A}(\omega_{j+1}^n - \omega_{j-1}^n) \\ &+ \frac{\lambda}{2}|\mathrm{A}|(\omega_{j+1}^n - 2\omega_j^n + \omega_{j-1}^n),\end{aligned} \quad (8.39)$$

where $|\mathrm{A}| = \mathrm{T}|\Lambda|\mathrm{T}^{-1}$ and $|\Lambda|$ is the diagonal matrix of the moduli of the eigenvalues of A;

2. the *Lax-Wendroff method*

$$w_j^{n+1} = w_j^n - \frac{\lambda}{2}A(w_{j+1}^n - w_{j-1}^n)$$
$$+ \frac{\lambda^2}{2}A^2(w_{j+1}^n - 2w_j^n + w_{j-1}^n). \quad (8.40)$$

The upwind method is first order accurate (in time and in space), while the Lax-Wendroff scheme is second order.

About stability, since all these schemes are explicit, they can only be conditionally stable. In particular, the upwind and the Lax-Wendroff schemes satisfy $\|w^n\|_\Delta \leq \|w^0\|_\Delta$, where

$$\|\mathbf{v}\|_\Delta = \sqrt{h \sum_{j=-\infty}^{\infty} v_j^2}, \quad \mathbf{v} = (v_j),$$

is a discrete norm under the following condition

$$\Delta t < \frac{h}{\rho(A)}, \quad (8.41)$$

known as the CFL or Courant, Friedrichs and Lewy condition. As usual $\rho(A)$ denotes the spectral radius of A. For the proof, see, e.g., [QV94], [LeV02], [GR96], [QSS06, Chapter 13].

See Exercises 8.9-8.10.

8.4 What we haven't told you

We could simply say that we have told you almost nothing, since the field of numerical analysis which is devoted to the numerical approximation of partial differential equations is so broad and multifaceted to deserve an entire monograph simply for addressing the most essential concepts (see, e.g., [TW98], [EEHJ96]).

We would like to mention that the finite element method is nowadays probably the most widely diffused method for the numerical solution of partial differential equations (see, e.g., [QV94], [Bra97], [BS01]). As already mentioned the MATLAB toolbox pde allows the solution of a broad family of partial differential equations by the linear finite element method.

Other popular techniques are the spectral methods (see, [CHQZ06], [Fun92], [BM92], [KS99]) and the finite volume method (see, [Krö98], [Hir88] and [LeV02]).

Octave 8.1 Neither Octave nor Octave-forge feature a pde toolbox. However, several Octave programs for partial differential equations can be found surfing on the web. ∎

8.5 Exercises

Exercise 8.1 Verify that matrix (8.15) is positive definite.

Exercise 8.2 Verify that the eigenvalues of the matrix $A \in \mathbb{R}^{(N-1)\times(N-1)}$, defined in (8.15), are

$$\lambda_j = 2(1 - \cos(j\theta)), \quad j = 1, \ldots, N-1,$$

while the corresponding eigenvectors are

$$\mathbf{q}_j = (\sin(j\theta), \sin(2j\theta), \ldots, \sin((N-1)j\theta))^T,$$

where $\theta = \pi/N$. Deduce that $K(A)$ is proportional to h^{-2}.

Exercise 8.3 Prove that the quantity (8.12) provides a second order approximation of $u''(\bar{x})$ with respect to h.

Exercise 8.4 Compute the matrix and the right-hand side of the numerical scheme that we have proposed to approximate problem (8.17).

Exercise 8.5 Use the finite difference method to approximate the boundary-value problem

$$\begin{cases} -u'' + \dfrac{k}{T}u = \dfrac{w}{T} & \text{in } (0,1), \\ u(0) = u(1) = 0, \end{cases}$$

where $u = u(x)$ represents the vertical displacement of a string of length 1, subject to a transverse load of intensity w per unit length. T is the tension and k is the elastic coefficient of the string. For the case in which $w = 1 + \sin(4\pi x)$, $T = 1$ and $k = 0.1$, compute the solution corresponding to $h = 1/i$, $i = 10, 20, 40$, and deduce the order of accuracy of the method.

Exercise 8.6 We consider problem (8.17) on the interval $(0,1)$ with $\gamma = 0$, $f = 0$, $\alpha = 0$ and $\beta = 1$. Using the Program 8.1 find the maximum value h_{crit} of h for which the numerical solution is monotone (as is the exact solution) when $\delta = 100$. What happens if $\delta = 1000$? Suggest an empirical formula for $h_{crit}(\delta)$ as a function of δ, and verify it for several values of δ.

Exercise 8.7 Use the finite difference method to solve problem (8.17) in the case where the following *Neumann* boundary conditions are prescribed at the endpoints

$$u'(a) = \alpha, \ u'(b) = \beta.$$

Use the formulae given in (4.11) to discretize $u'(a)$ and $u'(b)$.

Exercise 8.8 Verify that, when using a uniform grid, the right-hand side of the system associated with the centered finite difference scheme coincides with that of the finite element scheme provided that the composite trapezoidal formula is used to compute the integrals on the elements I_{k-1} and I_k.

Exercise 8.9 Verify that $\text{div}\nabla\phi = \Delta\phi$, where ∇ is the gradient operator that associates to a function u the vector whose components are the first order partial derivatives of u.

Exercise 8.10 (Thermodynamics) Consider a square plate whose side length is 20 cm and whose thermal conductivity is $k = 0.2$ cal/sec·cm·C. Denote by $Q = 5$ cal/cm^3·sec the heat production rate per unit area. The temperature $T = T(x,y)$ of the plate satisfies the equation $-\Delta T = Q/k$. Assuming that T is null on three sides of the plate and is equal to 1 on the fourth side, determine the temperature T at the center of the plate.

9
Solutions of the exercises

9.1 Chapter 1

Solution 1.1 Only the numbers of the form $\pm 0.1 a_2 \cdot 2^e$ with $a_2 = 0, 1$ and $e = \pm 2, \pm 1, 0$ belong to the set $\mathbb{F}(2, 2, -2, 2)$. For a given exponent, we can represent in this set only the two numbers 0.10 and 0.11, and their opposites. Consequently, the number of elements belonging to $\mathbb{F}(2, 2, -2, 2)$ is 20. Finally, $\epsilon_M = 1/2$.

Solution 1.2 For any fixed exponent, each of the digits $a_2, \ldots, a_t$ can assume β different values, while a_1 can assume only $\beta-1$ values. Therefore $2(\beta-1)\beta^{t-1}$ different numbers can be represented (the 2 accounts for the positive and negative sign). On the other hand, the exponent can assume $U - L + 1$ values. Thus, the set $\mathbb{F}(\beta, t, L, U)$ contains $2(\beta - 1)\beta^{t-1}(U - L + 1)$ different elements.

Solution 1.3 Thanks to the Euler formula $i = e^{i\pi/2}$; we obtain $i^i = e^{-\pi/2}$, that is, a real number. In MATLAB

```
>> exp(-pi/2)
ans =
    0.2079
>> i^i
ans =
    0.2079
```

Solution 1.4 Use the instruction `U=2*eye(10)-3*diag(ones(8,1),2)` (respectively, `L=2*eye(10)-3*diag(ones(8,1),-2)`).

Solution 1.5 We can interchange the third and seventh rows of the previous matrix using the instructions: `r=[1:10]; r(3)=7; r(7)=3; Lr=L(r,:)`. Notice that the character : in `L(r,:)` ensures that all columns of L are spanned in the usual increasing order (from the first to the last). To interchange the fourth column with the eighth column we can write `c=[1:10]; c(8)=4; c(4)=8; Lc=L(:,c)`. Similar instructions can be used for the upper triangular matrix.

`L(r,:)`

9 Solutions of the exercises

Solution 1.6 We can define the matrix A = [v1;v2;v3;v4] where v1, v2, v3 and v4 are the 4 given row vectors. They are linearly independent iff the determinant of A is different from 0, which is not true in our case.

Solution 1.7 The two given functions f and g have the symbolic expression:

```
>> syms x
>> f=sqrt(x^2+1); pretty(f)
```

$$(x^2+1)^{1/2}$$

```
>> g=sin(x^3)+cosh(x); pretty(g)
```

$$\sin(x^3) + \cosh(x)$$

pretty The command **pretty(f)** prints the symbolic expression f in a format that resembles type-set mathematics. At this stage, the symbolic expression of the first and second derivatives and the integral of f can be obtained with the following instructions:

```
>> diff(f,x)
ans =
1/(x^2+1)^(1/2)*x
>> diff(f,x,2)
ans =
-1/(x^2+1)^(3/2)*x^2+1/(x^2+1)^(1/2)
>> int(f,x)
ans =
1/2*x*(x^2+1)^(1/2)+1/2*asinh(x)
```

Similar instructions can be used for the function g.

Solution 1.8 The accuracy of the computed roots downgrades as the polynomial degree increases. This experiment reveals that the accurate computation of the roots of a polynomial of high degree can be troublesome.

Solution 1.9 Here is a possible program to compute the sequence:

```
function I=sequence(n)
I = zeros(n+2,1);  I(1) = (exp(1)-1)/exp(1);
for i = 0:n, I(i+2) = 1 - (i+1)*I(i+1); end
```

The sequence computed from this program doesn't tend to zero (as n increases), but it diverges with alternating sign.

Solution 1.10 The anomalous behavior of the computed sequence is due to the propagation of roundoff errors from the innermost operation. In particular, when $4^{1-n}z_n^2$ is less than $\epsilon_M/2$, the elements of the sequence are equal to 0. This happens for $n \geq 29$.

Solution 1.11 The proposed method is a special instance of the Monte Carlo method and is implemented by the following program:

9.1 Chapter 1

```
function mypi=pimontecarlo(n)
x = rand(n,1); y = rand(n,1);
z = x.^2+y.^2;
v = (z <= 1);
m=sum(v); mypi=4*m/n;
```

The command **rand** generates a sequence of pseudo-random numbers. The instruction v = (z <= 1) is a shortand version of the following procedure: we check whether z(k) <= 1 for any component of the vector z. If the inequality is satisfied for the k-th component of z (that is, the point (x(k),y(k)) belongs to the interior of the unit circle) v(k) is set equal to 1, and to 0 otherwise. The command sum(v) computes the sum of all components of v, that is, the number of points falling in the interior of the unit circle.   sum

By launching the program as mypi=pimontecarlo(n) for different values of n, when n increases, the approximation mypi of π becomes more accurate. For instance, for n=1000 we obtain mypi=3.1120, whilst for n=300000 we have mypi=3.1406.

Solution 1.12 To answer the question we can use the following *function*:

```
function pig=bbpalgorithm(n)
pig = 0;
for m=0:n
  m8 = 8*m;
  pig = pig + (1/16)^m*(4/(m8+1)-(2/(m8+4)+ ...
        1/(m8+5)+1/(m8+6)));
end
return
```

For n=10 we obtain an approximation **pig** of π that coincides (in the MATLAB precision) with the persistent MATLAB variable **pi**. In fact, this algorithm is extremely efficient and allows the rapid computation of hundreds of significant digits of π.

Solution 1.13 The binomial coefficient can be computed by the following program (see also the MATLAB function **nchoosek**):   nchoosek

```
function bc=bincoeff(n,k)
k = fix(k); n = fix(n);
if k > n, disp('k must be between  0 and n');
  break; end
if k > n/2, k = n-k; end
if k <= 1,  bc = n^k; else
  num = (n-k+1):n; den = 1:k; el = num./den;
  bc = prod(el);
end
```

The command **fix(k)** rounds k to the nearest integer smaller than k.   fix
The command **disp(string)** displays the string, without printing its name.   disp
In general, the command **break** terminates the execution of **for** and **while**   break
loops. If break is executed in an **if**, it terminates the statement at that point.
Finally, **prod(el)** computes the product of all elements of the vector **el**.   prod

Solution 1.14 The following *functions* compute f_n using the form $f_i = f_{i-1} + f_{i-2}$ (**fibrec**) or using the form (1.14) (**fibmat**):

```
function f=fibrec(n)
if n == 0
    f = 0;
elseif n == 1
    f = 1;
else
    f = fibrec(n-1)+fibrec(n-2);
end
return

function f=fibmat(n)
f = [0;1];
A = [1 1; 1 0];
f = A^n*f;
f = f(1);
return
```

For n=20 we obtain the following results:

```
>> t=cputime; fn=fibrec(20), cpu=cputime-t
fn =
        6765
cpu =
    1.3400
>> t=cputime; fn=fibmat(20), cpu=cputime-t
fn =
        6765
cpu =
    0
```

The recursive *function* fibrec requires much more CPU time than fibmat. The latter requires to compute only the power of a matrix, an easy operation in MATLAB.

9.2 Chapter 2

Solution 2.1 The command fplot allows us to study the graph of the given function f for various values of γ. For $\gamma = 1$, the corresponding function does not have real zeros. For $\gamma = 2$, there is only one zero, $\alpha = 0$, with multiplicity equal to four (that is, $f(\alpha) = f'(\alpha) = f''(\alpha) = f'''(\alpha) = 0$, while $f^{(4)}(\alpha) \neq 0$). Finally, for $\gamma = 3$, f has two distinct zeros, one in the interval $(-3, -1)$ and the other one in $(1, 3)$. In the case $\gamma = 2$, the bisection method cannot be used since it is impossible to find an interval (a, b) in which $f(a)f(b) < 0$. For $\gamma = 3$, starting from the interval $[a, b] = [-3, -1]$, the bisection method (Program 2.1) converges in 34 iterations to the value $\alpha = -1.85792082914850$ (with $f(\alpha) \simeq -3.6 \cdot 10^{-12}$), using the following instructions:

```
>> f=inline('cosh(x)+cos(x)-3'); a=-3; b=-1; tol=1.e-10; nmax=200;
>> [zero,res,niter]=bisection(f,a,b,tol,nmax)
zero =
    -1.8579
```

```
res =
    -3.6872e-12
niter =
    34
```

Similarly, choosing a=1 and b=3, for $\gamma = 3$ the bisection method converges after 34 iterations to the value $\alpha = 1.8579208291485$ with $f(\alpha) \simeq -3.6877 \cdot 10^{-12}$.

Solution 2.2 We have to compute the zeros of the function $f(V) = pV + aN^2/V - abN^3/V^2 - pNb - kNT$. Plotting the graph of f, we see that this function has just a simple zero in the interval $(0.01, 0.06)$ with $f(0.01) < 0$ and $f(0.06) > 0$. We can compute this zero using the bisection method as follows:

```
>> f=inline('35000000*x+401000./x-17122.7./x.^2-1494500');
>> [zero,res,niter]=bisection(f,0.01,0.06,1.e-12,100)
zero =
    0.0427
res =
    -6.3814e-05
niter =
    35
```

Solution 2.3 The unknown value of ω is the zero of the function $f(\omega) = s(1,\omega) - 1 = 9.8[\sinh(\omega) - \sin(\omega)]/(2\omega^2) - 1$. From the graph of f we conclude that f has a unique real zero in the interval $(0.5, 1)$. Starting from this interval, the bisection method computes the value $\omega = 0.61214447021484$ with the desired tolerance in 15 iterations as follows:

```
>> f=inline('9.8/2*(sinh (omega)- sin(omega))./omega.^2 -1','omega');
>> [zero,res,niter]=bisection(f,0.5,1,1.e-05,100)
zero =
    6.1214e-01
res =
    3.1051e-06
niter =
    15
```

Solution 2.4 The inequality (2.6) can be derived by observing that $|e^{(k)}| < |I^{(k)}|/2$ with $|I^{(k)}| < \frac{1}{2}|I^{(k-1)}| < 2^{-k-1}(b-a)$. Consequently, the error at the iteration k_{min} is less than ε if k_{min} is such that $2^{-k_{min}-1}(b-a) < \varepsilon$, that is, $2^{-k_{min}-1} < \varepsilon/(b-a)$, which proves (2.6).

Solution 2.5 The first formula is less sensitive to the roundoff error.

Solution 2.6 In Solution 2.1 we have analyzed the zeros of the given function with respect to different values of γ. Let us consider the case when $\gamma = 2$. Starting from the initial guess $x^{(0)} = 1$, the Newton method (Program 2.2) converges to the value $\bar{\alpha} = 0.0056$ in 18 iterations with tol=1.e-10 while the exact zero of f is equal to 0. This discrepancy is due to the fact that f is almost a constant in a neighborhood of its zero. Actually, the corresponding residual computed by MATLAB is 0. Let us set now $\gamma = 3$. The Newton method with tol=1.e-16 converges to the value 1.85792082915020 in 9 iterations starting from $x^{(0)} = 1$, while if $x^{(0)} = -1$ after 10 iterations it converges to the value -1.85792082915020 (in both cases the residuals are zero in MATLAB).

Solution 2.7 The square and the cube roots of a number a are the solutions of the equations $x^2 = a$ and $x^3 = a$, respectively. Thus, the corresponding algorithms are: for a given $x^{(0)}$ compute

$$x^{(k+1)} = \frac{1}{2}\left(x^{(k)} + \frac{a}{x^{(k)}}\right), \quad k \geq 0 \quad \text{for the square root,}$$

$$x^{(k+1)} = \frac{1}{3}\left(2x^{(k)} + \frac{a}{(x^{(k)})^2}\right), \quad k \geq 0 \text{ for the cube root.}$$

Solution 2.8 Setting $\delta x^{(k)} = x^{(k)} - \alpha$, from the Taylor expansion of f we find:

$$0 = f(\alpha) = f(x^{(k)}) - \delta x^{(k)} f'(x^{(k)}) + \frac{1}{2}(\delta x^{(k)})^2 f''(x^{(k)}) + \mathcal{O}((\delta x^{(k)})^3). \quad (9.1)$$

The Newton method yields

$$\delta x^{(k+1)} = \delta x^{(k)} - f(x^{(k)})/f'(x^{(k)}). \quad (9.2)$$

Combining (9.1) with (9.2), we have

$$\delta x^{(k+1)} = \frac{1}{2}(\delta x^{(k)})^2 \frac{f''(x^{(k)})}{f'(x^{(k)})} + \mathcal{O}((\delta x^{(k)})^3).$$

After division by $(\delta x^{(k)})^2$ and letting $k \to \infty$ we prove the convergence result.

Solution 2.9 For certain values of β the equation (2.2) can have two roots that correspond to different configurations of the rods system. The two initial values that are suggested have been chosen conveniently to allow the Newton method to converge toward one or the other root, respectively. We solve the problem for $\beta = k\pi/100$ with $k = 0, \ldots, 80$ (if $\beta > 2.6389$ the Newton method does not converge since the system has no admissible configuration). We use the following instructions to obtain the solution of the problem (shown in Figure 9.1):

```
>> a1=10; a2=13; a3=8; a4=10;
>> ss = num2str((a1^2 + a2^2 - a3^2+ a4^2)/(2*a2*a4),15);
>> n=100; x01=-0.1; x02=2*pi/3; nmax=100;
>> for i=0:80
    w = i*pi/n; k=i+1; beta(k) = w;
    ws = num2str(w,15);
    f  = inline(['10/13*cos(',ws,')-cos(x)
         -cos(',ws,'-x)+',ss],'x');
    df = inline(['sin(x)-sin(',ws,'-x)'],'x');
    [zero,res,niter]=newton(f,df,x01,1e-12,nmax);
```

```
        alpha1(k) = zero; niter1(k) = niter;
        [zero,res,niter]=newton(f,df,x02,1e-12,nmax);
        alpha2(k) = zero; niter2(k) = niter;
end
```

The components of the vectors alpha1 and alpha2 are the angles computed for different values of β, while the components of the vectors niter1 and niter2 are the number of Newton iterations (5-7) necessary to compute the zeros with the requested tolerance.

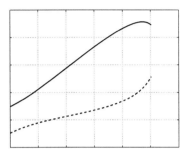

Fig. 9.1. The two curves representing the two possible configurations which correspond to the choice of the parameter $\beta \in [0, 2\pi/3]$

Solution 2.10 From an inspection of its graph we see that f has two positive real zeros ($\alpha_2 \simeq 1.5$ and $\alpha_3 \simeq 2.5$) and one negative ($\alpha_1 \simeq -0.5$). The Newton method converges in 4 iterations (having set $x^{(0)} = -0.5$ and tol = 1.e-10) to the value α_1:

```
>> f=inline('exp(x)-2*x^2'); df=inline('exp(x)-4*x');
>> x0=-0.5; tol=1.e-10; nmax=100;
>> format long; [zero,res,niter]=newton(f,df,x0,tol,nmax)
zero =
  -0.53983527690282
res =
     0
niter =
     4
```

The given function has a maximum at $\bar{x} \simeq 0.3574$ (which can be obtained by applying the Newton method to the function f'): for $x^{(0)} < \bar{x}$ the method converges to the negative zero. If $x^{(0)} = \bar{x}$ the Newton method cannot be applied since $f'(\bar{x}) = 0$. For $x^{(0)} > \bar{x}$ the method converges to the positive zero.

Solution 2.11 Let us set $x^{(0)} = 0$ and tol= 10^{-17}. The Newton method converges in 39 iterations to the value 0.64118239763649, which we identify with the exact zero α. We can observe that the (approximate) errors $x^{(k)} - \alpha$, for $k = 0, 1, \ldots, 29$, decrease only linearly when k increases. This behavior is due to the fact that α has multiplicity greater than 1 (see Figure 9.2). To recover a second-order method we can use the modified Newton method.

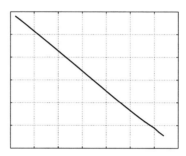

Fig. 9.2. Error vs iteration number of the Newton method for the computation of the zero of the function $f(x) = x^3 - 3x^2 2^{-x} + 3x 4^{-x} - 8^{-x}$

Solution 2.12 We should compute the zero of the function $f(x) = \sin(x) - \sqrt{2gh/v_0^2}$. From an inspection of its graph, we can conclude that f has one zero in the interval $(0, \pi/2)$. The Newton method with $x^{(0)} = \pi/4$ and tol$= 10^{-10}$ converges in 5 iterations to the value 0.45862863227859.

Solution 2.13 Using the data given in the exercise, the solution can be obtained with the following instructions:

```
>> f=inline('6000-1000*(1+x).*((1+x).^5 - 1)./x');
>> df=inline('1000*((1+x).^5.*(1-5*x) - 1)./(x.^2)');
>> [zero,res,niter]=bisection(f,0.01,0.1,1.e-12,4);
>> [zero,res,niter]=newton(f,df,zero,1.e-12,100);
```

The Newton method converges to the desired result in 3 iterations.

Solution 2.14 By a graphical study, we see that (2.32) is satisfied for a value of α in $(\pi/6, \pi/4)$. Using the following instructions:

```
>> f=inline('-12*cos(g+a)/sin(g+a)^2-11*cos(a)/sin(a)^2',...
            'a','g','11','12');
>> df=inline('12/sin(g+a)+2*12*cos(g+a)^2/sin(g+a)^3+...
   11/sin(a)+2*11*cos(a)^2/sin(a)^3','a','g','11','12')
>> [zero,res,niter]=newton(f,df,pi/4,1.e-15,100,3*pi/5,8,10);
```

the Newton method provides the approximate value 0.59627992746547 in 6 iterations, starting from $x^{(0)} = \pi/4$. We deduce that the maximum length of a rod that can pass in the corridor is $L = 30.84$.

Solution 2.15 If α is a zero of f with multiplicity m, then there exists a function h such that $h(\alpha) \neq 0$ and $f(x) = h(x)(x-\alpha)^m$. By computing the first derivative of the iteration function of the Newton method, we have

$$\phi'_N(x) = 1 - \frac{[f'(x)]^2 - f(x)f''(x)}{[f'(x)]^2} = \frac{f(x)f''(x)}{[f'(x)]^2}.$$

By replacing f, f' and f'' with the corresponding expressions as functions of $h(x)$ and $(x-\alpha)^m$, we obtain $\lim_{x \to \alpha} \phi'_N(x) = 1 - 1/m$, hence $\phi'_N(\alpha) = 0$ if and only if $m = 1$. Consequently, if $m = 1$ the method converges at least quadratically, according to (2.9). If $m > 1$ the method converges with order 1 following Proposition 2.1.

Solution 2.16 Let us inspect the graph of f by using the following commands:

```
>> f= 'x.^3+4*x.^2-10'; fplot(f,[-10,10]); grid on;
>> fplot(f,[-5,5]); grid on;
>> fplot(f,[0,5]); grid on
```

We can see that f has only one real zero, equal approximately to 1.36 (see Figure 9.3). The iteration function and its derivative are:

$$\phi(x) = \frac{2x^3 + 4x^2 + 10}{3x^2 + 8x} = -\frac{f(x)}{3x^2 + 8x} + x,$$

$$\phi'(x) = \frac{(6x^2 + 8x)(3x^2 + 8x) - (6x + 8)(2x^3 + 4x^2 + 10)}{(3x^2 + 8x)^2},$$

and $\phi(\alpha) = \alpha$. We easily deduce that $\phi'(\alpha) = 0$ by noting that $\phi'(x) = (6x+8)f(x)/(3x^2+8x)^2$. Consequently, the proposed method converges (at least) quadratically.

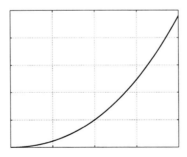

Fig. 9.3. Graph of $f(x) = x^3 + 4x^2 - 10$ for $x \in [0, 2]$

Solution 2.17 The proposed method is convergent at least with order 2 since $\phi'(\alpha) = 0$.

Solution 2.18 By keeping the remaining parameters unchanged, the method converges after only 3 iterations to the value 0.64118573649623 which differs by less than 10^{-9} from the result previously computed. However, the behavior of the function, which is quite flat near $x = 0$, suggests that the result computed previously could be more accurate. In Figure 9.4 we show the graph of f in $(0.5, 0.7)$, obtained with the following instructions:

```
>> f='x^3-3*x^2*2^(-x) + 3*x*4^(-x) - 8^(-x)';
>> fplot(f,[0.5 0.7]); grid on
```

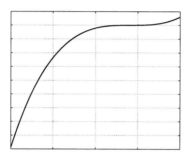

Fig. 9.4. Graph of $f(x) = x^3 - 3x^2 2^{-x} + 3x4^{-x} - 8^{-x}$ for $x \in [0.5, 0.7]$

9.3 Chapter 3

Solution 3.1 Since $x \in (x_0, x_n)$, there exists an interval $I_i = (x_{i-1}, x_i)$ such that $x \in I_i$. We can easily see that $\max_{x \in I_i} |(x - x_{i-1})(x - x_i)| = h^2/4$. If we bound $|x - x_{i+1}|$ above by $2h$, $|x - x_{i-2}|$ by $3h$ and so on, we obtain the inequality (3.6).

Solution 3.2 In all cases we have $n = 4$ and thus we should estimate the fifth derivative of each function in the given interval. We find: $\max_{x \in [-1,1]} |f_1^{(5)}| <$ 1.18, $\max_{x \in [-1,1]} |f_2^{(5)}| < 1.54$, $\max_{x \in [-\pi/2, \pi/2]} |f_3^{(5)}| < 1.41$. The corresponding errors are therefore bounded by 0.0018, 0.0024 and 0.0211, respectively.

Solution 3.3 Using the command `polyfit` we compute the interpolating polynomials of degree 3 in the two cases:

```
>> years=[1975 1980 1985 1990];
>> east=[70.2 70.2 70.3 71.2];
>> west=[72.8 74.2 75.2 76.4];
>> ceast=polyfit(years,east,3);
>> cwest=polyfit(years,west,3);
>> esteast=polyval(ceast,[1970 1983 1988 1995])
```

```
esteast =
   69.6000    70.2032    70.6992    73.6000
>> estwest=polyval(cwest,[1970 1983 1988 1995])
estwest =
   70.4000    74.8096    75.8576    78.4000
```

Thus, for Western Europe the life expectation in the year 1970 is equal to 70.4 years (`estwest(1)`), with a discrepancy of 1.4 years from the real value. The symmetry of the graph of the interpolating polynomial suggests that the estimation for the life expectation of 78.4 years for the year 1995, can be overestimated by the same quantity (in fact, the real life expectation is equal to 77.5 years). A different conclusion holds concerning Eastern Europe. Indeed, in that case the estimation for 1970 coincides exactly with the real value, while the estimation for 1995 is largely overestimated (73.6 years instead of 71.2).

Solution 3.4 We choose the month as time-unit. The initial time $t_0 = 1$ corresponds to November 1987, while $t_7 = 157$ to November 2000. With the following instructions we compute the coefficients of the polynomial interpolating the given prices:

```
>> time = [1 14 37 63 87 99 109 157];
>> price = [4.5 5 6 6.5 7 7.5 8 8];
>> [c] = polyfit(time,price,7);
```

Setting `[price2002]= polyval(c,181)` we find that the estimated price of the magazine in November 2002 is approximately 11.2 euros.

Solution 3.5 The interpolatory cubic spline, computed by the command `spline` in this special case, coincides with the interpolating polynomial. This wouldn't be true for the natural interpolating cubic spline.

Solution 3.6 We use the following instructions:

```
>> T = [4:4:20];
>> rho=[1000.7794,1000.6427,1000.2805,999.7165,998.9700];
>> Tnew = [6:4:18]; format long e;
>> rhonew = spline(T,rho,Tnew)
rhonew =
  Columns 1 through 2
     1.000740787500000e+03     1.000488237500000e+03
  Columns 3 through 4
     1.000022450000000e+03     9.993649250000000e+02
```

The comparison with the further measures shows that the approximation is extremely accurate. Note that the state equation for the sea-water (UNESCO, 1980) assumes a fourth-order dependence of the density on the temperature. However, the coefficient of the fourth power of T is of order of 10^{-9}.

Solution 3.7 We compare the results computed using the interpolatory cubic spline obtained using the MATLAB command `spline` (denoted with `s3`), the

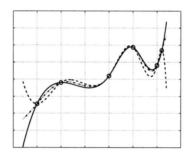

Fig. 9.5. The cubic splines s3 (*continuous line*), s3d (*dashed line*) and s3n (*dotted line*) for the data of Exercise 3.7. The circles denote the values used in the interpolation

interpolatory natural spline (s3n) and the interpolatory spline with null first derivatives at the endpoints of the interpolatory interval (s3d) (computed with Program 3.1). We use the following instructions:

```
>> year=[1965 1970 1980 1985 1990 1991];
>> production=[17769 24001 25961 34336 29036 33417];
>> z=[1962:0.1:1992];
>> s3  = spline(year,production,z);
>> s3n = cubicspline(year,production,z);
>> s3d = cubicspline(year,production,z,0,[0 0]);
```

In the following table we resume the computed values (expressed in thousands of tons of goods):

year	1962	1977	1992
s3	514.6	2264.2	4189.4
s3n	1328.5	2293.4	3779.8
s3d	2431.3	2312.6	2216.6

The comparison with the real data (1238, 2740.3 and 3205.9 thousands of tons, respectively) shows that the values predicted by the natural spline are accurate also outside the interpolation interval (see Figure 9.5). On the contrary, the interpolating polynomial introduces large oscillations near this end-point and underestimates the production of as many as -7768.5×10^6 Kg for 1962.

Solution 3.8 The interpolating polynomial p and the spline s3 can be evaluated by the following instructions:

```
>> pert = 1.e-04;
>> x=[-1:2/20:1]; y=sin(2*pi*x)+(-1).^[1:21]*pert; z=[-1:0.01:1];
>> c=polyfit(x,y,20); p=polyval(c,z); s3=spline(x,y,z);
```

When we use the unperturbed data (pert=0) the graphs of both p and s3 are indistinguishable from that of the given function. The situation changes dramatically when the perturbed data are used (pert=1.e-04). In particular, the interpolating polynomial shows strong oscillations at the end-points of the interval, whereas the spline remains practically unchanged (see Figure 9.6). This example shows that approximation by splines is in general more stable with respect to perturbation errors.

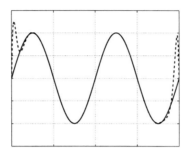

Fig. 9.6. The interpolating polynomial (*dotted line*) and the interpolatory cubic spline (*continuous line*) corresponding to the perturbed data. Note the severe oscillations of the interpolating polynomial near the end-points of the interval

Solution 3.9 If $n = m$, setting $\tilde{f} = \Pi_n f$ we find that the first member of (3.21) is null. Thus in this case $\Pi_n f$ is the solution of the least-squares problem. Since the interpolating polynomial is unique, we deduce that this is the only solution to the least-squares problem.

Solution 3.10 The coefficients (obtained by the command `polyfit`) of the requested polynomials are (only the first 4 significant digits are shown):

$K = 0.67$, $a_4 = 6.301 \ 10^{-8}$, $a_3 = -8.320 \ 10^{-8}$, $a_2 = -2.850 \ 10^{-4}$, $a_1 = 9.718 \ 10^{-4}$, $a_0 = -3.032$;
$K = 1.5$, $a_4 = -4.225 \ 10^{-8}$, $a_3 = -2.066 \ 10^{-6}$, $a_2 = 3.444 \ 10^{-4}$, $a_1 = 3.364 10^{-3}$, $a_0 = 3.364$;
$K = 2$, $a_4 = -1.012 \ 10^{-7}$, $a_3 = -1.431 \ 10^{-7}$, $a_2 = 6.988 \ 10^{-4}$, $a_1 = -1.060 \ 10^{-4}$, $a_0 = 4.927$;
$K = 3$, $a_4 = -2.323 \ 10^{-7}$, $a_3 = 7.980 \ 10^{-7}$, $a_2 = 1.420 \ 10^{-3}$, $a_1 = -2.605 \ 10^{-3}$, $a_0 = 7.315$.

In Figure 9.7 we show the graph of the polynomial computed using the data in the column with $K = 0.67$ of Table 3.1.

Solution 3.11 By repeating the first 3 instructions reported in Solution 3.7 and using the command `polyfit`, we find the following values (in 10^5 Kg): 15280.12 in 1962; 27407.10 in 1977; 32019.01 in 1992, which represent good approximations to the real ones (12380, 27403 and 32059, respectively).

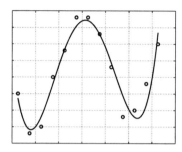

Fig. 9.7. Least-squares polynomial of degree 4 (*continuous line*) compared with the data in the first column of Table 3.1

Solution 3.12 We can rewrite the coefficients of the system (3.23) in terms of mean and variance by noting that the variance can be expressed as $v = \frac{1}{n+1}\sum_{i=0}^{n} x_i^2 - M^2$.

Solution 3.13 The desired property is deduced from the first equation of the system that provides the coefficients of the least-squares straight line.

Solution 3.14 We can use the command `interpft` as follows:

```
>> discharge = [0 35 0.125 5 0 5 1 0.5 0.125 0];
>> y =interpft(discharge,100);
```

The graph of the obtained solution is reported in Figure 9.8.

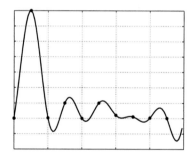

Fig. 9.8. The trigonometric interpolant obtained using the instructions in Solution 3.14. Dots refer to the experimental data available

9.4 Chapter 4

Solution 4.1 Using the following third-order Taylor expansions of f at the point x_0, we obtain

$$f(x_1) = f(x_0) + hf'(x_0) + \frac{h^2}{2}f''(x_0) + \frac{h^3}{6}f'''(\xi_1),$$
$$f(x_2) = f(x_0) + 2hf'(x_0) + 2h^2 f''(x_0) + \frac{4h^3}{3}f'''(\xi_2),$$

with $\xi_1 \in (x_0, x_1)$ and $\xi_2 \in (x_0, x_2)$ as two suitable points. Summing this two expressions yields

$$\frac{1}{2h}[-3f(x_0) + 4f(x_1) - f(x_2)] = f'(x_0) + \frac{h^2}{3}[f'''(\xi_1) - 2f'''(\xi_2)],$$

then the thesis follows for a suitable $\xi_0 \in (x_0, x_2)$. A similar procedure can be used for the formula at x_n.

Solution 4.2 Taylor expansions yield

$$f(\bar{x} + h) = f(\bar{x}) + hf'(\bar{x}) + \frac{h^2}{2}f''(\bar{x}) + \frac{h^3}{6}f'''(\xi),$$
$$f(\bar{x} - h) = f(\bar{x}) - hf'(\bar{x}) + \frac{h^2}{2}f''(\bar{x}) - \frac{h^3}{6}f'''(\eta),$$

where ξ and η are suitable points. Subtracting these two expressions and dividing by $2h$ we obtain the result (4.10).

Solution 4.3 Assuming that $f \in C^4$ and proceeding as in Solution 4.2 we obtain the following errors (for suitable points ξ_1, ξ_2 and ξ_3):

a. $-\frac{1}{4}f^{(4)}(\xi_1)h^3$, b. $-\frac{1}{12}f^{(4)}(\xi_2)h^3$, c. $\frac{1}{30}f^{(4)}(\xi_3)h^4$.

Solution 4.4 Using the approximation (4.9), we obtain the following values:

t (months)	0	0.5	1	1.5	2	2.5	3
δn	--	78	45	19	7	3	--
n'	--	77.91	39.16	15.36	5.91	1.99	--

By comparison with the exact values of $n'(t)$ we can conclude that the computed values are sufficiently accurate.

Solution 4.5 The quadrature error can be bounded by

$$(b-a)^3/(24M^2) \max_{x \in [a,b]} |f''(x)|,$$

where $[a, b]$ is the integration interval and M the (unknown) number of subintervals.

The function f_1 is infinitely differentiable. From the graph of f_1'' we infer that $|f_1''(x)| \leq 2$ in the integration interval. Thus the integration error for f_1 is less than 10^{-4} provided that $5^3/(24M^2)2 < 10^{-4}$, that is $M > 322$.

Also the function f_2 is differentiable to any order. Since $\max_{x \in [0,\pi]} |f_2''(x)| = \sqrt{2}e^{3/4\pi}$, the integration error is less than 10^{-4} provided that $M > 439$. These inequalities actually provide an over estimation of the integration errors. Indeed, the (effective) minimum number of intervals which ensures that the error is below the fixed tolerance of 10^{-4} is much lower than that predicted by our result (for instance, for the function f_1 this number is 51). Finally, we note that since f_3 is not differentiable in the integration interval, our theoretical error estimate doesn't hold.

Solution 4.6 On each interval I_k, $k = 1, \ldots, M$, the error is equal to $H^3/24 f''(\xi_k)$ with $\xi_k \in (x_{k-1}, x_k)$ and hence the global error will be $H^3/24 \sum_{k=1}^{M} f''(\xi_k)$. Since f'' is a continuous function in (a,b) there exists a point $\xi \in (a,b)$ such that $f''(\xi) = \frac{1}{M} \sum_{k=1}^{M} f''(\xi_k)$. Using this result and the fact that $MH = b - a$, we derive equation (4.14).

Solution 4.7 This effect is due to the accumulation of local errors on each sub-interval.

Solution 4.8 By construction, the mid-point formula integrates exactly the constants. To verify that the linear polynomials also are exactly integrated, it is sufficient to verify that $I(x) = I_{PM}(x)$. As a matter of fact we have

$$I(x) = \int_a^b x \, dx = \frac{b^2 - a^2}{2}, \quad I_{PM}(x) = (b-a)\frac{b+a}{2}.$$

Solution 4.9 For the function f_1 we find $M = 71$ if we use the trapezoidal formula and only $M = 7$ for the Gauss formula. Indeed, the computational advantage of this latter formula is evident.

Solution 4.10 Equation (4.18) states that the quadrature error for the composite trapezoidal formula with $H = H_1$ is equal to CH_1^2, with $C = -\frac{b-a}{12} f''(\xi)$. If f'' does not vary "too much", we can assume that also the error with $H = H_2$ behaves like CH_2^2. Then, by equating the two expressions

$$I(f) \simeq I_1 + CH_1^2, \quad I(f) \simeq I_2 + CH_2^2, \tag{9.3}$$

we obtain $C = (I_1 - I_2)/(H_2^2 - H_1^2)$. Using this value in one of the expressions (9.3), we obtain equation (4.32), that is, a better approximation than the one produced by I_1 or I_2.

Solution 4.11 We seek the maximum positive integer p such that $I_{approx}(x^p) = I(x^p)$. For $p = 0, 1, 2, 3$ we find the following nonlinear system with 4 equations in the 4 unknowns α, β, $\bar{x}$ and $\bar{z}$:

$$p = 0 \to \alpha + \beta = b - a,$$
$$p = 1 \to \alpha\bar{x} + \beta\bar{z} = \frac{b^2 - a^2}{2},$$
$$p = 2 \to \alpha\bar{x}^2 + \beta\bar{z}^2 = \frac{b^3 - a^3}{3},$$
$$p = 3 \to \alpha\bar{x}^3 + \beta\bar{z}^3 = \frac{b^4 - a^4}{4}.$$

From the first two equations we can eliminate α and $\bar{z}$ and reduce the system to a new one in the unknowns β and $\bar{x}$. In particular, we find a second-order equation in β from which we can compute β as a function of $\bar{x}$. Finally, the nonlinear equation in $\bar{x}$ can be solved by the Newton method, yielding two values of $\bar{x}$ that are the abscissae of the Gauss quadrature points.

Solution 4.12 Since

$$f_1^{(4)}(x) = \frac{24}{(1+(x-\pi)^2)^5(2x-2\pi)^4} - \frac{72}{(1+(x-\pi)^2)^4(2x-2\pi)^2}$$
$$+ \frac{24}{(1+(x-\pi)^2)^3},$$
$$f_2^{(4)}(x) = -4e^x \cos(x),$$

we find that the maximum of $|f_1^{(4)}(x)|$ is bounded by $M_1 \simeq 25$, while that of $|f_2^{(4)}(x)|$ by $M_2 \simeq 93$. Consequently, from (4.22) we obtain $H < 0.21$ in the first case and $H < 0.16$ in the second case.

Solution 4.13 Using the command int('exp(-x^2/2)',0,2) we obtain for the integral at hand the value 1.19628801332261.

The Gauss formula applied to the same interval would provide the value 1.20278027622354 (with an absolute error equal to 6.4923e-03), while the Simpson formula gives 1.18715264069572 with a slightly larger error (equal to 9.1354e-03).

Solution 4.14 We note that $I_k > 0 \; \forall k$, since the integrand is non-negative. Therefore, we expect that all the values produced by the recursive formula should be non-negative. Unfortunately, the recursive formula is unstable to the propagation of roundoff errors and produces negative elements:

```
>> I(1)=1/exp(1); for k=2:20, I(k)=1-k*I(k-1); end
>> I(20)
    -30.1924
```

Using the composite Simpson formula, with $H < 0.25$, we can compute the integral with the desired accuracy.

Solution 4.15 For the Simpson formula we obtain

$$I_1 = 1.19616568040561, \; I_2 = 1.19628173356793, \; \Rightarrow \; I_R = 1.19628947044542,$$

with an absolute error in I_R equal to -1.4571e-06 (we gain two orders of magnitude with respect to I_1 and a factor $1/4$ with respect to I_2). Using the Gauss formula we obtain (the errors are reported between parentheses):

$$I_1 = 1.19637085545393 \; (-8.2842e-05),$$
$$I_2 = 1.19629221796844 \; (-4.2046e-06),$$
$$I_R = 1.19628697546941 \; (1.0379e-06).$$

The advantage of using the Richardson extrapolation method is evident.

Solution 4.16 We must compute by the Simpson formula the values $j(r) = \sigma/(\varepsilon_0 r^2) \int_0^r f(\xi) d\xi$ with $r = k/10$, for $k = 1, \ldots, 10$ and $f(\xi) = e^\xi \xi^2$.

In order to estimate the integration error we need the fourth derivative $f^{(4)}(\xi) = e^\xi(\xi^2 + 8\xi + 12)$. The maximum of $f^{(4)}$ in the integration interval $(0,r)$ is attained at $\xi = r$ since $f^{(4)}$ is monotonically increasing. Then we obtain the following values:

```
>> r=[0.1:0.1:1];
>> maxf4=exp(r).*(r.^2+8*r+12);
maxf4 =
  Columns 1 through 6
   14.1572   16.6599   19.5595   22.9144   26.7917   31.2676
  Columns 7 through 10
   36.4288   42.3743   49.2167   57.0839
```

For a given r the error is below 10^{-10} provided that $H_r^4 < 10^{-10} 2880/(r f^{(4)}(r))$. For $r = k/10$ with $k = 1, \ldots, 10$ by the following instructions we can compute the minimum numbers of subintervals which ensure that the previous inequalities are satisfied. The components of the vector M contain these numbers:

```
>> x=[0.1:0.1:1]; f4=exp(x).*(x.^2+8*x+12);
>> H=(10^(-10)*2880./(x.*f4)).^(1/4); M=fix(x./H)
M =
    4    11    20    30    41    53    67    83   100   118
```

Therefore, the values of $j(r)$ are:

```
>> sigma=0.36; epsilon0 = 8.859e-12;
   f = inline('exp(x).*x.^2');
   for k = 1:10
      r = k/10;
      j(k)=simpsonc(0,r,M(k),f);
      j(k) = j(k)*sigma/r*epsilon0;
   end
```

Solution 4.17 We compute $E(213)$ using the Simpson composite formula by increasing the number of intervals until the difference between two consecutive approximations (divided by the last computed value) is less than 10^{-11}:

```
>> f=inline('2.39e-11./((x.^5).*(exp(1.432./(T*x))-1))','x','T');
>> a=3.e-04; b=14.e-04; T=213;
>> i=2; err = 1; Iold = 0; while err >= 1.e-11
I=simpsonc(a,b,i,f,T);
err = abs(I-Iold)/abs(I);
Iold=I;
i=i+1;
end
```

The procedure returns the value $i = 59$. Therefore, using 58 equispaced intervals we can compute the integral $E(213)$ with ten exact significant digits. The same result could be obtained by the Gauss formula using 53 intervals. Note that as many as 1609 intervals would be nedeed if using the composite trapezoidal formula.

Solution 4.18 On the whole interval the given function is not regular enough to allow the application of the theoretical convergence result (4.22). One possibility is to decompose the integral into the sum of two intervals, $(0, 0.5)$ and $(0.5, 1)$, in which the function is regular (it is actually a polynomial of degree 3). In particular, if we use the Simpson rule on each interval we can even integrate f exactly.

9.5 Chapter 5

Solution 5.1 The number r_k of algebraic operations (sums, subtractions and multiplications) required to compute a determinant of a matrix of order $k \geq 2$ with the Laplace rule (1.8), satisfies the following difference equation:

$$r_k - kr_{k-1} = 2k - 1,$$

with $r_1 = 0$. Multiplying both side of this equation by $1/k!$, we obtain

$$\frac{r_k}{k!} - \frac{r_{k-1}}{(k-1)!} = \frac{2k-1}{k!}.$$

Summing both sides from 2 to n gives the solution:

$$r_n = n! \sum_{k=2}^{n} \frac{2k-1}{k!} = n! \sum_{k=1}^{n-1} \frac{2k+1}{(k+1)!}, \qquad n \geq 1.$$

Solution 5.2 We use the following MATLAB commands to compute the determinants and the corresponding CPU-times:

```
>> t = [ ]; for i = 3:500
     A = magic(i); tt = cputime; d=det(A); t=[t, cputime-tt];
   end
```

The coefficients of the cubic least-squares polynomial that approximate the data n=[3:500] and t are

```
>> format long; c=polyfit(n,t,3)
c =
  Columns 1 through 3
    0.00000002102187   0.00000171915661  -0.00039318949610
  Column 4
    0.01055682398911
```

The first coefficient (that multiplies n^3), is small, but not small enough with respect to the second one to be neglected. Indeed, if we compute the fourth degree least-squares polynomial we obtain the following coefficients:

```
>> c=polyfit(i,t,4)
c =
  Columns 1 through 3
   -0.00000000000051   0.00000002153039   0.00000155418071
  Columns 4 through 6
   -0.00037453657810  -0.00037453657810   0.01006704351509
```

From this result, we can conclude that the computation of a determinant of a matrix of dimension n requires approximately n^3 operations.

Solution 5.3 We have: $\det A_1 = 1$, $\det A_2 = \varepsilon$, $\det A_3 = \det A = 2\varepsilon + 12$. Consequently, if $\varepsilon = 0$ the second principal submatrix is singular and the Proposition 5.1 cannot be applied. The matrix is singular if $\varepsilon = -6$. In this case the Gauss factorization yields

286   9 Solutions of the exercises

$$L = \begin{bmatrix} 1 & 0 & 0 \\ 2 & 1 & 0 \\ 3 & 1.25 & 1 \end{bmatrix}, U = \begin{bmatrix} 1 & 7 & 3 \\ 0 & -12 & -4 \\ 0 & 0 & 0 \end{bmatrix}.$$

Note that U is singular (as we could have predicted since A is singular).

Solution 5.4 At step 1, $n-1$ divisions were used to calculate the l_{1k} entries for $i = 2,\ldots,n$. Then $(n-1)^2$ multiplications and $(n-1)^2$ additions were used to create the new entries $a_{ij}^{(2)}$, for $j = 2,\ldots,n$. At step 2, the numbers of divisions is $(n-2)$, while the numbers of multiplications and additions will be $(n-2)^2$. At final step $n-1$ only 1 addition, 1 multiplication and 1 division is required. Thus, using the identies

$$\sum_{s=1}^{q} s = \frac{q(q+1)}{2}, \sum_{s=1}^{q} s^2 = \frac{q(q+1)(2q+1)}{6}, \quad q \geq 1,$$

we can conclude that to complete the Gaussian factorization $2(n-1)n(n+1)/3+n(n-1)$ operations are required. Neglecting the lower order terms, we can state that the Gaussian factorization process has a cost of $2n^3/3$ operations.

Solution 5.5 By definition, the inverse X of a matrix $A \in \mathbb{R}^{n \times n}$ satisfies $XA = AX = I$. Therefore, for $j = 1,\ldots,n$ the column vector $\mathbf{y}_j$ of X is the solution of the linear system $A\mathbf{y}_j = \mathbf{e}_j$, where $\mathbf{e}_j$ is the j-th vector of the canonical basis of $\mathbb{R}^n$ with all components equal to zero except the j-th that is equal to 1. After computing the LU factorization of A, the computation of the inverse of A requires the solution of n linear systems with the same matrix and different right-hand sides.

Solution 5.6 Using the Program 5.1 we compute the L and U factors:

$$L = \begin{bmatrix} 1 & 0 & 0 \\ 2 & 1 & 0 \\ 3 & -3.38 \cdot 10^{15} & 1 \end{bmatrix}, U = \begin{bmatrix} 1 & 1 & 3 \\ 0 & -8.88 \cdot 10^{-16} & 14 \\ 0 & 0 & 4.73 \cdot 10^{-16} \end{bmatrix}.$$

If we compute their product we obtain the matrix

```
>> L*U
ans =
    1.0000    1.0000    3.0000
    2.0000    2.0000   20.0000
    3.0000    6.0000   -2.0000
```

which differs from A since the entry in position (3,3) is equal to -2 while in A it is equal to 4.

Solution 5.7 Usually, only the triangular (upper or lower) part of a symmetric matrix is stored. Therefore, any operation that does not respect the symmetry of the matrix is not optimal in view of the memory storage. This is the case when row pivoting is carried out. A possibility is to exchange simultaneously rows and columns having the same index, limiting therefore the choice of the pivot only to the diagonal elements. More generally, a pivoting strategy involving exchange of rows and columns is called *complete pivoting* (see, e.g., [QSS06, Chap. 3]).

Solution 5.8 The L and U factors are:

$$L = \begin{bmatrix} 1 & 0 & 0 \\ (\varepsilon - 2)/2 & 1 & 0 \\ 0 & -1/\varepsilon & 1 \end{bmatrix}, \quad U = \begin{bmatrix} 2 & -2 & 0 \\ 0 & \varepsilon & 0 \\ 0 & 0 & 3 \end{bmatrix}.$$

When $\varepsilon \to 0$ $l_{32} \to \infty$. In spite of that, the solution of the system is accurate also when ε tends to zero as confirmed by the following instructions:

```
>> e=1; for k=1:10
b=[0; e; 2];
L=[1 0 0; (e-2)*0.5 1 0; 0 -1/e 1]; U=[2 -2 0; 0 e 0; 0 0 3];
y=L\b; x=U\y; err(k)=max(abs(x-ones(3,1))); e=e*0.1;
end
>> err
err =
     0    0    0    0    0    0    0    0    0    0
```

Solution 5.9 The computed solutions become less and less accurate when i increases. Indeed, the error norms are equal to $2.63 \cdot 10^{-14}$ for $i = 1$, to $9.89 \cdot 10^{-10}$ for $i = 2$ and to $2.10 \cdot 10^{-6}$ for $i = 3$. This can be explained by observing that the condition number of A_i increases as i increases. Indeed, using the command cond we find that the condition number of A_i is $\simeq 10^3$ for $i = 1$, $\simeq 10^7$ for $i = 2$ and $\simeq 10^{11}$ for $i = 3$.

Solution 5.10 If $(\lambda, \mathbf{v})$ are an eigenvalue-eigenvector pair of a matrix A, then λ^2 is an eigenvalue of A^2 with the same eigenvector. Indeed, from $A\mathbf{v} = \lambda \mathbf{v}$ follows $A^2\mathbf{v} = \lambda A\mathbf{v} = \lambda^2 \mathbf{v}$. Consequently, if A is symmetric and positive definite $K(A^2) = (K(A))^2$.

Solution 5.11 The iteration matrix of the Jacobi method is:

$$B_J = \begin{bmatrix} 0 & 0 & -\alpha^{-1} \\ 0 & 0 & 0 \\ -\alpha^{-1} & 0 & 0 \end{bmatrix}.$$

Its eigenvalues are $\{0, \alpha^{-1}, -\alpha^{-1}\}$. Thus the method converges if $|\alpha| > 1$.

The iteration matrix of the Gauss-Seidel method is

$$B_{GS} = \begin{bmatrix} 0 & 0 & -\alpha^{-1} \\ 0 & 0 & 0 \\ 0 & 0 & \alpha^{-2} \end{bmatrix}$$

with eigenvalues $\{0, 0, \alpha^{-2}\}$. Therefore, the method converges if $|\alpha| > 1$. In particular, since $\rho(B_{GS}) = [\rho(B_J)]^2$, the Gauss-Seidel converges more rapidly than the Jacobi method.

Solution 5.12 A sufficient condition for the convergence of the Jacobi and the Gauss-Seidel methods is that A is strictly diagonally dominant. The second row of A satisfies the condition of diagonal dominance provided that $|\beta| < 5$. Note that if we require directly that the spectral radii of the iteration matrices are less than 1 (which is a sufficient and necessary condition for convergence), we find the (less restrictive) limitation $|\beta| < 25$ for both methods.

288   9 Solutions of the exercises

Solution 5.13 The relaxation method in vector form is

$$(I - \omega D^{-1}E)\mathbf{x}^{(k+1)} = [(1 - \omega)I + \omega D^{-1}F]\mathbf{x}^{(k)} + \omega D^{-1}\mathbf{b}$$

where $A = D - E - F$, D being the diagonal of A, and E and F the lower (resp. upper) part of A. The corresponding iteration matrix is

$$B(\omega) = (I - \omega D^{-1}E)^{-1}[(1 - \omega)I + \omega D^{-1}F].$$

If we denote by λ_i the eigenvalues of $B(\omega)$, we obtain

$$\left| \prod_{i=1}^{n} \lambda_i \right| = \left| \det \left[(1 - \omega)I + \omega D^{-1}F \right] \right| = |1 - \omega|^n.$$

Therefore, at least one eigenvalue must satisfy the inequality $|\lambda_i| \geq |1 - \omega|$. Thus, a necessary condition to ensure convergence is that $|1 - \omega| < 1$, that is, $0 < \omega < 2$.

Solution 5.14 The given matrix is symmetric. To verify whether it is also definite positive, that is, $\mathbf{z}^T A \mathbf{z} > 0$ for all $\mathbf{z} \neq \mathbf{0}$ of $\mathbb{R}^2$, we use the following instructions:

```
>> syms z1 z2 real
>> z=[z1;z2]; A=[3 2; 2 6];
>> pos=z'*A*z; simple(pos)
   ans =
3*z1^2+4*z1*z2+6*z2^2
```

The command `syms z1 z2 real` is necessary to declare that the symbolic variables z1 and z2 are real numbers, while the command `simple(pos)` tries several algebraic simplifications of pos and returns the shortest. It is easy to see that the computed quantity is positive since it can be rewritten as `2*(z1+z2)^2 +z1^2+4*z2^2`. Thus, the given matrix is symmetric and positive definite, and the Gauss-Seidel method is convergent.

Solution 5.15 We find:

for the Jacobi method: $\begin{cases} x_1^{(1)} = \frac{1}{2}(1 - x_2^{(0)}), \\ x_2^{(1)} = -\frac{1}{3}(x_1^{(0)}); \end{cases} \Rightarrow \begin{cases} x_1^{(1)} = \frac{1}{4}, \\ x_2^{(1)} = -\frac{1}{3}; \end{cases}$

for the Gauss-Seidel method: $\begin{cases} x_1^{(1)} = \frac{1}{2}(1 - x_2^{(0)}), \\ x_2^{(1)} = -\frac{1}{3}x_1^{(1)}, \end{cases} \Rightarrow \begin{cases} x_1^{(1)} = \frac{1}{4}, \\ x_2^{(1)} = -\frac{1}{12}; \end{cases}$

for the gradient method, we first compute the initial residual

$$\mathbf{r}^{(0)} = \mathbf{b} - A\mathbf{x}^{(0)} = \begin{bmatrix} 1 \\ 0 \end{bmatrix} - \begin{bmatrix} 2 & 1 \\ 1 & 3 \end{bmatrix} \mathbf{x}^{(0)} = \begin{bmatrix} -3/2 \\ -5/2 \end{bmatrix}.$$

Then, since

$$P^{-1} = \begin{bmatrix} 1/2 & 0 \\ 0 & 1/3 \end{bmatrix},$$

we have $\mathbf{z}^{(0)} = P^{-1}\mathbf{r}^{(0)} = (-3/4, -5/6)^T$. Therefore

$$\alpha_0 = \frac{(\mathbf{z}^{(0)})^T \mathbf{r}^{(0)}}{(\mathbf{z}^{(0)})^T A \mathbf{z}^{(0)}} = \frac{77}{107},$$

and

$$\mathbf{x}^{(1)} = \mathbf{x}^{(0)} + \alpha_0 \mathbf{z}^{(0)} = (197/428, -32/321)^T.$$

Solution 5.16 In the stationary case, $\rho(B_\alpha) = \min_\lambda |1 - \alpha\lambda|$, where λ are the eigenvalues of $P^{-1}A$. The optimal value of α is obtained solving the equation $|1 - \alpha\lambda_{min}| = |1 - \alpha\lambda_{max}|$, that is $1 - \alpha\lambda_{min} = -1 + \alpha\lambda_{max}$, which yields (5.48). Since,

$$\rho(B_\alpha) = 1 - \alpha\lambda_{min} \ \forall \alpha \leq \alpha_{opt},$$

for $\alpha = \alpha_{opt}$ we obtain (5.59).

Solution 5.17 In this case the matrix associated to the Leontieff model is not positive definite. Indeed, using the following instructions:

```
>> for i=1:20; for j=1:20; c(i,j)=i+j; end; end; A=eye(20)-c;
>> min(eig(A))
ans =
  -448.5830
>> max(eig(A))
ans =
   30.5830
```

we can see that the minimum eigenvalue is a negative number and the maximum eigenvalue is a positive number. Therefore, the convergence of the gradient method is not guaranteed. However, since A is nonsingular, the given system is equivalent to the system $A^T A \mathbf{x} = A^T \mathbf{b}$, where $A^T A$ is symmetric and positive definite. We solve the latter by the gradient method requiring that the norm of the residual be less than 10^{-10} and starting from the initial data $\mathbf{x}^{(0)} = 0$:

```
>> b = [1:20]'; aa=A'*A; b=A'*b; x0 = zeros(20,1);
>> [x,iter]=itermeth(aa,b,x0,100,1.e-10);
```

The method converges in 15 iterations. A drawback of this approach is that the condition number of the matrix $A^T A$ is, in general, larger than the condition number of A.

9.6 Chapter 6

Solution 6.1 A_1: the power method converges in 34 iterations to the value 2.00000000004989. A_2: starting from the same initial vector, the power method requires now 457 iterations to converge to the value 1.99999999990611. The slower convergence rate can be explained by observing that the two largest eigenvalues are very close one another. Finally, for the matrix A_3 the method doesn't converge since A_3 features two distinct eigenvalues (i and $-i$) of maximum modulus.

Solution 6.2 The Leslie matrix associated with the values in the table is

$$A = \begin{bmatrix} 0 & 0.5 & 0.8 & 0.3 \\ 0.2 & 0 & 0 & 0 \\ 0 & 0.4 & 0 & 0 \\ 0 & 0 & 0.8 & 0 \end{bmatrix}.$$

Using the power method we find $\lambda_1 \simeq 0.5353$. The normalized distribution of this population for different age intervals is given by the components of the corresponding unitary eigenvector, that is, $\mathbf{x}_1 \simeq (0.8477, 0.3167, 0.2367, 0.3537)^T$.

Solution 6.3 We rewrite the initial guess as

$$\mathbf{y}^{(0)} = \beta^{(0)} \left(\alpha_1 \mathbf{x}_1 + \alpha_2 \mathbf{x}_2 + \sum_{i=3}^{n} \alpha_i \mathbf{x}_i \right),$$

with $\beta^{(0)} = 1/\|\mathbf{x}^{(0)}\|$. By calculations similar to those carried out in Section 6.1, at the generic step k we find:

$$\mathbf{y}^{(k)} = \gamma^k \beta^{(k)} \left(\alpha_1 \mathbf{x}_1 e^{ik\vartheta} + \alpha_2 \mathbf{x}_2 e^{-ik\vartheta} + \sum_{i=3}^{n} \alpha_i \frac{\lambda_i^k}{\gamma^k} \mathbf{x}_i \right).$$

The first two terms don't vanish and, due to the opposite sign of the exponents, the sequence of the $\mathbf{y}^{(k)}$ oscillates and cannot converge.

Solution 6.4 From the eigenvalue equation $A\mathbf{x} = \lambda \mathbf{x}$, we deduce $A^{-1}A\mathbf{x} = \lambda A^{-1}\mathbf{x}$, and therefore $A^{-1}\mathbf{x} = (1/\lambda)\mathbf{x}$.

Solution 6.5 The power method applied to the matrix A generates an oscillating sequence of approximations of the maximum modulus eigenvalue (see, Figure 9.9). This behavior is due to the fact that this eigenvalue is not unique.

Solution 6.6 To compute the eigenvalue of maximum modulus of A we use Program 6.1:

```
>> A=wilkinson(7);
>> x0=ones(7,1); tol=1.e-15; nmax=100;
>> [lambda,x,iter]=eigpower(A,tol,nmax,x0);
```

After 35 iterations we obtain `lambda=3.76155718183189`. To find the largest negative eigenvalue of A, we can use the power method with shift and, in particular, we can choose a shift equal to the largest positive eigenvalue that we have just computed. We find:

```
>> [lambda2,x,iter]=eigpower(A-lambda*eye(7),tol,nmax,x0);
>> lambda2+lambda
ans =
    -1.12488541976457
```

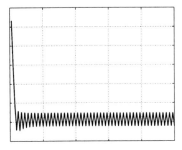

Fig. 9.9. The approximations of the maximum modulus eigenvalue of the matrix of Solution 6.5 computed by the power method

after iter = 33 iterations. These results are satisfactory approximations of the largest (positive and negative) eigenvalues of A.

Solution 6.7 Since all the coefficients of A are real, eigenvalues occur in conjugate pairs. Note that in this situation conjugate eigenvalues must belong to the same Gershgorin circle. The matrix A presents 2 column circles isolated from the others (see Figure 9.10 on the left). Each of them must contain only one eigenvalue that must therefore be real. Then A admits at least 2 real eigenvalues.

Let us consider now the matrix B that admits only one isolated column circle (see Figure 9.10 on the right). Then, thanks to the previous consideration the corresponding eigenvalue must be real. The remaining eigenvalues can be either all real, or one real and 2 complex.

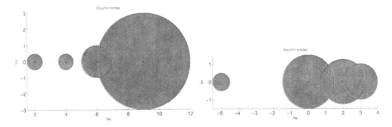

Fig. 9.10. *On the left,* column circles of the matrix A of Solution 6.7. *On the right,* column circles of the matrix B of Solution 6.7

Solution 6.8 The row circles of A feature an isolated circle of center 5 and radius 2 the maximum modulus eigenvalue must belong to. Therefore, we can set the value of the shift equal to 5. The comparison between the number of iterations and the computational cost of the power method with and without shift can be found using the following commands:

292   9 Solutions of the exercises

```
A=[5 0 1 -1; 0 2 0 -1/2; 0 1 -1 1; -1 -1 0 0];
tol=1e-14; x0=[1 2 3 4]';   nmax=1000;
tic; [lambda1,x1,iter1]=eigpower(A,tol,nmax,x0);
toc, iter1

Elapsed time is 0.033607 seconds.
iter1 = 35

tic; [lambda2,x2,iter2]=invshift(A,5,tol,nmax,x0);
toc, iter2

Elapsed time is 0.018944 seconds.
iter2 = 12
```

The power method with shift requires in this case a lower number of iterations (1 versus 3) and almost half the cost than the usual power method (also accounting for the extra time needed to compute the Gauss factorization of A off-line).

Solution 6.9 Using the qr command we have immediately:

```
>> A=[2 -1/2 0 -1/2; 0 4 0 2; -1/2 0 6 1/2; 0 0 1 9];
>> [Q,R]=qr(A)
Q =
    -0.9701    0.0073   -0.2389   -0.0411
          0   -0.9995   -0.0299   -0.0051
     0.2425    0.0294   -0.9557   -0.1643
          0         0   -0.1694    0.9855
R =
    -2.0616    0.4851    1.4552    0.6063
          0   -4.0018    0.1764   -1.9881
          0         0   -5.9035   -1.9426
          0         0         0    8.7981
```

To verify that RQ is similar to A, we observe that

$$Q^T A = Q^T QR = R$$

thanks to the orhogonality of Q. Thus $C = Q^T AQ = RQ$, since $Q^T = Q^{-1}$, and we conclude that C is similar to A.

Solution 6.10 We can use the command eig in the following way: [X,D]=eig (A), where X is the matrix whose columns are the unit eigenvectors of A and D is a diagonal matrix whose elements are the eigenvalues of A. For the matrices A and B of Exercise 6.7 we should execute the following instructions:

```
>> A=[2 -1/2 0 -1/2; 0 4 0 2; -1/2 0 6 1/2; 0 0 1 9];
>> sort(eig(A))
ans =
    2.0000
    4.0268
```

```
      5.8003
      9.1728
>> B=[-5 0 1/2 1/2; 1/2 2 1/2 0; 0 1 0 1/2; 0 1/4 1/2 3];
>> sort(eig(B))
ans =
   -4.9921
   -0.3038
    2.1666
    3.1292
```

9.7 Chapter 7

Solution 7.1 Let us approximate the exact solution $y(t) = \frac{1}{2}[e^t - \sin(t) - \cos(t)]$ of the Cauchy problem (7.72) by the forward Euler method using different values of h: $1/2, 1/4, 1/8, \ldots, 1/512$. The associated error is computed by the following instructions:

```
>> y0=0; f=inline('sin(t)+y','t','y');
>> y='0.5*(exp(t)-sin(t)-cos(t))';
>> tspan=[0 1]; N=2; for k=1:10
 [tt,u]=feuler(f,tspan,y0,N);t=tt(end);e(k)=abs(u(end)-eval(y));
 N=2*N;end
>> e
e =
  Columns 1 through 6
    0.4285    0.2514    0.1379    0.0725    0.0372    0.0189
  Columns 7 through 10
    0.0095    0.0048    0.0024    0.0012
```

Now we apply formula (1.12) to estimate the order of convergence:

```
>>  p=log(abs(e(1:end-1)./e(2:end)))/log(2)
p =
  Columns 1 through 6
    0.7696    0.8662    0.9273    0.9620    0.9806    0.9902
  Columns 7 through 9
    0.9951    0.9975    0.9988
```

As expected the order of convergence is one. With the same instructions (substituting the program `feuler` with the program `beuler`) we obtain an estimate of the convergence order of the backward Euler method:

```
>>  p=log(abs(e(1:end-1)./e(2:end)))/log(2)
p =
  Columns 1 through 6
    1.5199    1.1970    1.0881    1.0418    1.0204    1.0101
  Columns 7 through 9
    1.0050    1.0025    1.0012
```

Solution 7.2 The numerical solution of the given Cauchy problem by the forward Euler method can be obtained as follows:

```
>> tspan=[0 1]; N=100;f=inline('-t*exp(-y)','t','y');y0=0;
>> [t,u]=feuler(f,tspan,y0,N);
```

To compute the number of exact significant digits we can estimate the constants L and M which appear in (7.13). Note that, since $f(t, y(t)) < 0$ in the given interval, $y(t) = \log(1 - t^2/2)$ is a monotonically decreasing function, vanishing at $t = 0$. Since f is continuous together with its first derivative, we can approximate L as $L = \max_{0 \le t \le 1} |L(t)|$ with $L(t) = \partial f/\partial y = te^{-y}$. Note that $L(0) = 0$ and $L(t) > 0$ for all $t \in (0, 1]$. Thus, $L = e$.

Similarly, in order to compute $M = \max_{0 \le t \le 1} |y''(t)|$ with $y'' = -e^{-y} - t^2 e^{-2y}$, we can observe that this function has its maximum at $t = 1$, and then $M = e + e^2$. From (7.13) we deduce

$$|u_{100} - y(1)| \le \frac{e^L - 1}{L} \frac{M}{200} = 0.26.$$

Therefore, there is no guarantee that more than one significant digit be exact. Indeed, we find u(end)=-0.6785, while the exact solution at $t = 1$ is $y(1) = -0.6931$.

Solution 7.3 The iteration function is $\phi(u) = u - ht_{n+1}e^{-u}$ and the fixed-point iteration converges if $|\phi'(u)| < 1$. This property is ensured if $h(t_0 + (n+1)h) < e^u$. If we substitute u with the exact solution, we can provide an a priori estimate of the value of h. The most restrictive situation occurs when $u = -1$ (see Solution 7.2). In this case the solution of the inequality $(n+1)h^2 < e^{-1}$ is $h < \sqrt{e^{-1}/(n+1)}$.

Solution 7.4 We repeat the same set of instructions of Solution 7.1, however now we use the program cranknic (Program 7.3) instead of feuler. According to the theory, we obtain the following result that shows second-order convergence:

```
>> p=log(abs(e(1:end-1)./e(2:end)))/log(2)
p =
  Columns 1 through 6
    2.0379    2.0092    2.0023    2.0006    2.0001    2.0000
  Columns 7 through 9
    2.0000    2.0000    2.0000
```

Solution 7.5 Consider the integral formulation of the Cauchy problem (7.5) in the interval $[t_n, t_{n+1}]$:

$$y(t_{n+1}) - y(t_n) = \int_{t_n}^{t_{n+1}} f(\tau, y(\tau)) d\tau$$
$$\simeq \frac{h}{2} \left[f(t_n, y(t_n)) + f(t_{n+1}, y(t_{n+1})) \right],$$

where we have approximated the integral by the trapezoidal formula (4.19). By setting $u_0 = y(t_0)$ and replacing $y(t_n)$ by the approximate value u_n and the symbol $\simeq$ by $=$, we obtain

$$u_{n+1} = u_n + \frac{h}{2}[f(t_n, u_n) + f(t_{n+1}, u_{n+1})], \quad \forall n \geq 0,$$

which is the Crank-Nicolson method.

Solution 7.6 We must impose the limitation $|1 - h + ih| < 1$, which yields $0 < h < 1$.

Solution 7.7 Let us rewrite the Heun method in the following (Runge-Kutta like) form:

$$u_{n+1} = u_n + \frac{1}{2}(k_1 + k_2), \quad k_1 = hf(t_n, u_n), \quad k_2 = hf(t_{n+1}, u_n + k_1). \quad (9.4)$$

We have $h\tau_{n+1}(h) = y(t_{n+1}) - y(t_n) - (\widehat{k}_1 + \widehat{k}_2)/2$, with $\widehat{k}_1 = hf(t_n, y(t_n))$ and $\widehat{k}_2 = hf(t_{n+1}, y(t_n) + \widehat{k}_1)$. Therefore, the method is consistent since

$$\lim_{h \to 0} \tau_{n+1} = y'(t_n) - \frac{1}{2}[f(t_n, y(t_n)) + f(t_n, y(t_n))] = 0.$$

The Heun method is implemented in Program 9.1. Using this program, we can verify the order of convergence as in Solution 7.1. By the following instructions, we find that the Heun method is second-order with respect to h

```
>> p=log(abs(e(1:end-1)./e(2:end)))/log(2)
p =
  Columns 1 through 6
    1.7642    1.8796    1.9398    1.9700    1.9851    1.9925
  Columns 7 through 9
    1.9963    1.9981    1.9991
```

Program 9.1. rk2: Heun method

```
function [t,u]=rk2(odefun,tspan,y0,Nh,varargin)
h=(tspan(2)-tspan(1)-t0)/Nh;  tt=[tspan(1):h:tspan(2)];
u(1)=y0;
for s=tt(1:end-1)
    t = s;   y = u(end);
    k1=h*feval(odefun,t,y,varargin{:});
    t = t + h;
    y = y + k1;  k2=h*feval(odefun,t,y,varargin{:});
    u = [u, u(end) + 0.5*(k1+k2)];
end
t=tt;
return
```

Solution 7.8 Applying the method (9.4) to the model problem (7.28) we obtain $k_1 = h\lambda u_n$ and $k_2 = h\lambda u_n(1 + h\lambda)$. Therefore $u_{n+1} = u_n[1 + h\lambda + (h\lambda)^2/2] = u_n p_2(h\lambda)$. To ensure absolute stability we must require that $|p_2(h\lambda)| < 1$, which is equivalent to $0 < p_2(h\lambda) < 1$, since $p_2(h\lambda)$ is positive. Solving the latter inequality, we obtain $-2 < h\lambda < 0$, that is, $h < 2/|\lambda|$.

Solution 7.9 Note that

$$u_n = u_{n-1}(1 + h\lambda_{n-1}) + hr_{n-1}.$$

Then proceed recursively on n.

Solution 7.10 The inequality (7.38) follows from (7.37) by setting

$$\varphi(\lambda) = \left|1 + \frac{1}{\lambda}\right| + \left|\frac{1}{\lambda}\right|.$$

The conclusion follows easily.

Solution 7.11 From (7.35) we have

$$|z_n - u_n| \le \rho_{max} a^n + h\rho_{max} \sum_{k=0}^{n-1} \delta(h)^{n-k-1}.$$

The result follows using (7.36).

Solution 7.12 We have

$$h\tau_{n+1}(h) = y(t_{n+1}) - y(t_n) - \frac{1}{6}(\widehat{k}_1 + 4\widehat{k}_2 + \widehat{k}_3),$$

$$\widehat{k}_1 = hf(t_n, y(t_n)), \quad \widehat{k}_2 = hf(t_n + \tfrac{h}{2}, y(t_n) + \tfrac{\widehat{k}_1}{2}),$$

$$\widehat{k}_3 = hf(t_{n+1}, y(t_n) + 2\widehat{k}_2 - \widehat{k}_1).$$

This method is consistent since

$$\lim_{h\to 0} \tau_{n+1} = y'(t_n) - \frac{1}{6}[f(t_n, y(t_n)) + 4f(t_n, y(t_n)) + f(t_n, y(t_n))] = 0.$$

This method is an explicit Runge-Kutta method of order 3 and is implemented in Program 9.2. As in Solution 7.7, we can derive an estimate of its order of convergence by the following instructions:

```
>> p=log(abs(e(1:end-1)./e(2:end)))/log(2)
p =
  Columns 1 through 6
    2.7306    2.8657    2.9330    2.9666    2.9833    2.9916
  Columns 7 through 9
    2.9958    2.9979    2.9990
```

Solution 7.13 From Solution 7.8 we obtain the relation

$$u_{n+1} = u_n[1 + h\lambda + \frac{1}{2}(h\lambda)^2 + \frac{1}{6}(h\lambda)^3] = u_n p_3(h\lambda).$$

By inspection of the graph of p_3, obtained with the instruction

```
>> c=[1/6 1/2 1 1]; z=[-3:0.01:1]; p=polyval(c,z); plot(z,abs(p))
```

we deduce that $|p_3(h\lambda)| < 1$ for $-2.5 < h\lambda < 0$.

Program 9.2. rk3: explicit Runge-Kutta method of order 3

```
function [t,u]=rk3(odefun,tspan,y0,Nh,varargin)
h=(tspan(2)-tspan(1))/Nh; tt=[tspan(1):h:tspan(2)];
u(1)=y0;
for s=tt(1:end-1)
   t = s; y = u(end);
   k1=h*feval(odefun,t,y,varargin{:});
   t = t + h*0.5;  y = y + 0.5*k1;
   k2=h*feval(odefun,t,y,varargin{:});
   t = s + h;      y = u(end) + 2*k2-k1;
   k3=h*feval(odefun,t,y,varargin{:});
   u = [u, u(end) + (k1+4*k2+k3)/6];
end
t=tt;
```

Solution 7.14 The method (7.74) applied to the model problem (7.28) gives the equation $u_{n+1} = u_n(1 + h\lambda + (h\lambda)^2)$. From the graph of $1 + z + z^2$ with $z = h\lambda$, we deduce that the method is absolutely stable if $-1 < h\lambda < 0$.

Solution 7.15 To solve Problem 7.1 with the given values, we repeat the following instructions with N=10 and N=20:

```
>> f=inline('-1.68*10^(-9)*y^4+2.6880','t','y');
>> [t,uc]=cranknic(f,[0,200],180,N);
>> [t,u]=predcor(f,[0 200],180,N,'feonestep','cnonestep');
```

The graphs of the computed solutions are shown in Figure 9.11. The solutions obtained by the Crank-Nicolson method are more accurate than those obtained by the Heun method.

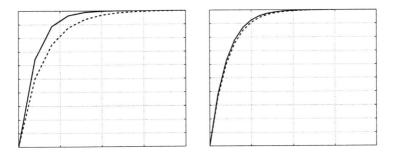

Fig. 9.11. Computed solutions with N = 10 (*left*) and N = 20 (*right*) for the Cauchy problem of Solution 7.15: the solutions computed by the Crank-Nicolson method (*continuous line*), and by the Heun method (*dashed line*)

Solution 7.16 Heun method applied to the model problem (7.28), gives

$$u_{n+1} = u_n \left(1 + h\lambda + \frac{1}{2}h^2\lambda^2\right).$$

In the complex plane the boundary of its region of absolute stability satisfies $|1 + h\lambda + h^2\lambda^2/2|^2 = 1$, having set $h\lambda = x + iy$. This equation is satisfied by the pairs (x, y) such that $f(x, y) = x^4 + y^4 + 2x^2y^2 + 4x^3 + 4xy^2 + 8x^2 + 8x = 0$. We can represent this curve as the level curve $f(x, y) = z$ (corresponding to the level $z = 0$). This can be done by means of the following instructions:

```
>> f='x.^4+y.^4+2*(x.^2).*(y.^2)+4*x.*y.^2+4*x.^3+8*x.^2+8*x';
>> [x,y]=meshgrid([-2.1:0.1:0.1],[-2:0.1:2]);
>> contour(x,y,eval(f),[0 0])
```

meshgrid The command **meshgrid** draws in the rectangle $[-2.1, 0.1] \times [-2, 2]$ a grid with 23 equispaced nodes in the x-direction, and 41 equispaced nodes in the **contour** y-direction. With the command **contour** we plot the level curve of $f(x, y)$ (evaluated with the command **eval(f)**) corresponding to the value $z = 0$ (made precise in the input vector [0 0] of **contour**). In Figure 9.12 the continuous line delimitates the region of absolute stability of the Heun method. This region is larger than the corresponding region of the forward Euler method (which corresponds to the interior of the dashed circle). Both curves are tangent to the imaginary axis at the origin $(0, 0)$.

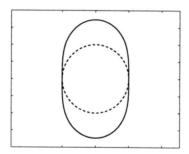

Fig. 9.12. Boundaries of the regions of absolute stability for the Heun method (*continuous line*) and the forward Euler method (*dashed line*). The corresponding regions lie at the interior of the boundaries

Solution 7.17 We use the following instructions:

```
>> tspan=[0 1]; y0=0; f=inline('cos(2*y)','t','y');
>> y='0.5*asin((exp(4*t)-1)./(exp(4*t)+1))';
>> N=2; for k=1:10
   [tt,u]=predcor(f,tspan,y0,N,'feonestep','cnonestep');
   t=tt(end); e(k)=abs(u(end)-eval(y)); N=2*N; end
>> p=log(abs(e(1:end-1)./e(2:end)))/log(2)
p =
   Columns 1 through 6
```

2.4733   2.2507   2.1223   2.0601   2.0298   2.0148
Columns 7 through 9
2.0074   2.0037   2.0018

As expected, we find that the order of convergence of the method is 2. However, the computational cost is comparable with that of the forward Euler method, which is first-order accurate only.

Solution 7.18 The second-order differential equation of this exercise is equivalent to the following first-order system:

$$x' = z, \quad z' = -5z - 6x,$$

with $x(0) = 1$, $z(0) = 0$. We use the Heun method as follows:

```
>> tspan=[0 5]; y0=[1 0];
>> [tt,u]=predcor('fspring',tspan,y0,N,'feonestep','cnonestep');
```

where N is the number of nodes and fspring.m is the following function:

```
function y=fspring(t,y)
b=5; k=6;
yy=y; y(1)=yy(2); y(2)=-b*yy(2)-k*yy(1);
```

In Figure 9.13 we show the graphs of the two components of the solution, computed with N=20,40 and compare them with the graph of the exact solution $x(t) = 3e^{-2t} - 2e^{-3t}$ and that of its first derivative.

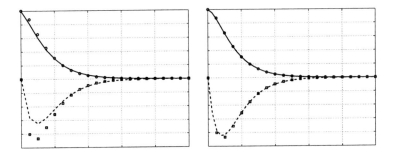

Fig. 9.13. Approximations of $x(t)$ (*continuous line*) and $x'(t)$ (*dashed line*) computed with N=20 (*thin line*) and N=40 (*thick line*). Small circles and squares refer to the exact functions $x(t)$ and $x'(t)$, respectively

Solution 7.19 The second-order system of differential equations is reduced to the following first-order system:

$$\begin{cases} x' = z, \\ y' = v, \\ z' = 2\omega \sin(\Psi) - k^2 x, \\ v' = -2\omega \sin(\Psi)z - k^2 y. \end{cases} \quad (9.5)$$

If we suppose that the pendulum at the initial time $t_0 = 0$ is at rest in the position $(1, 0)$, the system (9.5) must be given the following initial conditions:

$$x(0) = 1, \ y(0) = 0, \ z(0) = 0, \ v(0) = 0.$$

Setting $\Psi = \pi/4$, which is the average latitude of the Northern Italy, we use the forward Euler method as follows:

```
>> [t,y]=feuler('ffocault',[0 300],[1 0 0 0],Nh);
```

where Nh is the number of steps and ffocault.m is the following function:

```
function y=ffocault(t,y)
l=20;    k2=9.8/l;    psi=pi/4; omega=7.29*1.e-05;
yy=y;    y(1)=yy(3);  y(2)=yy(4);
y(3)=2*omega*sin(psi)*yy(4)-k2*yy(1);
y(4)=-2*omega*sin(psi)*yy(3)-k2*yy(2);
```

By some numerical experiments we conclude that the forward Euler method cannot produce acceptable solutions for this problem even for very small h. For instance, on the left of Figure 9.14 we show the graph, in the phase plane (x, y), of the motion of the pendulum computed with N=30000, that is, $h = 1/100$. As expected, the rotation plane changes with time, but also the amplitude of the oscillations increases. Similar results can be obtained for smaller h and using the Heun method. In fact, the model problem corresponding to the problem at hand has a coefficient λ that is purely imaginary. The corresponding solution (a sinusoid) is bounded for t that tends to infinity, however it doesn't tend to zero.

Unfortunately, both the forward Euler and Heun methods feature a region of absolute stability that doesn't include any point of the imaginary axis (with the exception of the origin). Thus, to ensure the absolute stability one should choose the prohibited value $h = 0$.

To get an acceptable solution we should use a method whose region of absolute stability includes a portion of the imaginary axis. This is the case, for instance, for the adaptive Runge-Kutta method of order 3, implemented in the MATLAB function ode23. We can invoke it by the following command:

```
>> [t,u]=ode23('ffocault',[0 300],[1 0 0 0]);
```

In Figure 9.14 (*right*) we show the solution obtained using only 1022 integration steps. Note that the numerical solution is in good agreement with the analytical one.

Octave 7.1 In Octave, ode23 returns after 1419 iterations. Moreover ode23 returns a different final result. ∎

Solution 7.20 We fix the right hand side of the problem in the following *function*

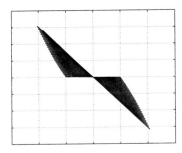 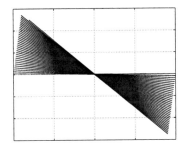

Fig. 9.14. Trajectories on the phase plane for the Foucault pendulum of Solution 7.19 computed by the forward Euler method (*left*) and the third-order adaptive Runge-Kutta method (*right*)

```
function y=baseball(t,y)
phi = 0;   omega = 1800*1.047198e-01;
B = 4.1*1.e-4;  yy=y;
g = 9.8;
vmodulo = sqrt(y(4)^2+y(5)^2+y(6)^2);
Fv = 0.0039+0.0058/(1+exp((vmodulo-35)/5));
y(1)=yy(4);
y(2)=yy(5);
y(3)=yy(6);
y(4)=-Fv*vmodulo*y(4)+B*omega*(yy(6)*sin(phi)-yy(5)
    *cos(phi));
y(5)=-Fv*vmodulo*y(5)+B*omega*yy(4)*cos(phi);
y(6)=-g-Fv*vmodulo*y(6)-B*omega*yy(4)*sin(phi);
return
```

At this point we only need to recall **ode23** as follows:

```
>> [t,u]=ode23('baseball',[0 0.4],...
    [0 0 0 38*cos(1*pi/180) 0 38*sin(1*pi/180)]);
```

Using command **find** we approximately compute the time at which the altitude becomes negative, which corresponds to the exact time of impact with the ground:

```
>> n=max(find(u(:,3)>=0));
t(n)
ans = 0.1066
```

In Figure 7.1 we report the trajectories of the baseball with an inclination of 1 and 3 degrees represented on the plane $x_1 x_3$ and on the $x_1 x_2 x_3$ space.

9.8 Chapter 8

Solution 8.1 We can verify directly that $\mathbf{x}^T A \mathbf{x} > 0$ for all $\mathbf{x} \neq \mathbf{0}$. Indeed,

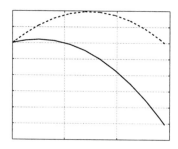 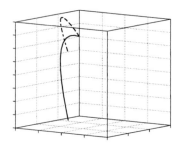

Fig. 9.15. The trajectories followed by a baseball launched with an initial angle of 1 degree (*solid line*), respectively, 3 degrees (*dashed line*)

$$[x_1\ x_2\ \ldots\ x_{N-1}\ x_N] \begin{bmatrix} 2 & -1 & 0 & \ldots & 0 \\ -1 & 2 & \ddots & & \vdots \\ 0 & \ddots & \ddots & -1 & 0 \\ \vdots & & -1 & 2 & -1 \\ 0 & \ldots & 0 & -1 & 2 \end{bmatrix} \begin{bmatrix} x_1 \\ x_2 \\ \vdots \\ x_{N-1} \\ x_N \end{bmatrix}$$

$$= 2x_1^2 - 2x_1x_2 + 2x_2^2 - 2x_2x_3 + \ldots - 2x_{N-1}x_N + 2x_N^2.$$

The last expression is equivalent to $(x_1-x_2)^2+\ldots+(x_{N-1}-x_N)^2+x_1^2+x_N^2$, which is, positive provided that at least one x_i is non-null.

Solution 8.2 We verify that $A\mathbf{q}_j = \lambda_j \mathbf{q}_j$. Computing the matrix-vector product $\mathbf{w} = A\mathbf{q}_j$ and requiring that $\mathbf{w}$ is equal to the vector $\lambda_j \mathbf{q}_j$, we find:

$$\begin{cases} 2\sin(j\theta) - \sin(2j\theta) = 2(1 - \cos(j\theta))\sin(j\theta), \\ -\sin(jk\theta) + 2\sin(j(k+1)\theta) - \sin(j(k+2)\theta) = 2(1 - \cos(j\theta))\sin(2j\theta), \\ \qquad k = 1,\ldots, N-2 \\ 2\sin(Nj\theta) - \sin((N-1)j\theta) = 2(1 - \cos(j\theta))\sin(Nj\theta). \end{cases}$$

The first equation is an identity since $\sin(2j\theta) = 2\sin(j\theta)\cos(j\theta)$. The other equations can be simplified since

$$\sin(jk\theta) = \sin((k+1)j\theta)\cos(j\theta) - \cos((k+1)j\theta)\sin(j\theta),$$

$$\sin(j(k+2)\theta) = \sin((k+1)j\theta)\cos(j\theta) + \cos((k+1)j\theta)\sin(j\theta).$$

Since A is symmetric and positive definite, its condition number is $K(A) = \lambda_{max}/\lambda_{min}$, that is, $K(A) = \lambda_1/\lambda_N = (1-\cos(N\pi/(N+1)))/(1-\cos(\pi/(N+1)))$. Using the Taylor expansion of order 2 of the cosine function, we obtain $K(A) \simeq N^2$, that is, $K(A) \simeq h^{-2}$.

Solution 8.3 We note that

$$u(\bar{x}+h) = u(\bar{x}) + hu'(\bar{x}) + \frac{h^2}{2}u''(\bar{x}) + \frac{h^3}{6}u'''(\bar{x}) + \frac{h^4}{24}u^{(4)}(\xi_+),$$

$$u(\bar{x}-h) = u(\bar{x}) - hu'(\bar{x}) + \frac{h^2}{2}u''(\bar{x}) - \frac{h^3}{6}u'''(\bar{x}) + \frac{h^4}{24}u^{(4)}(\xi_-),$$

where $\xi_+ \in (x, x+h)$ and $\xi_- \in (x-h, x)$. Summing the two expression we obtain

$$u(\bar{x}+h) + u(\bar{x}-h) = 2u(\bar{x}) + h^2 u''(\bar{x}) + \frac{h^4}{24}(u^{(4)}(\xi_+) + u^{(4)}(\xi_-)),$$

which is the desired property.

Solution 8.4 The matrix is again tridiagonal with entries $a_{i,i-1} = -1 - h\frac{\delta}{2}$, $a_{ii} = 2 + h^2\gamma$, $a_{i,i+1} = -1 + h\frac{\delta}{2}$. The right-hand side, accounting for the boundary conditions, becomes $\mathbf{f} = (f(x_1) + \alpha(1+h\delta/2)/h^2, f(x_2), \ldots, f(x_{N-1}), f(x_N) + \beta(1-h\delta/2)/h^2)^T$.

Solution 8.5 With the following instructions we compute the corresponding solutions to the three given values of h:

```
>> fbvp=inline('1+sin(4*pi*x)','x');
>> [z,uh10]=bvp(0,1,9,0,0.1,fbvp,0,0);
>> [z,uh20]=bvp(0,1,19,0,0.1,fbvp,0,0);
>> [z,uh40]=bvp(0,1,39,0,0.1,fbvp,0,0);
```

Since we don't know the exact solution, to estimate the convergence order we compute an approximate solution on a very fine grid (for instance $h = 1/1000$), then we use this latter as a surrogate for the exact solution. We find:

```
>> [z,uhex]=bvp(0,1,999,0,0.1,fbvp,0,0);
>> max(abs(uh10-uhex(1:100:end)))
ans =
    8.6782e-04
>> max(abs(uh20-uhex(1:50:end)))
ans =
    2.0422e-04
>> max(abs(uh40-uhex(1:25:end)))
ans =
    5.2789e-05
```

Halving h, the error is divided by 4, proving that the convergence order with respect to h is 2.

Solution 8.6 To find the largest h_{crit} which ensures a monotonic solution (as the analytical one) we execute the following cycle:

```
>> fbvp=inline('1+0.*x','x'); for k=3:1000
   [z,uh]=bvp(0,1,k,100,0,fbvp,0,1); if sum(diff(uh)>0)==length(uh)
   -1, break, end, end
```

We let $h(= 1/(\text{k}+1))$ vary till the forward incremental ratios of the numerical solution uh are all positive. Then we compute the vector diff(uh) whose components are 1 if the corresponding incremental ratio is positive, 0 otherwise. If the sum of all components equals the vector length of uh diminished by 1, then all incremental ratios are positive. The cycle stops when k=499, that is, when $h = 1/500$ if $\delta = 1000$, and when $h = 1/1000$ if $\delta = 2000$. We can therefore guess that one should require $h < 2/\delta = h_{crit}$ in order to get a monotonically increasing numerical solution. Indeed, this restriction on h is precisely what can be proven theoretically (see, for instance, [QV94]). In Figure 9.16 we show the numerical solutions obtained when $\delta = 100$ for two values of h.

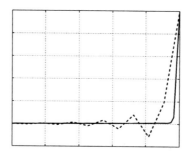

Fig. 9.16. Numerical solution for Problem 8.6 obtained for $h = 1/10$ (*dashed line*) and $h = 1/60$ (*continuous line*)

Solution 8.7 We should modify the Program 8.1 in order to impose Neumann boundary conditions. In the Program 9.3 we show one possible implementation.

Program 9.3. neumann: approximation of a Neumann boundary-value problem

```
function [x,uh]=neumann(a,b,N,delta,gamma,bvpfun, ...
     ua,ub,varargin)
h = (b-a)/(N+1);   x = [a:h:b];   e = ones(N+2,1);
A = spdiags([-e-0.5*h*delta 2*e+gamma*h^2 ...
     -e+0.5*h*delta], -1:1, N+2, N+2);
f = h^2*feval(bvpfun,'x',varargin{:});   f=f';
A(1,1)=-3/2*h;   A(1,2)=2*h;   A(1,3)=-1/2*h;
f(1)=h^2*ua;
A(N+2,N+2)=3/2*h;   A(N+2,N+1)=-2*h;   A(N+2,N)=1/2*h;
f(N+2)=h^2*ub;
uh = A\f;
return
```

Solution 8.8 The trapezoidal integration formula, used on the two subintervals I_{k-1} and I_k, produces the following approximation

$$\int_{I_{k-1} \cup I_k} f(x)\varphi_k(x)\, dx \simeq \frac{h}{2}f(x_k) + \frac{h}{2}f(x_k) = hf(x_k),$$

since $\varphi_k(x_j) = \delta_{jk}$, $\forall j,k$. Thus, we obtain the same right-hand side of the finite difference method.

Solution 8.9 We have $\nabla \phi = (\partial \phi/\partial x, \partial \phi/\partial y)^T$ and therefore $\text{div} \nabla \phi = \partial^2 \phi/\partial x^2 + \partial^2 \phi/\partial y^2$, that is, the Laplacian of ϕ.

Solution 8.10 To compute the temperature at the center of the plate, we solve the corresponding Poisson problem for various values of $\Delta_x = \Delta_y$, using the following instructions:

```
>> k=0; fun=inline('25','x','y'); bound=inline('(x==1)','x','y');
>> for N = [10,20,40,80,160],
   [u,x,y]=poissonfd(0,0,1,1,N,N,fun,bound);
   k=k+1; uc(k) = u(N/2+1,N/2+1); end
```

The components of the vector uc are the values of the computed temperature at the center of the plate as the step-size h of the grid decreases. We have

```
>> uc
   2.0168    2.0616    2.0789    2.0859    2.0890
```

We can therefore conclude that at the center of the plate the temperature is about $2.08°C$. In Figure 9.17 we show the isolines of the temperature for two different values of h.

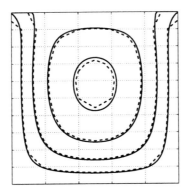

Fig. 9.17. The isolines of the computed temperature for $\Delta_x = \Delta_y = 1/10$ (*dashed lines*) and for $\Delta_x = \Delta_y = 1/80$ (*continuous lines*)

References

[ABB+99] Anderson E., Bai Z., Bischof C., Blackford S., Demmel J., Dongarra J., Croz J. D., Greenbaum A., Hammarling S., McKenney A., and Sorensen D. (1999) *LAPACK User's Guide*. SIAM, Philadelphia, 3rd edition.

[Ada90] Adair R. (1990) *The physics of baseball*. Harper and Row, New York.

[Arn73] Arnold V. (1973) *Ordinary Differential Equations*. The MIT Press, Cambridge.

[Atk89] Atkinson K. (1989) *An Introduction to Numerical Analysis*. John Wiley, New York.

[Axe94] Axelsson O. (1994) *Iterative Solution Methods*. Cambridge University Press, New York.

[BB96] Brassard G. and Bratley P. (1996) *Fundamentals of Algorithmics, 1/e*. Prentice Hall, New York.

[BM92] Bernardi C. and Maday Y. (1992) *Approximations Spectrales des Problémes aux Limites Elliptiques*. Springer-Verlag, Paris.

[Bra97] Braess D. (1997) *Finite Elements: Theory, Fast Solvers and Applications in Solid Mechanics*. Cambridge University Press, Cambridge.

[BS01] Babuska I. and Strouboulis T. (2001) *The Finite Element Method and its Reliability*. Oxford University Press, Padstow.

[But87] Butcher J. (1987) *The Numerical Analysis of Ordinary Differential Equations: Runge-Kutta and General Linear Methods*. Wiley, Chichester.

[CHQZ06] Canuto C., Hussaini M. Y., Quarteroni A., and Zang T. A. (2006) *Spectral Methods: Fundamentals in Single Domains*. Springer-Verlag, Berlin Heidelberg.

[CLW69] Carnahan B., Luther H., and Wilkes J. (1969) *Applied Numerical Methods*. John Wiley ans Sons, Inc., New York.

[Dav63] Davis P. (1963) *Interpolation and Approximation*. Blaisdell Pub., New York.

[DD99] Davis T. and Duff I. (1999) A combined unifrontal/multifrontal method for unsymmetric sparse matrices. *ACM Transactions on Mathematical Software* 25(1): 1–20.

References

[Dem97] Demmel J. (1997) *Applied Numerical Linear Algebra.* SIAM, Philadelphia.

[Deu04] Deuflhard P. (2004) *Newton Methods for Nonlinear Problems. Affine Invariance and Adaptive Algorithms.* Springer Series in Computational Mathematics, 35: Springer-Verlag, Berlin.

[Die93] Dierckx P. (1993) *Curve and Surface Fitting with Splines.* Claredon Press, New York.

[DL92] DeVore R. and Lucier J. (1992) Wavelets. *Acta Numerica* 1: 1–56.

[DR75] Davis P. and Rabinowitz P. (1975) *Methods of Numerical Integration.* Academic Press, New York.

[DS83] Dennis J. and Schnabel R. (1983) *Numerical Methods for Unconstrained Optimization and Nonlinear Equations.* Prentice-Hall, Englewood Cliffs, New York.

[dV89] der Vorst H. V. (1989) High Performance Preconditioning. *SIAM J. Sci. Stat. Comput.* 10: 1174–1185.

[Eat02] Eaton J. (2002) *GNU Octave manual.* Network Theory Ltd., Bristol.

[EEHJ96] Eriksson K., Estep D., Hansbo P., and Johnson C. (1996) *Computational Differential Equations.* Cambridge Univ. Press, Cambridge.

[EKM05] Etter D., Kuncicky D., and Moore H. (2005) *Introduction to MATLAB 7.* Prentice Hall, Englewood Cliffs.

[Fun92] Funaro D. (1992) *Polynomial Approximation of Differential Equations.* Springer-Verlag, Berlin Heidelberg.

[Gau97] Gautschi W. (1997) *Numerical Analysis. An Introduction.* Birkhäuser, Berlin.

[Gea71] Gear C. (1971) *Numerical Initial Value Problems in Ordinary Differential Equations.* Prentice-Hall, Upper Saddle River NJ.

[Gio97] Giordano N. (1997) *Computational physics.* Prentice-Hall, Upper Saddle River NJ.

[GL96] Golub G. and Loan C. V. (1996) *Matrix Computations.* The John Hopkins Univ. Press, Baltimore and London, 3rd edition.

[GR96] Godlewski E. and Raviart P.-A. (1996) *Hyperbolic Systems of Conservations Laws*, volume 118. Springer-Verlag, New York.

[Hac85] Hackbusch W. (1985) *Multigrid Methods and Applications.* Springer-Verlag, Berlin Heidelberg.

[Hac94] Hackbusch W. (1994) *Iterative Solution of Large Sparse Systems of Equations.* Springer-Verlag, New York.

[HH05] Higham D. and Higham N. (2005) *MATLAB Guide. Second edition.* SIAM, Philadelphia.

[Hig02] Higham N. (2002) *Accuracy and Stability of Numerical Algorithms. Second edition.* SIAM Publications, Philadelphia, PA.

[Hir88] Hirsh C. (1988) *Numerical Computation of Internal and External Flows*, volume 1. John Wiley and Sons, Chichester.

[HLR01] Hunt B., Lipsman R., and Rosenberg J. (2001) *A guide to MATLAB: for Beginners and Experienced Users.* Cambridge University Press.

[IK66] Isaacson E. and Keller H. (1966) *Analysis of Numerical Methods.* Wiley, New York.

[Krö98]	Kröner D. (1998) *Finite volume schemes in multidimensions*. Pitman Res. Notes Math. Ser., 380. Longman, Harlow.
[KS99]	Karniadakis G. and Sherwin S. (1999) *Spectral/hp Element Methods for CFD*. Oxford University Press, Padstow.
[Lam91]	Lambert J. (1991) *Numerical Methods for Ordinary Differential Systems*. John Wiley and Sons, Chichester.
[Lan03]	Langtangen H. (2003) *Advanced Topics in Computational Partial Differential Equations: Numerical Methods and Diffpack Programming*. Springer-Verlag, Berlin Heidelberg.
[LeV02]	LeVeque R. (2002) *Finite Volume Methods for Hyperbolic Problems*. Cambridge University Press, Cambridge.
[Mei67]	Meinardus G. (1967) *Approximation of Functions: Theory and Numerical Methods*. Springer-Verlag, Berlin Heidelberg.
[MH03]	Marchand P. and Holland O. (2003) *Graphics and Guis With Matlab*. CRC Press.
[Pal04]	Palm W. (2004) *Introduction to Matlab 7 for Engineers*. McGraw-Hill, New York.
[Pan92]	Pan V. (1992) Complexity of Computations with Matrices and Polynomials. *SIAM Review* 34: 225–262.
[PBP02]	Prautzsch H., Boehm W., and Paluszny M. (2002) *Bezier and B-Spline Techniques*. Springer-Verlag, Berlin Heidelberg.
[PdDKÜK83]	Piessens R., de Doncker-Kapenga E., Überhuber C., and Kahaner D. (1983) *QUADPACK: A Subroutine Package for Automatic Integration*. Springer-Verlag, Berlin Heidelberg.
[Pra02]	Pratap R. (2002) *Getting Started with MATLAB: A Quick Introduction for Scientists and Engineers*. Oxford University Press, Padstow.
[QSS06]	Quarteroni A., Sacco R., and Saleri F. (2006) *Numerical Mathematics*, volume 37 of *Texts in Applied Mathematics*. Springer-Verlag, New York, 2nd edition.
[QV94]	Quarteroni A. and Valli A. (1994) *Numerical Approximation of Partial Differential Equations*. Springer-Verlag, Berlin Heidelberg.
[RR85]	Ralston A. and Rabinowitz P. (1985) *A First Course in Numerical Analysis*. McGraw-Hill, Singapore.
[Saa92]	Saad Y. (1992) *Numerical Methods for Large Eigenvalue Problems*. Halstead Press, New York.
[Saa96]	Saad Y. (1996) *Iterative Methods for Sparse Linear Systems*. PWS Publishing Company, Boston.
[SM03]	Süli E. and Mayers D. (2003) *An Introduction to Numerical Analysis*. Cambridge University Press, Cambridge.
[TW98]	Tveito A. and Winther R. (1998) *Introduction to Partial Differential Equations. A Computational Approach*. Springer-Verlag, Berlin Heidelberg.
[Übe97]	Überhuber C. (1997) *Numerical Computation: Methods, Software, and Analysis*. Springer-Verlag, Berlin Heidelberg.
[Urb02]	Urban K. (2002) *Wavelets in Numerical Simulation*. Springer Verlag, Berlin Heidelberg.
[vdV03]	van der Vorst H. (2003) *Iterative Krylov Methods for Large Linear systems*. Cambridge University Press, Cambridge.

[Wes04]   Wesseling P. (2004) *An Introduction to Multigrid Methods.* R.T. Edwards, Inc., Philadelphia.
[Wil65]   Wilkinson J. (1965) *The Algebraic Eigenvalue Problem.* Clarendon Press, Oxford.

Index

... 33
= 30

Abel's theorem 60
`abs` 7
absolute
   error 4
   stability 202
   stability region 204, 215, 235
Adams-Bashforth methods 213, 215
Adams-Moulton methods 214, 215
adaptive
   formula 115
   forward Euler method 203
adaptivity 87, 115
`aitken` 59
Aitken's
   convergence 58
   extrapolation 57
   method 56
algorithm 26
   Gauss 129
   Hörner 61
   Strassen 27
   synthetic division 61
   Thomas 140, 241
   Winograd and Coppersmith 27
aliasing 85
`angle` 7
`ans` 30
approximation 74
   least-squares 93

`arpackc` 183
average 99
`axis` 177

backward
   difference formula 214
   finite difference 104
   substitutions 128
Bairstow's method 66
baseball trajectory 188, 236
basis 3
biomechanics 72, 94
`bisection` 43
bisection method 41, 53
Bogacki and Shampine pair 213
boundary conditions 240, 264
boundary-value problem 237
   Dirichlet 240
   Neumann 240
`break` 269
Broyden method 66
Butcher array 212, 213

cancellation 5
Cauchy
   problem 190
   theorem 61
characteristic
   polynomial 167, 201
   variables 259
Chebyshev
   interpolation 80
   nodes 80

chol 133
Cholesky factorization 133
cholinc 164
clear 30
climatology 71, 76
communications 239
compass 7
complex 6
complex numbers 6
complexity 26
computational cost 26
cond 138
condest 138
condition number 138, 249
conj 7
consistency 196, 251
consistent method 195, 252
conv 20
convergence 25
 Euler method 194
 finite differences 251
 Gauss-Seidel method 149
 interpolation 79
 iterative method 145
 Newton method 46
 order 25
 power method 173
 Richardson method 151
cos 30
cputime 27
Cramer rule 125
Crank-Nicolson method 197, 255, 257
cross 15
cubicspline 90
cumtrapz 109
cyclic composite methods 216

Dahlquist barrier 214, 215
dblquad 119
deconv 20
deflation 61, 63, 183
degree of exactness 107
Dekker-Brent method 65
demography 102, 110, 120
derivative
 partial 49
Descartes's rule 61
det 131

determinant 131
diag 12
diff 22
direct methods 126
discretization step 191
disp 269
divergence operator 238
domain of dependence 259
Dormand-Prince pair 213
dot 14

economy 125
eig 179
eigenvalue 15, 167
 extremal 170
 problem 167
eigenvector 15, 167
eigs 181
elastic membrane 250
elastic springs 168
electrical circuits 189, 221, 224
electromagnetism 102, 121
elliptic operator 260
end 27
eps 3, 5
ϵ_M 3
equation
 heat 253
 Poisson 240
 telegraph 239
 wave 257
error
 absolute 4, 24
 computational 24
 estimator 25, 47, 56, 115, 139
 interpolation 77
 local truncation 194
 perturbation 206
 relative 4, 24
 roundoff 3
 truncation 24, 194, 252
etime 27
Euler
 backward method 191, 257
 formula 7
 forward method 191, 202
 improved method 217
exit 29
exp 30

explicit method 192
exponent 3
extrapolation, Richardson method 121
eye 9

factorization
  Cholesky 133, 175
  Gauss 129
  LU 127, 136, 175
  QR 142
feuler 192
feval 16, 35
fft 83
Fibonacci sequence 32
figure 177
finance 71, 92, 94
find 43
finite
  difference method 241, 245
  element method 162, 243, 245
finite difference
  backward 104
  centered 104
  forward 103
fix 269
fixed point 52
  iterations
    convergence 58
  convergence 53, 55
  iteration function 53
  iterations 53
floating-point
  numbers 2, 3
  operations 26
format 3
formula
  adaptive 115
  backward difference 214
  Euler 7
forward
  Euler adaptive method 203
  Euler method 191
  finite difference 103
  substitutions 127
Foucault pendulum 236
Fourier
  discrete series 83
  fast transform 83

inverse fast transform 83
  law 239
fplot 15, 41, 87
fsolve 66, 67, 193
function
  derivative 21
  graph 15
  primitive 21
function 33
funtool 23
fzero 17, 65, 67

Gauss
  algorithm 129
  factorization 129, 132
  quadrature formulae 119
Gauss-Legendre formula 113
Gauss-Seidel method 149, 158
Gershgorin circles 176, 178, 184
gmres 156
grid 16
griddata 97
griddata3 97
griddatan 97

heat equation 238, 253
help 30
Heun method 217, 218, 235
Hilbert matrix 136, 139, 155, 157
hold off 177
hold on 177
horner 62
hydraulic network 123
hydraulics 101, 105
hydrogeology 238
hyperbolic
  operator 260
  system 258
Hörner's algorithm 61

if 27
ifft 83
imag 7
image compression 169, 182
implicit method 192
improved Euler method 217
increment 157
Inf 4
inline 35

int 22
interp1 87
interp1q 87
interp2 97
interp3 97
interpft 84
interpolant 74
interpolation
  Chebyshev 80
  composite 86, 97
  error 77
  Lagrangian polynomial 75
  nodes 74
  piecewise linear 86
  polynomial 74
  rational 75
  spline 88
  trigonometric 74, 81
interurban viability 169, 172
inv 10
investment fund 39, 68
invshift 175
iteration function 57
iterative methods 126, 144
itermeth 147

Jacobi method 146, 158

Kirchoff law 189
Kronecker symbol 75
Krylov methods 156, 164

Lagrange
  characteristic polynomials 76
  form 76
Lanczos method 183
Laplace
  operator 237, 247
  rule 11
law
  Fourier 239
  Kirchoff 189
  Ohm 189
leap-frog method 223
least-squares
  method 92
  solution 141, 142
Legendre polynomials 112
Leontief model 125

lexicographic order 247
Lipschitz continuous function 191
load 30
loglog 25
Lotka and Leslie model 169
Lotka-Volterra equations 188
LU
  factorization 127, 136
    incomplete 161
lu 131
lugauss 131
luinc 161, 164

magic 164
mantissa 3
matrix 8
  bidiagonal 140
  companion 66
  determinant 10
  diagonal 12
  diagonally dominant 133, 147
  Hankel 161
  hermitian 13
  Hilbert 136, 139, 157, 161
  ill conditioned 138
  inverse 10
  iteration 145
  Leslie 169, 184
  permutation 134
  positive definite 133, 149
  product 10
  Riemann 162
  similar 179
  sparse 141, 143
  square 9
  strictly diagonal 149
  sum 9
  symmetric 13, 133, 149
  transpose 13
  triangular 12
  tridiagonal 140, 149, 241
  Vandermonde 130, 161
  well conditioned 138
  Wilkinson 184
mesh 249
  contour 298
meshgrid 97, 298
method
  $\theta-$ 254

Adams-Bashforth 213
Adams-Moulton 214
Aitken 56
backward Euler 191, 257
Bairstow 66
Bi-CGStab 156
bisection 41
Broyden 66
CGS 156
conjugate gradient 153
consistent 195
Crank-Nicolson 197, 255, 257
Dekker-Brent 65
dynamic Richardson 150
explicit 192
finite elements 162, 263
forward Euler 191
forward Euler/uncentred 262
Gauss-Seidel 149
GMRES 156, 161
gradient 151
Heun 217, 218, 235
implicit 192
improved Euler 217
Jacobi 146
Krylov 156, 164
Lax-Wendroff 263
leap-frog 223
least-squares 92
Monte Carlo 268
multifrontal 164
multigrid 164
multistep 200, 213
Müller 66
Newmark 222, 223, 260
Newton 45
Newton-Hörner 63
power 171
predictor-corrector 216
QR 179
quasi-Newton 66
relaxation 149, 166, 288
Runge-Kutta 212
SOR 166
spectral 263
stationary Richardson 150
upwind 262
midpoint formula 106
composite 106

mkpp 90
model
Leontief 125
problem 202
generalized 205
multistep method 200, 213
Müller's method 66

NaN 5
nargin 35
nchoosek 269
Neumann boundary conditions 264
newmark 223
newmark 315
Newmark method 222, 223, 260
newton 48
Newton method 45, 56
adaptive 46
for systems 49
modified 46
Newton-Cotes formulae 119
Newton-Hörner, method 63
newtonhorner 64
newtonsys 50
nodes
Chebyshev 80
Gauss-Legendre-Lobatto 113
norm 15
normal equations 96, 141
not a number 5
not-a-knot condition 90
numerical integration 105

ode 213
ode113 218
ode15s 216
ode23 213, 221
ode23tb 213
ode45 213, 221
Ohm law 189
one-step method 192
ones 14
optics 102, 122
ordinary differential equation 187
overflow 4, 5

parabolic operator 260
partial
derivative 237

differential equation 187
patch 177
path 32
pattern of a matrix 141
pcg 155
pchip 91
pde 251
pdetool 97, 263
permutation matrix 134
phase plane 220
piecewise linear interpolation 86
pivot elements 129
pivoting 134
  by row 134
  complete 286
$\mathbb{P}_n$ 17
Poisson equation 237
poly 38, 79
polyder 20, 79
polyderiv 21
polyfit 21, 76, 94
polyint 20
polyinteg 21
polynomial 18
  characteristic 167
  division 20, 62
  Lagrangian interpolation 75
  product 20
  Taylor 22
polyval 18, 76
population dynamics 41, 54, 168, 184, 188, 219
  Malthus model 41
  predator/prey model 41, 54
  Verhulst model 41, 54
eigpower 171
power method 171
  inverse 174
  with shift 175
ppval 90
preconditioner 146, 150
  incomplete LU factorization 164
predator/prey model 41
predictor-corrector method 216
pretty 268
problem
  boundary-value 237
  Cauchy 190
  stiff 230, 231

prod 269
ptomedioc 107

QR
  factorization 142
  method 179
qrbasic 180
quad2dc 119
quad2dg 119
quadl 114
quadrature
  nodes 111
  weights 111
quadratures
  Gauss 113
  interpolatory 111
quasi-Newton methods 66
quit 29
quiver 15
quiver3 15

rand 27
rank 141
  full 141
Rayleigh quotient 167
rcond 138
real 7
realmax 4
realmin 4
rectangle formula 106
  composite 106
region of absolute stability 204, 235
regression line 94
relaxation method 149, 166, 288
residual 47, 139, 156
return 34
robotics 73, 90
rods system 40, 68
root 16
  condition 201
roots 19, 66
roundoff error 3, 135, 136
rpmak 97
rsmak 97
rule
  Cramer 125
  Descartes 61
  Laplace 11
Runge's function 79, 81

Runge-Kutta method   212

save   30
scalar product   14
semi-discretization   254
shape function   244
shift   175
significant digits   3
simpadpt   117, 118
simple   23, 288
Simpson
  adaptive formula   116
  composite formula   109
  formula   109
simpsonc   110
sin   30
singular value decomposition   167
singular values   167
sparse matrix   141, 249
spdemos   97
spdiags   140
spectral radius   145
spectrometry   124, 129
spectrum   170
spherical pendulum   225
spline   88
  natural cubic   88
spline   90
spy   249
sqrt   30
stability
  absolute   205
  asymptotical   254
  conditioned absolute   204
  region of absolute   235
  unconditioned absolute   204
Steffensen's method   58
stencil   247
stiff problems   230
stopping test
  for fixed point iterations   55
  for iterative methods   156
Strassen algorithm   27
Sturm, sequences   66, 183
successive over-relaxation method
  166
sum   269
svd   181
syms   22, 288

synthetic division algorithm   61
system
  hyperbolic   258
  linear   123
  nonlinear   49
  overdetermined   141
  triangular   127
  tridiagonal   140
  underdetermined   128, 141

taylor   22
Taylor polynomial   22, 73
taylortool   73
theorem
  Abel   60
  Cauchy   61
  Descartes   61
  equivalence   201
  first mean-value   21
  of integration   21
thermodynamics   187, 235, 239, 265
Thomas algorithm   140, 241
three-body problem   228
title   177
toolbox   18, 28, 30
trapezoidal formula   109
  composite   108
trapz   109
trigonometric interpolant   82
tril   12
triu   12
truncation error   194, 252, 255

underflow   4

Van der Pol equation   232
vander   130
variance   99, 280
vector   14
  column   9
  component   14
  linearly independent   14
  norm   15
  row   9

wave equation   238
wavelet   98
wavelets   98
weak formulation   243

wilkinson 184
xlabel 177
ylabel 177
zero
    multiple 16
    of a function 16
    simple 16
zero-stability 199
zeros 9, 14

Editorial Policy

1. Textbooks on topics in the field of computational science and engineering will be considered. They should be written for courses in CSE education. Both graduate and undergraduate textbooks will be published in TCSE. Multidisciplinary topics and multidisciplinary teams of authors are especially welcome.

2. Format: Only works in English will be considered. They should be submitted in camera-ready form according to Springer-Verlag's specifications. Electronic material can be included if appropriate. Please contact the publisher. Technical instructions and/or TEX macros are available via http://www.springer.com/sgw/cda/frontpage/0,11855,5-40017-2-71391-0,00.html

3. Those considering a bookwhichmight be suitable for the series are strongly advised to contact the publisher or the series editors at an early stage.

General Remarks

TCSE books are printed by photo-offset from the master-copy delivered in camera-ready form by the authors. For this purpose Springer-Verlag provides technical instructions for the preparation of manuscripts. See also *Editorial Policy*.

Careful preparation of manuscripts will help keep production time short and ensure a satisfactory appearance of the finished book.

The following terms and conditions hold:
Regarding free copies and royalties, the standard terms for Springer mathematics monographs and textbooks hold. Please write to martin.peters@springer.com for details.

Authors are entitled to purchase further copies of their book and other Springer books for their personal use, at a discount of 33.3% directly from Springer-Verlag.

Series Editors

Timothy J. Barth
NASA Ames Research Center
NAS Division
Moffett Field, CA 94035, USA
e-mail: barth@nas.nasa.gov

Michael Griebel
Institut für Numerische Simulation
der Universität Bonn
Wegelerstr. 6
53115 Bonn, Germany
e-mail: griebel@ins.uni-bonn.de

David E. Keyes
Department of Applied Physics
and Applied Mathematics
Columbia University
200 S. W. Mudd Building
500 W. 120th Street
New York, NY 10027, USA
e-mail: david.keyes@columbia.edu

Risto M. Nieminen
Laboratory of Physics
Helsinki University of Technology
02150 Espoo, Finland
e-mail: rni@fyslab.hut.fi

Dirk Roose
Department of Computer Science
Katholieke Universiteit Leuven
Celestijnenlaan 200A
3001 Leuven-Heverlee, Belgium
e-mail: dirk.roose@cs.kuleuven.ac.be

Tamar Schlick
Department of Chemistry
Courant Institute of Mathematical
Sciences
New York University
and Howard Hughes Medical Institute
251 Mercer Street
New York, NY 10012, USA
e-mail: schlick@nyu.edu

Editor at Springer: Martin Peters
Springer-Verlag, Mathematics Editorial IV
Tiergartenstrasse 17
D-69121 Heidelberg, Germany
Tel.: *49 (6221) 487-8185
Fax: *49 (6221) 487-8355
e-mail: martin.peters@springer.com

Texts in Computational Science and Engineering

Vol. 1 H. P. Langtangen, *Computational Partial Differential Equations. Numerical Methods and Diffpack Programming.* 2nd Edition 2003. XXVI, 855 pp. Hardcover. ISBN 3-540-43416-X

Vol. 2 A. Quarteroni, F. Saleri, *Scientific Computing with MATLAB and Octave.* 2nd Edition 2006. XIV, 318 pp. Hardcover. ISBN 3-540-32612-X

Vol. 3 H. P. Langtangen, *Python Scripting for Computational Science.* 2nd Edition 2006. XXIV, 736 pp. Hardcover. ISBN 3-540-29415-5

For further information on these books, please have a look at our mathematics catalogue at the following URL: www.springer.com/series/5151

Monographs in Computational Science and Engineering

Vol. 1 J. Sundnes, G.T. Lines, X. Cai, B.F. Nielsen, K.-A. Mardal, A. Tveito. *Computing the Electrical Activity in the Heart.* 2006. XI, 318 pp. Hardcover. ISBN 3-540-33432-7

For further information on this book, please have a look at our mathematics catalogue at the following URL: www.springer.com/series/7417

Lecture Notes in Computational Science and Engineering

Vol. 1 D. Funaro, *Spectral Elements for Transport-Dominated Equations.* 1997. X, 211 pp. Softcover. ISBN 3-540-62649-2

Vol. 2 H. P. Langtangen, *Computational Partial Differential Equations. Numerical Methods and Diffpack Programming.* 1999. XXIII, 682 pp. Hardcover. ISBN 3-540-65274-4

Vol. 3 W. Hackbusch, G. Wittum (eds.), *Multigrid Methods V.* Proceedings of the Fifth European Multigrid Conference held in Stuttgart, Germany, October 1-4, 1996. 1998. VIII, 334 pp. Softcover. ISBN 3-540-63133-X

Vol. 4 P. Deuflhard, J. Hermans, B. Leimkuhler, A. E. Mark, S. Reich, R. D. Skeel (eds.), *Computational Molecular Dynamics: Challenges, Methods, Ideas.* Proceedings of the 2nd International Symposium on Algorithms for Macromolecular Modelling, Berlin, May 21-24, 1997. 1998. XI, 489 pp. Softcover. ISBN 3-540-63242-5

Vol. 5 D. Kröner, M. Ohlberger, C. Rohde (eds.), *An Introduction to Recent Developments in Theory and Numerics for Conservation Laws.* Proceedings of the International School on Theory and Numerics for Conservation Laws, Freiburg / Littenweiler, October 20-24, 1997. 1998. VII, 285 pp. Softcover. ISBN 3-540-65081-4

Vol. 6 S. Turek, *Efficient Solvers for Incompressible Flow Problems. An Algorithmic and Computational Approach.* 1999. XVII, 352 pp, with CD-ROM. Hardcover. ISBN 3-540-65433-X

Vol. 7 R. von Schwerin, *Multi Body System SIMulation.* Numerical Methods, Algorithms, and Software. 1999. XX, 338 pp. Softcover. ISBN 3-540-65662-6

Vol. 8 H.-J. Bungartz, F. Durst, C. Zenger (eds.), *High Performance Scientific and Engineering Computing.* Proceedings of the International FORTWIHR Conference on HPSEC, Munich, March 16-18, 1998. 1999. X, 471 pp. Softcover. ISBN 3-540-65730-4

Vol. 9 T. J. Barth, H. Deconinck (eds.), *High-Order Methods for Computational Physics.* 1999. VII, 582 pp. Hardcover. ISBN 3-540-65893-9

Vol. 10 H. P. Langtangen, A. M. Bruaset, E. Quak (eds.), *Advances in Software Tools for Scientific Computing.* 2000. X, 357 pp. Softcover. ISBN 3-540-66557-9

Vol. 11 B. Cockburn, G. E. Karniadakis, C.-W. Shu (eds.), *Discontinuous Galerkin Methods.* Theory, Computation and Applications. 2000. XI, 470 pp. Hardcover. ISBN 3-540-66787-3

Vol. 12 U. van Rienen, *Numerical Methods in Computational Electrodynamics.* Linear Systems in Practical Applications. 2000. XIII, 375 pp. Softcover. ISBN 3-540-67629-5

Vol. 13 B. Engquist, L. Johnsson, M. Hammill, F. Short (eds.), *Simulation and Visualization on the Grid.* Parallelldatorcentrum Seventh Annual Conference, Stockholm, December 1999. Proceedings. 2000. XIII, 301 pp. Softcover. ISBN 3-540-67264-8

Vol. 14 E. Dick, K. Riemslagh, J. Vierendeels (eds.), *Multigrid Methods VI.* Proceedings of the Sixth European Multigrid Conference Held in Gent, Belgium, September 27-30, 1999. 2000. IX, 293 pp. Softcover. ISBN 3-540-67157-9

Vol. 15 A. Frommer, T. Lippert, B. Medeke, K. Schilling (eds.), *Numerical Challenges in Lattice Quantum Chromodynamics.* Joint Interdisciplinary Workshop of John von Neumann Institute for Computing, Jülich and Institute of Applied Computer Science, Wuppertal University, August 1999. 2000. VIII, 184 pp. Softcover. ISBN 3-540-67732-1

Vol. 16 J. Lang, *Adaptive Multilevel Solution of Nonlinear Parabolic PDE Systems.* Theory, Algorithm, and Applications. 2001. XII, 157 pp. Softcover. ISBN 3-540-67900-6

Vol. 17 B. I. Wohlmuth, *Discretization Methods and Iterative Solvers Based on Domain Decomposition.* 2001. X, 197 pp. Softcover. ISBN 3-540-41083-X

Vol. 18 U. van Rienen, M. Günther, D. Hecht (eds.), *Scientific Computing in Electrical Engineering.* Proceedings of the 3rd International Workshop, August 20-23, 2000, Warnemünde, Germany. 2001. XII, 428 pp. Softcover. ISBN 3-540-42173-4

Vol. 19 I. Babuška, P. G. Ciarlet, T. Miyoshi (eds.), *Mathematical Modeling and Numerical Simulation in Continuum Mechanics.* Proceedings of the International Symposium on Mathematical Modeling and Numerical Simulation in Continuum Mechanics, September 29 - October 3, 2000, Yamaguchi, Japan. 2002. VIII, 301 pp. Softcover. ISBN 3-540-42399-0

Vol. 20 T. J. Barth, T. Chan, R. Haimes (eds.), *Multiscale and Multiresolution Methods.* Theory and Applications. 2002. X, 389 pp. Softcover. ISBN 3-540-42420-2

Vol. 21 M. Breuer, F. Durst, C. Zenger (eds.), *High Performance Scientific and Engineering Computing.* Proceedings of the 3rd International FORTWIHR Conference on HPSEC, Erlangen, March 12-14, 2001. 2002. XIII, 408 pp. Softcover. ISBN 3-540-42946-8

Vol. 22 K. Urban, *Wavelets in Numerical Simulation.* Problem Adapted Construction and Applications. 2002. XV, 181 pp. Softcover. ISBN 3-540-43055-5

Vol. 23 L. F. Pavarino, A. Toselli (eds.), *Recent Developments in Domain Decomposition Methods.* 2002. XII, 243 pp. Softcover. ISBN 3-540-43413-5

Vol. 24 T. Schlick, H. H. Gan (eds.), *Computational Methods for Macromolecules: Challenges and Applications.* Proceedings of the 3rd International Workshop on Algorithms for Macromolecular Modeling, New York, October 12-14, 2000. 2002. IX, 504 pp. Softcover. ISBN 3-540-43756-8

Vol. 25 T. J. Barth, H. Deconinck (eds.), *Error Estimation and Adaptive Discretization Methods in Computational Fluid Dynamics.* 2003. VII, 344 pp. Hardcover. ISBN 3-540-43758-4

Vol. 26 M. Griebel, M. A. Schweitzer (eds.), *Meshfree Methods for Partial Differential Equations.* 2003. IX, 466 pp. Softcover. ISBN 3-540-43891-2

Vol. 27 S. Müller, *Adaptive Multiscale Schemes for Conservation Laws*. 2003. XIV, 181 pp. Softcover. ISBN 3-540-44325-8

Vol. 28 C. Carstensen, S. Funken, W. Hackbusch, R. H. W. Hoppe, P. Monk (eds.), *Computational Electromagnetics*. Proceedings of the GAMM Workshop on "Computational Electromagnetics", Kiel, Germany, January 26-28, 2001. 2003. X, 209 pp. Softcover. ISBN 3-540-44392-4

Vol. 29 M. A. Schweitzer, *A Parallel Multilevel Partition of Unity Method for Elliptic Partial Differential Equations*. 2003. V, 194 pp. Softcover. ISBN 3-540-00351-7

Vol. 30 T. Biegler, O. Ghattas, M. Heinkenschloss, B. van Bloemen Waanders (eds.), *Large-Scale PDE-Constrained Optimization*. 2003. VI, 349 pp. Softcover. ISBN 3-540-05045-0

Vol. 31 M. Ainsworth, P. Davies, D. Duncan, P. Martin, B. Rynne (eds.), *Topics in Computational Wave Propagation*. Direct and Inverse Problems. 2003. VIII, 399 pp. Softcover. ISBN 3-540-00744-X

Vol. 32 H. Emmerich, B. Nestler, M. Schreckenberg (eds.), *Interface and Transport Dynamics*. Computational Modelling. 2003. XV, 432 pp. Hardcover. ISBN 3-540-40367-1

Vol. 33 H. P. Langtangen, A. Tveito (eds.), *Advanced Topics in Computational Partial Differential Equations*. Numerical Methods and Diffpack Programming. 2003. XIX, 658 pp. Softcover. ISBN 3-540-01438-1

Vol. 34 V. John, *Large Eddy Simulation of Turbulent Incompressible Flows*. Analytical and Numerical Results for a Class of LES Models. 2004. XII, 261 pp. Softcover. ISBN 3-540-40643-3

Vol. 35 E. Bänsch (ed.), *Challenges in Scientific Computing - CISC 2002*. Proceedings of the Conference Challenges in Scientific Computing, Berlin, October 2-5, 2002. 2003. VIII, 287 pp. Hardcover. ISBN 3-540-40887-8

Vol. 36 B. N. Khoromskij, G. Wittum, *Numerical Solution of Elliptic Differential Equations by Reduction to the Interface*. 2004. XI, 293 pp. Softcover. ISBN 3-540-20406-7

Vol. 37 A. Iske, *Multiresolution Methods in Scattered Data Modelling*. 2004. XII, 182 pp. Softcover. ISBN 3-540-20479-2

Vol. 38 S.-I. Niculescu, K. Gu (eds.), *Advances in Time-Delay Systems*. 2004. XIV, 446 pp. Softcover. ISBN 3-540-20890-9

Vol. 39 S. Attinger, P. Koumoutsakos (eds.), *Multiscale Modelling and Simulation*. 2004. VIII, 277 pp. Softcover. ISBN 3-540-21180-2

Vol. 40 R. Kornhuber, R. Hoppe, J. Périaux, O. Pironneau, O. Wildlund, J. Xu (eds.), *Domain Decomposition Methods in Science and Engineering*. 2005. XVIII, 690 pp. Softcover. ISBN 3-540-22523-4

Vol. 41 T. Plewa, T. Linde, V.G. Weirs (eds.), *Adaptive Mesh Refinement – Theory and Applications*. 2005. XIV, 552 pp. Softcover. ISBN 3-540-21147-0

Vol. 42 A. Schmidt, K.G. Siebert, *Design of Adaptive Finite Element Software*. The Finite Element Toolbox ALBERTA. 2005. XII, 322 pp. Hardcover. ISBN 3-540-22842-X

Vol. 43 M. Griebel, M.A. Schweitzer (eds.), *Meshfree Methods for Partial Differential Equations II*. 2005. XIII, 303 pp. Softcover. ISBN 3-540-23026-2

Vol. 44 B. Engquist, P. Lötstedt, O. Runborg (eds.), *Multiscale Methods in Science and Engineering*. 2005. XII, 291 pp. Softcover. ISBN 3-540-25335-1

Vol. 45 P. Benner, V. Mehrmann, D.C. Sorensen (eds.), *Dimension Reduction of Large-Scale Systems*. 2005. XII, 402 pp. Softcover. ISBN 3-540-24545-6

Vol. 46 D. Kressner (ed.), *Numerical Methods for General and Structured Eigenvalue Problems*. 2005. XIV, 258 pp. Softcover. ISBN 3-540-24546-4

Vol. 47 A. Boriçi, A. Frommer, B. Joó, A. Kennedy, B. Pendleton (eds.), *QCD and Numerical Analysis III*. 2005. XIII, 201 pp. Softcover. ISBN 3-540-21257-4

Vol. 48 F. Graziani (ed.), *Computational Methods in Transport*. 2006. VIII, 524 pp. Softcover. ISBN 3-540-28122-3

Vol. 49 B. Leimkuhler, C. Chipot, R. Elber, A. Laaksonen, A. Mark, T. Schlick, C. Schütte, R. Skeel (eds.), *New Algorithms for Macromolecular Simulation*. 2006. XVI, 376 pp. Softcover. ISBN 3-540-25542-7

Vol. 50 M. Bücker, G. Corliss, P. Hovland, U. Naumann, B. Norris (eds.), *Automatic Differentiation: Applications, Theory, and Implementations.* 2006. XVIII, 362 pp. Softcover. ISBN 3-540-28403-6

Vol. 51 A.M. Bruaset, A. Tveito (eds.), *Numerical Solution of Partial Differential Equations on Parallel Computers* 2006. XII, 482 pp. Softcover. ISBN 3-540-29076-1

Vol. 52 K.H. Hoffmann, A. Meyer (eds.), *Parallel Algorithms and Cluster Computing.* 2006. X, 374 pp. Softcover. ISBN 3-540-33539-0

Vol. 53 H.-J. Bungartz, M. Schäfer (eds.), *Fluid-Structure Interaction.* 2006. VII, 388 pp. Softcover. ISBN 3-540-34595-7

For further information on these books please have a look at our mathematics catalogue at the following URL: www.springer.com/series/3527